International Conference on Advanced Materials and Technology (ICAMT 2019)

Innovation of Advanced Materials Technology and Multidisciplinary Research for a Better Life

IOP Conference Series: Materials Science and Engineering
Volume 924

Bogor, Indonesia
8 - 9 October 2019

Editors:

Rida Nurul Shelni Rofika
Shinta Widyaningrum
Kartika Sari

ISBN: 978-1-7138-3191-4
ISSN: 1757-8981

Printed from e-media with permission by:

Curran Associates, Inc.
57 Morehouse Lane
Red Hook, NY 12571

Some format issues inherent in the e-media version may also appear in this print version.

Institute of Physics
Dirac House, Temple Back
Bristol BS1 6BE UK

Phone: 44 1 17 929 7481
Fax: 44 1 17 920 0979

techtracking@iop.org

Additional copies of this publication are available from:

Curran Associates, Inc.
57 Morehouse Lane
Red Hook, NY 12571 USA
Phone: 845-758-0400
Fax: 845-758-2633
Email: curran@proceedings.com
Web: www.proceedings.com

TABLE OF CONTENTS

ADVANCED FUNCTIONAL AND STRUCTURAL MATERIALS

MATERIALS PROCESSING AND CHARACTERIZATION

INFORMATIC COMPUTATIONAL TECHNOLOGY AND COMPUTATIONAL MATERIALS

ENERGY AND ENVIRONMENT MATERIALS

International Conference on Advance Materials and Technology 2019 (ICAMT 2019)

ASTON SENTUL LAKE RESORT & CONFERENCE CENTER
8-9 OCTOBER 2019,
BOGOR-INDONESIA

ORGANIZED BY

Accepted, peer reviewed papers from the ICAMT 2019

Edited by
Evvy Kartini *et.al*
Indonesia

2019

IOP Conf. Series: Materials Science and Engineering **924** (2020) 011001 doi:10.1088/1757-899X/924/1/011001

PREFACE

International Conference on Advance Materials and Technology 2019 (ICAMT 2019) was held in Bogor on 8 -9 October 2019. The theme of this conference was "***Innovation of Advanced Materials Technology and Multidisciplinary Research for a Better Life***".

The conference was organized by the Indonesian Material Research Society (MRS-INA) together with the University of Prof. Dr. Moestopo (Beragama). This conference aimed to provide an international forum of scientific results in the fields of material science by national scientists, academics, industry and government, regional and international. In this regard, the conference was attended by world-class speakers. There were 130 manuscripts submitted to the conference and the final acceptance were 59 manuscripts.

On behalf of the organizing committee, it is our pleasure to present the results of the conference in the IOP-MSE proceeding.

Prof. Dr. rer. nat. Evvy Kartini
President of MRS-INA

Prof. Dr. Rudy Harjanto, M.Sn.
Chair of the Committee

IOP Conf. Series: Materials Science and Engineering **924** (2020) 011001 doi:10.1088/1757-899X/924/1/011001

CONFERENCE ORGANIZERS

Steering Committee

Prof. Dr. Thomas Suyatno	Universitas Prof. Dr. Moestopo (Beragama)
Dr. Azwar Manaf	Universitas Indonesia (UI)
Prof. Dr. Kadarsah Suryadi	Institut Teknologi Bandung
Prof. Dr. Ir. Herman Yuwono	Universitas Indonesia (UI)
Prof. Dr. Anne Zulfia	Universitas Indonesia (UI)
Dr. Agus Purwanto	Universitas Sebelas Maret (UNS)
Drs. H. Hartono Laras	Universitas Prof. Dr. Moestopo (Beragama)
Prof. Dr. Ir. Djoko Santoso, M.Si.	Institut Teknologi Bandung
Prof. Dr. Darminto	Institut Teknologi Sepuluh Nopember (ITS)
Dr. Sunit Hendrana	The Indonesian Institute of Sciences (LIPI)
Dr. Yudi Darma	Institut Teknologi Bandung (ITB)
Dr. Triyana Kuwat	Universitas Gadjah Mada (UGM)
Prof. Dr. H. Wiboro, S.E., M.Phil	Universitas Dr. Mustopo (Beragama)

International Advisory

Prof. Dr. Yafang Han	President of IUMRS, China
Prof. Dr. Rodrigo Martins	Vice President of IUMRS, Portugal
Prof. Dr. rer. nat. Evvy Kartini	President of MRS-Indonesia, Indonesia
Prof. Dr. B.V.R. Chowdary	President of MRS-Singapore, NTU Singapore
Prof. Dr. Santi Maensiri	President of MRS-Thailand, SUT Thailand
Prof. Dr. Atsushi Takahara	President of MRS-Japan, Kyushu Univ. Japan
Prof. Dr. Alex Peng	President of MRS-Taiwan, ITRI Taiwan
Prof. Dr. Do Kyung Kim	President of MRS-Korea, KAIST Korea

IOP Conf. Series: Materials Science and Engineering **924** (2020) 011001 doi:10.1088/1757-899X/924/1/011001

Prof. Brendan Kennedy	President of AONSA, ANU, Australia
Prof. Dr. Toshiji Kanaya	Director of MLF J-PARC, Japan
Prof. Dr. Masatoshi Arai*	ESS Sweden
Prof. Dr. Takashi Kamiyama	KEK, Japan
Prof. Dr. Takashi Sakuma	Ibaraki University, Japan
Prof. Dr. Shahidan Radiman	UKM, Malaysia
Prof. Dr. Stefan Adams	NUS, Singapore
Dr. Jamie Schultz	Director of ANSC ANSTO, Australia
Dr. Maxeem Avdev	ANSTO, Australia

IOP Conf. Series: Materials Science and Engineering **924** (2020) 011001 doi:10.1088/1757-899X/924/1/011001

ORGANIZING COMMITTEE

Chair :

Prof. Dr. Rudy Harjanto, M.Sn.

Co Chairs :

Dr. Bambang Winarso, M.Sc.

Secretary :

Setya Ambar Pertiwi, SE., MA.

Yualina Riastuti,

Shinta Widyaningrum, S.Sos.

Treasury :

Noer'aida

Coordinator of Conference & Worksho :

Dr. Sudaryanto, M.Eng

Coordinator of Exhibition and Poster :

Yustinus Purwamargapratala, ST., M.Si.

Coordinator for Publication :

Rida Nurul Shelni Rofika, M.Si.,

Coordinator for Accommodation, Transportation and Tour :

Enggay Sugati, S.E., Agus Sujatno, Amd

Coordinator of Award/Sponsorhip :

Leli Deswindi, SE., M.T.

IOP Conf. Series: Materials Science and Engineering **924** (2020) 011001 doi:10.1088/1757-899X/924/1/011001

Adithya Kusuma Whardana, S.Kom., M.Kom.

Anggar Riskinanto, ST. MTI.

Prof. Dr. Ir. Anne Zulfia, MSc

Coordinator of Public Relations & Press Conference :

Dr. Maryono Basuki,M.Si, Hari Purnomo

Coordination for Collaboration:

Dewi Koraq

Webmaster :

Dr. Sudaryanto, M.Eng

Iman Sholihin

IOP Conf. Series: Materials Science and Engineering **924** (2020) 011001 doi:10.1088/1757-899X/924/1/011001

SCIENTIFIC COMMITTEE

Chair of the International Scientific Committee

Prof. Dr. rer. nat. Evvy Kartini
Center for Science and Technology of Advanced Materials
National Nuclear Energy Agency Indonesia
evvy.kartini@gmail.com

Members of the International Scientific Committee

Prof. Soo Wohn LEE
Research Center for Eco Multi-Functional Nanomaterials Global Research Laboratory
Soo Moon Unviersity, South Korea

Prof. Santi Maensiri
School of Physics, Institute of Science, Suranaree University of Technology 111 University Ave, Muang District, Nakhon Ratchsima, 30000 Thailand

Prof. Taku Sato
Institute for Solid State Physics
Tohoku University, Japan

Dr. Julwan Hendry Purba, ST., M.App.IT.
Center for Nuclear Science and Engineering Technology
National Nuclear Energy Agency Indonesia
julwan@batan.go.id

Ir. Chairul Hudaya, ST, M.Eng., Ph.D.
Department of Electrical Engineering
University of Indonesia
chairul.hudaya@gmail.com

Prof. Dr. Rudy Harjanto, M.Sn.
Communication Science Department
Universitas Prof. Dr. Moestopo (Beragama)
abrudyharjanto@gmail.com

Prof. Dr. H. Wibowo, M.Phil.
Economics and Business Department
Universitas Prof. Dr. Moestopo (Beragama)
wibowo303@yahoo.co.id

IOP Conf. Series: Materials Science and Engineering **924** (2020) 011001 doi:10.1088/1757-899X/924/1/011001

Komm Pechinthorn
International College
Rajamangala University of Technology Krungthep
komm.p@mail.rmutk.ac.th

Setya Ambar Pertiwi, SE, MA
International Relations Department
Universitas Prof. Dr. Moestopo (Beragama)
ambar.pertiwi@gmail.com

Assoc. Prof. Dr. M. Syafii Anwar (REVIEWER)
Department of International Relations Study Program
President University
syafii.anwar@gmail.com

Dr. Michael Adhi MBA (REVIEWER)
Post Graduate Studies
Universitas Prof. Dr. Moestopo (Beragama)
miki.nugroho@gmail.com

Assoc. Prof. Dr. Muhammad Muda
School of International Studies
Universiti Utara Malaysia
mmuda@uum.edu.my

Editors of the International Scientific Committee

Rida Nurul Shelni Rofika, M.Si.
Physics Department
Bandung Institute of Technology
rida.elsc@gmail.com

Shinta Widyaningrum, S.Sos.
International Relation Department
Universitas Prof. Dr. Moestopo (Beragama)
shintawi05@gmail.com

Dr. Kartika Sari, S.Si., M.Si.
Department of Physics, Faculty of Mathematics and Natural Sciences,
Universitas Jenderal Soedirman, Purwokerto 53123
kartika.sari@unsoed.ac.id

IOP Conf. Series: Materials Science and Engineering **924** (2020) 011002 doi:10.1088/1757-899X/924/1/011002

Peer review declaration

All conference organisers/editors are required to declare details about their peer review. Therefore, please provide the following information:

- **Type of peer review: Double-blind**

- **Conference submission management system:**
  First, participants had to submit their manuscripts to *Easychair*.
  Second, the manuscripts were reviewed and selected for presentation in the conference.
  Third, the manuscripts were checked for plagiarism by *Turnitin*.
  Fourth, for those manuscripts that have similarity less than 20% can be further reviewed by our referees.
  Fifth, some of those manuscripts that full fill the requirement as international publication are accepted. When necessary, the manuscripts have to be revised according to the referee's comments.
  Sixth, the editors checked all the layout form based on IOP format.
  Finally, proof reading of the manuscript were sent to the authors for approval, including the acceptance letter.
- **Number of submissions received:**
  130
- **Number of submissions sent for review:**
  87
- **Number of submissions accepted:**
  38
- **Acceptance Rate (Number of Submissions Accepted / Number of Submissions Received X 100):**
  29.23%
- **Average number of reviews per paper:**
  4
- **Total number of reviewers involved:**
  13
- **Any additional info on review process:**
  The manuscripts have been reviewed following the international journal procedures.
- **Contact person for queries:**
  Evvy.kartini@gmail.com

Please submit this form along with the rest of your files on the submission date written in your publishing agreement.

The information you provide will be published as part of your proceedings.

Aluminum Alloy Piston Fabrication Technology by Super Forging-KIRIU

Lydia Anggraini* and Adelya Anjany
Study Program of Mechanical Engineering, President University
Jl. Ki Hajar Dewantara, Jababeka Education Park, Cikarang, Bekasi 17550, Indonesia

*lydia.anggra@president.ac.id

Abstract. This research aims to produce pistons a lighter and stronger piston, even in high compression and high engine speed conditions. The piston fabrication process is carried out by super forging using Aluminum as a raw material with a compound choice by Silicon or Magnesium. In addition, the outer layer of the piston is covered by molybdenum with the coating process as nanotechnology to prevent corrosion. Furthermore, the finishing is controlled by the programming system machine named CNC (Computerized Numerical Control). In addition to CNC, the automation system is carried out by the KIRIU machine which is also used in this study. The results of this study show that the use of Al-Si alloy base material has better piston quality and able to withstand high compression or high engine speed. Then, the Al-Si superforging piston compared to the two final automation processes of CNC and KIRIU. The evaluation results have proved that the KIRIU machine provides high accuracy and the actual production increases by 20% compared to CNC.

Keywords **:** Piston, Casting, Superforging, CNC, KIRIU

1. Introduction

A piston is a main moving component contained in a cylinder. In a machine, the function is to transmit force generated from expansion of gas in the cylinder to the crankshaft through a connecting rod. In order to achieve a high-performance engine, power per step at higher speeds is necessary to withstand peak mechanical pressure.

Nowadays, the Al-Mg casting piston is very common to use in many industries which has the main problem of being susceptible to damage on high-compression and high-speed engines. The impact encountered was the high-cost maintenance [1]. If compared with the Al-Si piston forging which is lighter and stronger, it can be used in high-compression and high-speed engines. In this study, the molybdenum coating was examined to used to prevent the corrosion rate [1–4].

In the manufacturing of pistons, to produce cutting results in details has not been able to obtain by only CNC machines [5]. Therefore, to overcome this problem, KIRIU for an automated finishing system was chosen. The automation system by the KIRIU machining engine has a high degree of accuracy and can increase actual production [6]. The Al-Si piston forging is more efficient than the Al-Mg casting piston [7–9]. Efficiency is very important with respect to the costs incurred for maintenance, which will have a large influence on the graph of income budget production [10–13].

The aim of this study is to compare the fabrication technology of Al-alloy piston by casting-CNC and super forging-KIRIU. Through the experiment of the ideal piston speed, compression ratio, piston strength due to the influence of heat, and the volume of wear resistance are determined based on the experimental procedure. The difference results of piston finishing by both automation cutting systems are described in detail in this paper.

2. Experimental Procedure

The manufacturing of piston by casting process is started with foundry. Foundry is located on ingots of heated aluminum in the furnace. The molten aluminum is poured in a hydraulic mold then scooped up with a ladle from a crucible then allowed to cool. The process continued by dehorning, hardening and machining, in which the pistons are dehorned by a vertical milling machine, then hardened [6]. Furthermore, by a lathe machine, the rough edges are cut and the smooth profiles are obtained by means of a turning operation. In the casting process of piston, the CNC machining is used. CNC is the computer program to cut the ring grooves and create an accurate bore diameter by finishing the crown and making oil slots and pin bore. In addition, the grinding process is carried out as a final inspection to complete the size worked on the piston in the final stage.

The forging process begins from the solid aluminium rod type 2618. The slug can be obtained by a saw after cutting the rod into smaller pieces after heated up to 150°C to 200°C. The punch is purposed to heat up to 426°C and applies 2000 tons of pressure onto the slug, as shown in the Fig. 1. The wrist pinhole and oil control holes are large holes which obtained by drilling through both sides of the piston. The three ring grooves are created by a lathe machine. In the forging process, the milling machines are able to shave up to a couple of centimeters off of each side of the piston where the large holes are drilled for the wrist pin insertion. This is to reduce the overall weight of the piston, the step closer to its final form. As the type of press technique, super-forging pistons are forged using a backward extrusion. Backward extrusion is the process where the material from the slug flows back and around the descending punch become cup shape.

Next, the grinding process is carried out to give the crown the shape required according to the design. Then the last process is machining by KIRIU techno automation system which has high accuracy and increases actual production as the job finishes. KIRIU machine has higher precision compared with CNC. Detail parts up to millimicrons size are able to be shaved and increase the production to 20%. Steps to produce a part by KIRIU machine is start from reading drawing, programming, inputing program, and finalized with machining process. By input a programming language, the parameters determine the coordinates of the prefix and adjust the movement through coding in accordance with the KIRIU machine. Coordinate settings are the most important thing in this method. Import data from the part design has done by CAM (Computer Aided Manufacturing) program, continue with data import to the CAM program connected to the KIRIU machine. The final step in this process is applying the program to the every part which readed by KIRIU's machine, then continues to realize the raw material with the machining process. Figure 2 shows the KIRIU machine.

Figure 1. Slugs after cutting.

Figure 2. KIRIU Machine setup.

3. Results and Discussion

Piston speed is the relative speed of the piston against the cylinder wall, which if it exceeds the threshold. In order to prevent the damage in the engine, the amount of piston speed limitation must be considered. Therefore, the piston speed limit is adjusted to the type of engine and the purpose of its use. When the rotational engine speed reaches 15000 rpm, the other subsystem percentages are also gain such as crankshaft of 25.1%, the transmission of 7.9 %, pumps of 11.4 %, crankcase pumping of 8.5 %, pistons of 33.9 % and valve trains of 13.2 %. The ideal piston speed and friction loss is shown in the Table 1 and Figure 3.

Table 1. Typical Mean Piston Speed.

Engine Type	v_{mps} *@max power*
Small industrial engine	7 m/s
Marine & generator	8.5 m/s
Train & truck	11 m/s
Automobile (low speed)	14 m/s
Automobile (typical)	16 m/s
Automobile (hi-performance)	18 m/s
Racing (endurance)	22 m/s
Racing (short race)	25 m/s
Drag racing	30 m/s

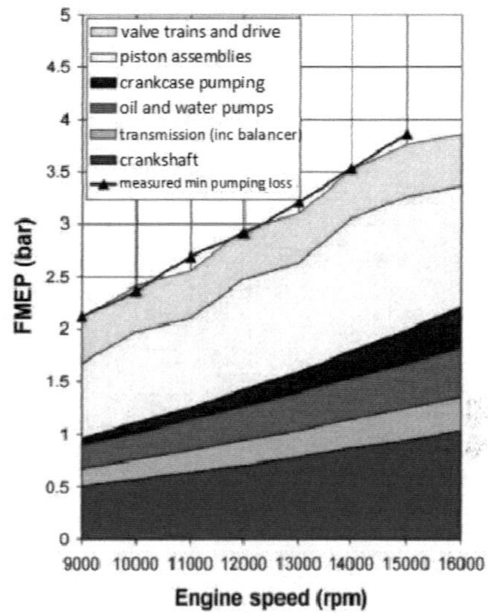

Figure 3. Friction loss of the piston [14].

The ideal piston speed can be calculated by the following equation:

$$\frac{2 \times stroke \times rpm}{60} \quad (1)$$

where 2 is the up and down motion of the piston when the engine rotates once a full rotation, stroke is the step of the piston (m), and rpm is the safe limit of the engine speed.

One of the causes of power loss is the piston friction and the ring with the cylinder (piston assembly), as the first thing that must be considered. This loss is caused by friction that occurs, which is affected by the amount of combustion chamber pressure and piston speed. The sudden increase in piston speed can increase the coefficient of friction between the ring and piston with the cylinder wall simultaneously, as shown in Figure 4.

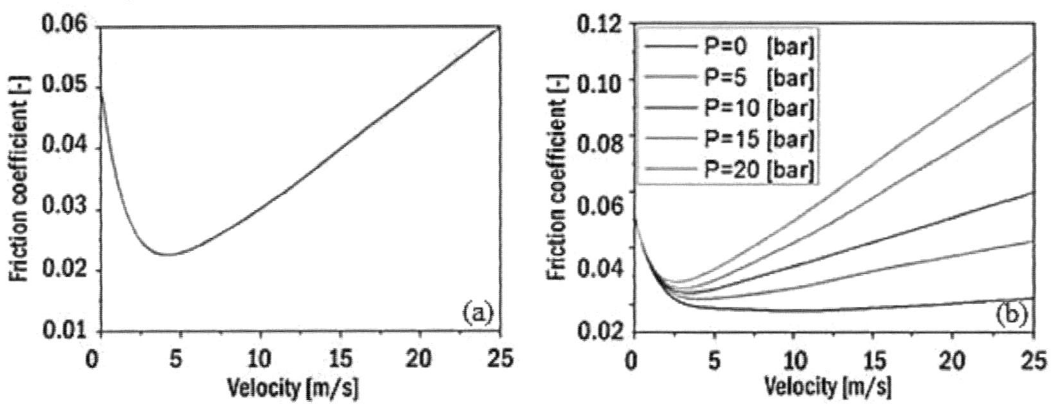

Figure 4. Coefficient of friction between the piston (a) and ring (b) with the cylinder wall.

Figure 5 shows the graph of piston-cylinder power loss on the velocity. The graph explains that the increase in combustion chamber pressure will also increase the pressure of the piston ring to the cylinder wall. Therefore, the coefficient of friction is increasing. As a result of the friction, the engine has a power loss, then it affects the engine room which will gain the temperature and become hot.

If the piston heat continues to increase with increasing piston speed (comparable to the engine speed/rpm), the overheating will occur. Furthermore, this condition to be unable cooled by oil, wind (engine fins) or even a radiator. The effect happens are as follows: the piston ring will expand, the gap with the cylinder wall will be narrower, the coefficient of friction rises and the piston heat is higher followed increased cylinder heat.

In the several engine cycles condition, the piston ring expands in such a way that the distance gap with the cylinder wall is 0, which means very tight and deformed. The piston is stuck and the piston handlebar is broken. If the fracture strikes the cylinder, the cylinder will lose or break. This worst possibility is the objective of the piston manufacturer, avoiding this condition is by giving a limiter at its rpm.

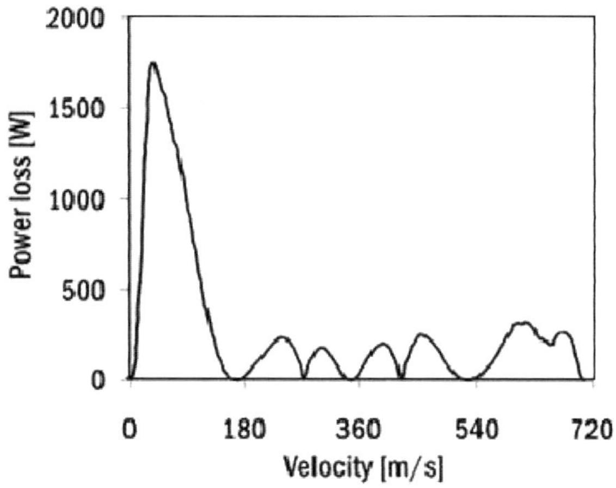

Figure 5. Graph of piston-cylinder power loss.

In addition, it increases the combustion and the piston ring pressures to the cylinder wall, so that the coefficient of friction increase is another thing to note too. As a result of the work of this machine is that it can suffer losses. Therefore, because of the loss of power produced, it can turn hot. Fig. 6 shows the graph of the relationship between piston temperature and engine speed.

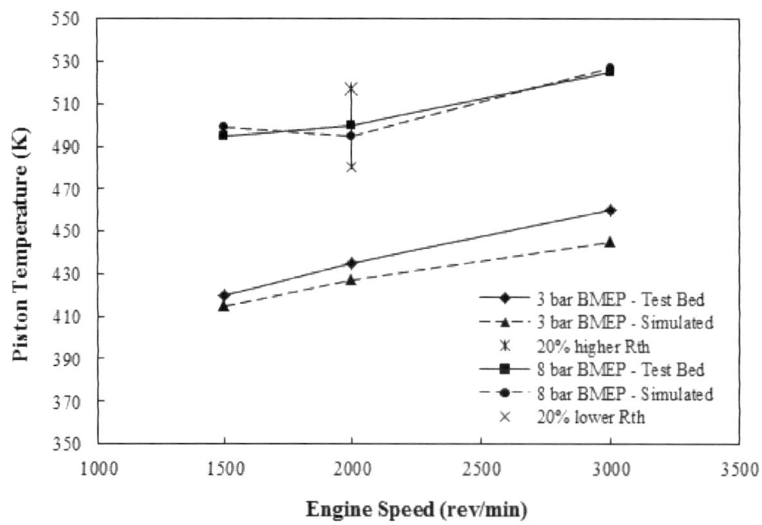

Figure 6. Graph of piston temperature and engine speed.

Besides the coefficient of friction, temperature and engine speed, the compression ratio should also be considered. The following equation is to obtain the compression ratio:

$$CR = \frac{V_1}{V_2} + 1 \tag{2}$$

where *CR* is the compression ratio, V_1 is the engine capacity or volume (cc) and V_2 is the combustion chamber volume (burette – the volume of the spark plug hole). Figure 7 shows the illustration of the combustion chamber with schematic changing of compression ratio. The physical differences between casting and super forging piston are shown in Figures 8 and 9, respectively.

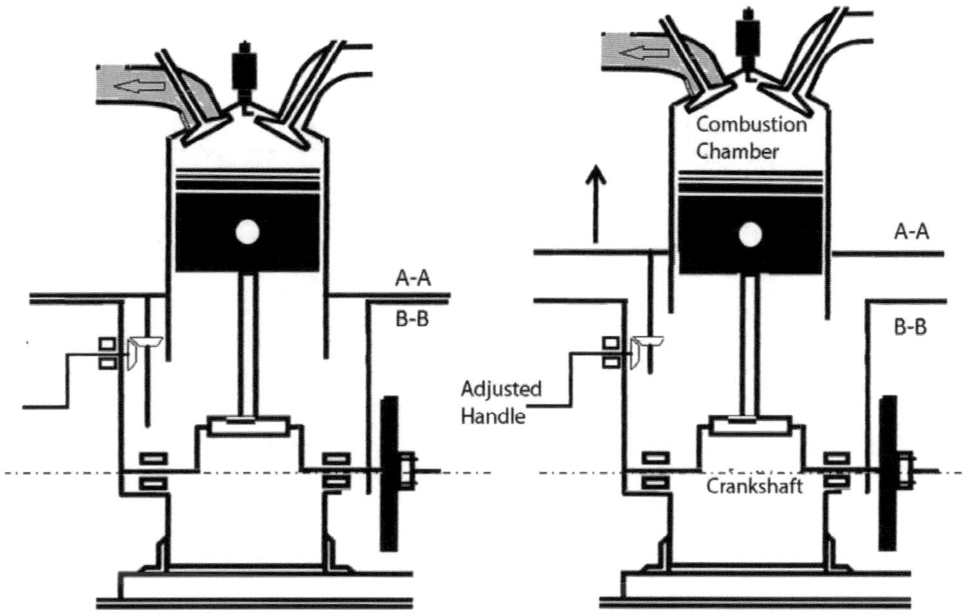

Figure 7. Illustration of a combustion chamber with schematic changing of compression ratio [15].

Figure 8. Casting Piston.

Figure 9. Superforging Piston.

The piston manufacturing process with the casting method is usually used for low engine speed (RPM) and pistons at affordable prices are easily obtained. This type of piston has poor heat resistance which can be realized. In addition, the physical shape of this type of piston is also thick and heavy. Therefore, this type of piston can increase the compression rate. To overcome this problem, a piston with an advanced manufacturing process was developed. The manufacturing process is through the more expensive super forging method, it is mostly used for high engine speed (RPM). Therefore, through the physical form of a lightweight piston, the compression level can be increased perfectly. By this method, the production successfully increases by 20%. Furthermore, this piston is not easy to expand on the cylinder wall of the combustion chamber. Table 2 and 3 show, the advantages and disadvantages of both manufacturing processes of casting - super forging pistons and the difference of the automation system by KIRIU compared to CNC are as follows.

Table 2. Comparison of casting and forging process.

Advantages		**Disadvantages**	
Casting	Forging	Casting	Forging
• Affordable price Rp. 100.000,- • Generally used for low speed	• Mostly used for high speed or high-speed engines • Resistance of heating higher compared to the casting process • Increases the compression rate • Lightweight 79.6 gr	• Rejecting (metallurgy's environmental effects) • Easy to expand on cylinder wall • Bad heat resistance • Thick and heavy 100.7 gr • Can't increase the compression rate	• Expensive Rp. 450.000,- • Tool path for processing is easy to break • Actual production is not maximum because the manufacturing process is quite complicated and requires a long time

Table 3. The difference of finishing the automation system by KIRIU compared to CNC machining.

KIRIU	**CNC**
• Only a few people are able to operate the KIRIU machine program in each company.	• Many people are able to operate the CNC machine with basic training and skills.
• Higher accuracy than CNC, able to shave a very detailed part into μm.	• Less accuracy compared to KIRIU.
• Able to increase an actual production by 20% with a constant production.	• Production of piston cannot exceed the KIRIU. The number of pistons produced is less than the KIRIU machining.

4. Conclusion

The manufacturing piston has been successfully produced by Si-Mg based Al. The quality of piston in which raw material made by Al-Si is higher compared with Al-Mg alloy. The durability of the piston is increased with super forging and lightweight piston can be obtained. By KIRIU, high accuracy piston can be achieved better than the CNC machining process. Furthermore, by the KIRIU machining process, a more detailed part up to millimicron size can be shaved and the production increases by 20%.

Acknowledgment
The authors gratefully acknowledge PT. Federal Izumi Manufacturing (FIM) and the riders who related to insightful discussions regarding this manuscript.

References

[1] D. Pierce, A. Haynes, J. Hughes, R. Graves, P. Maziasz, G. Muralidharan, A. Shyam, B. Wang, R. England, C. Daniel, "High Temperature Materials for Heavy Duty Diesel Engines: Historical and Future Trends," *Progress in Materials Science*, Vol. 103, pp. 109-179, 2019.

[2] G. Kozalka, A. Suchhecki, "Analysis of Design Parameters of Pistons and Piston Rings of a Combustion Engine," *MATEC Web of Conference*, Vol. 118, pp. 5, 2017.

[3] L. Anggraini, S. Sugeng, "Analysis of Porosity Defects in Aluminum as Part Handle Motor Vehicle Lever Processed by High-pressure Die Casting," In *IOP Conference Series: Materials Science and Engineering*, Vol. 367, No. 1, pp. 012039, 2018.

[4] E. Fracchia, S. Lombardo, M. Rosso, "Case Study of a Functionally Graded Aluminum Part," *Applied Sciences*, Vol. 8, No. 7, pp. 1113, 2018.

[5] A. Susilawati, N. Atmadioa, H. Siswantoa, "Tool Path Optimization and Cost Analysis for Manufacturing Process of Master Cylinder Piston of Motorcycle Brake," *Journal of Ocean, Mechanical and Aerospace - Science and Engineering*, Vol. 55, 2018.

[6] T. Okamura, "Effect of Material and Dimensional Homogeneity on Thermo-mechanical Deformation of Brake Discs during High-speed Braking," *SAE International Journal of Commercial Vehicles*, Vol. 8, No. 2015-01-2673, pp. 293-301, 2015.

[7] G.A. Sweet, B.S. Amirkhiz, B.W. Williams, A. Taylor, R.L. Hexemer, I.W. Donaldson, D.P. Bishop, "Microstructural Evolution of a Forged 2XXX Series Aluminum Powder Metallurgy Alloy," *Materials Characterization*, Vol. 151, pp. 342-350, 2019.

[8] B.A. Behrens, A. Chugreev, M. Kazhai, D. Yarcu, C. Büdenbender, R. Relge, "Fabrication of Piston Pins made of a Novel Aluminium-alloyed UHC Steel," *The International Journal of Advanced Manufacturing Technology*, Vol. 102, No. 9-12, pp. 3781-3789, 2019.

[9] D. Hashiguchi, D. Tricker, A. Tarrant, "Mechanically Alloyed Aluminum Metal Matrix Composites." *Material Technologies and Applications to Optics, Structures, Components, and Sub-Systems III*, Vol. 10372, pp. 1037203, 2017.

[10] F. Ma, C. Zhao, F. Zhang, Z. Zhao, Z. Zhang, Z. Xie, H. Wang, "An Experimental Investigation on the Combustion and Heat Release Characteristics of an Opposed-Piston Folded-Cranktrain Diesel Engine," *Energies*, Vol. 8, pp. 6365-6381, 2018.

[11] S. Jog, K. Anthony, M. Bhoinkar, K. Kadam, M.M. Patil, "Modelling and Analysis of IC Engine Piston with Composite Material ($AlSi_{17}Cu_5MgNi$)," In *International Conference on Reliability, Risk Maintenance and Engineering Management*, Springer, Singapore, pp. 159-169, 2019.

[12] V.K Rastogi, "Static Structural Analysis of a Forged Aluminum High-Performance Piston," *International Journal of Engineering Trends and Technology*, Vol. 47(6), pp. 356-360, 2017.

[13] A. Bedotti, M. Pastori, A. Lettini, P. Casoli, "Condition Monitoring Based on Thermodynamic Efficiency Method for an Axial Piston Pump," In *BATH/ASME 2018 Symposium on Fluid Power and Motion Control*, American Society of Mechanical Engineers Digital Collection, 2018.

[14] B. Jia, R. Mikalsen, A. Smallbone, A.P. Roskilly, "A Study and Comparison of Frictional Losses in Free-piston Engine and Crankshaft Engines." *Applied Thermal Engineering*, Vol. 140, pp.217-224, 2018.

[15] D. Feng, H. Wei, M. Pan, "Comparative Study on Combined Effects of Cooled EGR with Intake Boosting and Variable Compression Ratios on Combustion and Emissions Improvement in a SI Engine." *Applied Thermal Engineering*, Vol. 131, pp.192-200, 2018.

IOP Conf. Series: Materials Science and Engineering **924** (2020) 012002 doi:10.1088/1757-899X/924/1/012002

The Effect of Amorphous Polysilazane and Infiltration Method on The Densification of SiC_f/SiC Composites

Alvin Muhammad Habieb[1], Alfian Noviyanto[1,2,*], Galuh Sukmarani[1], Retno Kusumaningrum[1], Fiqhi Fauzi[1], Wahyu Bambang Widayatno[3], Agus Sukarto Wismogroho[3], Muhamad Ikhlasul Amal[4], and Nurul Taufiqu Rochman[4]

[1]Nano Center Indonesia, Jl. PUSPIPTEK, South Tangerang, Banten 15314 Indonesia
[2]Department of Mechanical Engineering, Mercu Buana University, Jl. Meruya Selatan, Kebun Jeruk, Jakarta 11650, Indonesia
[3]Research Center for Physics, Indonesian Institute of Sciences, PUSPIPTEK, South Tangerang, Banten 15314, Indonesia
[4]Research Center for Metallurgy and Materials, Indonesian Institute of Sciences, PUSPIPTEK, South Tangerang, Banten 15314, Indonesia

*a.noviyanto@nano.or.id

Abstract. In this study, we investigated the ability of polysilazane as starting materials to fabricate SiC_f/SiC composite prepared by slurry infiltration method, i.e., vacuum infiltration (VI) and electrophoretic infiltration (EPI). After infiltration, sintering was performed using hot pressing vacuum furnace at 1750 °C for 1 h in Ar atmosphere and applied pressure of 20 MPa. Phase analysis revealed the formation of α-SiC for EPI, while α-SiC and β-SiC were detected for VI. It seems the transformation of β- to α-SiC enhances significantly in the EPI, resulting in single-phase α-SiC. Moreover, the presence of excess carbon after sintering indicated that the amount of Si and C was not stoichiometry. The density of SiC_f/SiC composite infiltrated by VI and EPI was 2.68 and 2.92 g/cm^3, respectively, corresponding to 81.6 and 90.9% of theoretical density. Although the density is quite different, the average flexural strength for VI and EPI was 174 ± 5 and 200 ± 7 MPa, respectively, which does not differ significantly. However, the tail extension was observed for both samples, indicates the toughness increase compared to monolithic ceramic. This study indicates that polysilazane can be used as an alternative precursor for the fabrication of SiC_f/SiC composites using a slurry infiltration method.

***Keywords*:** Polysilazane, SiC_f/SiC composite, Electrophoretic infiltration, Vacuum infiltration

1. Introduction

Ceramic matrix composites (CMC), such as SiC_f/SiC composites, have been developed to overcome the catastrophic fracture behavior of monolithic SiC due to a weak interface between fiber and the matrix that improves the toughness of the composites. Moreover, SiC_f/SiC composites have recently considered as a structural component for fusion and advanced fission reactors due to their low induced radioactivity and excellent resistance under neutron irradiation [1]. However, the fabrication of SiC_f/SiC composites is very challenging due to the difficulty of sintering or infiltration methods.

Many technologies used to fabricate SiC_f/SiC composites, including chemical vapor infiltration (CVI) [2], polymer impregnation and pyrolysis (PIP) [3], reaction sintering (RS) [4], slurry infiltration

[5], and the combination of these technologies [6]. In terms of purity and highly crystalline of SiC_f/SiC composites, CVI is the best method that shows no degradation of strength under neutron irradiation. However, this process requires quite a long time and produced high porosity [1]. In terms of simplicity, PIP and RS are an alternative method in addition to the former process. However, the free Si existed in the composites might deteriorate the properties for high-temperature applications. Moreover, non-stoichiometry of SiC_f/SiC composites fabricated by PIP and RS can causes radiation instability under neutron irradiation [7]. In terms of density, the slurry infiltration method is possible to obtain SiC_f/SiC composites with close-to-zero porosity. Therefore, the mechanical properties of this composite, in general, higher compared to other methods. The key factor of this method is the dispersion of SiC powder in the suspension, which infiltrates the SiC_f/SiC composites due to capillary force. Several slurry infiltrations were developed, such as nano-infiltrated transient eutectoid (NITE) [8], vacuum infiltration (VI) [9], and electrophoretic infiltration (EPI) [10].

Recently, monolithic SiC can be fabricated using a polymer precursor such as polycarbosilane and polysilazane [11]. High density monolithic SiC is possible to fabricate without the sintering additives when the polymer precursor is used as starting material [12]. These polymer precursors are also used for the fabrication of SiC_f/SiC composite using the PIP method. To the best of author knowledge, there is no report about the fabrication of SiC_f/SiC composites prepared by slurry infiltration especially using EPI and VI methods which have low cost and easy to produce with the polysilazane as starting materials.

In this research, we investigated the ability of polysilazane as starting materials to fabricate the densification of SiC_f/SiC composite prepared by slurry infiltration method. Two infiltration methods were used in this study; VI and EPI. The density, phase, microstructural, and mechanical properties were evaluated and presented in this study.

2. Experimental method

Polysilazane (KiON Ceraset Polysilazane 20, USA) is used as a matrix phase for SiC_f/SiC composites. The as-received polysilazane, which is in liquid form, was heated at 200 °C for 90 min to obtain a cross-link polysilazane. The cross-linked polysilazane was ground by mortar and pestle to get the powder. The powder was pyrolyzed at 1000 °C for 2 h in Ar atmosphere to achieve the amorphous powder. The pyrolyzed powder added with 10% of sintering additives (Al_2O_3, D_m = 150 nm, >99.9% purity, Baikowski, Japan and Y_2O_3, D_m = 200 nm, >99.9% purity, Acros Organics, USA with mass ratio of 60:40) and mixed with ball milling for 24 h in wet condition using ethanol as medium. Before mixing, 3 wt% of Hypermer (KD1, ICI, UK) as a dispersant and 10 wt% of polyvinyl butyral-98 (Butvar B-98, Solutia, USA) as a binder was added to the suspension to improve the properties of the suspension. The suspension used to infiltrate Tyranno SA3 fabrics (Ube Industries LTD., Japan) that have a woven structure and coated with pyrolytic carbon (PyC) with a thickness of 400 nm. This experiment used two infiltration methods i.e., vacuum infiltration (VI) and electrophoretic infiltration (EPI). For VI, five layers of SiC fabrics were stacked and infiltrated in the special VI apparatus. Details about VI method were further explained by Yonathan et al [13]. The dual system electrode was used in EPI, whereas the distance between fabrics and the electrode was 20 mm, with 10 V as an applied voltage for 30 min. The surface sealing effect during EPI was minimized by using ultrasonicator (HD 2070, Bandelin, Germany) for the first 20 min. After drying the infiltrate fabrics both for VI and EPI, lamination of 15 layers of fabrics was performed under an applied pressure of 10 MPa at 80°C. Binder burn-out was carried out at 350°C for 2 h in air with the heating rate was 0.5°C/min to remove volatile materials in the infiltrate fabrics. SiC_f/SiC composites were sintered in the hot-pressing furnace at 1750°C for 1 h in Ar atmosphere.

The density of SiC_f/SiC composites measured with the Archimedes principle using distilled water as a medium. The scanning electron microscopy (SEM, S-4800, Hitachi, Japan) used to observe and analyze the microstructure of the sintered SiC_f/SiC composites. A three-point bending test was performed using the ultimate testing machine (UTM AG-50E, Shimadzu, Japan). The composites cut

IOP Conf. Series: Materials Science and Engineering **924** (2020) 012002 doi:10.1088/1757-899X/924/1/012002

into 4 × 2 × 20 mm and polished using 1 μm diamond paste to remove the scratch from the samples for bending test purposes.

3. Results and discussion

Figure 1 (a) shows the amorphous polysilazane after pyrolysis, showing the coarse particle size. Milling was performed in order to decrease the particle size of the pyrolyzed powder; hence it could be comparable with the size of sintering additives. Beads mill was performed at 3000 rpm for 1 h, resulting in excellent particle size of amorphous polysilazane powder (Figure 1 (b)). The initial particle size of pyrolyzed powder before and after milling was > 10 μm and < 1 μm, respectively.

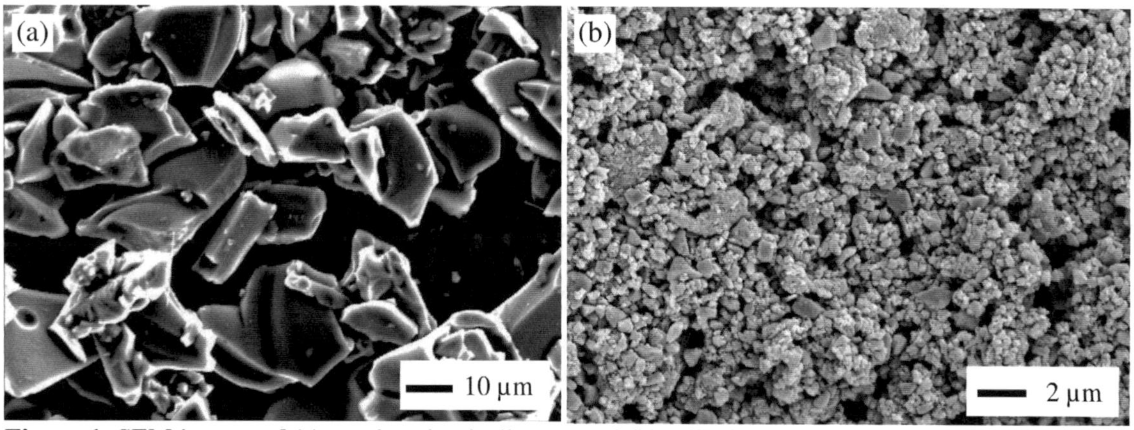

Figure 1. SEM images of (a) pyrolyzed polysilazane and (b) milled pyrolyzed polysilazane. Different magnifications were used for clarity.

Zeta potential of each constituent in the suspension is imperative for the infiltration of SiC fabrics using EPI due to its stabilization and the direction of the particle. Figure 2 shows the zeta potential of the suspension that contains amorphous polysilazane and sintering additives. According to the manufacturer, the polysilazane comprises Si, C, N, and O. Therefore, the zeta potential of Si, C, Al_2O_3, and Y_2O_3 was determined for the electrophoretic infiltration. As shown in Figure 2, positive zeta potential for each powder was obtained at pH < 10, while the isoelectric point for all fine particles was at pH 9.5-9.7. The highest zeta potential was Y_2O_3, i.e., 35 mV, and the lowest was ~ 15 mV. Thus, pH 1 was chosen for the electrophoretic infiltration.

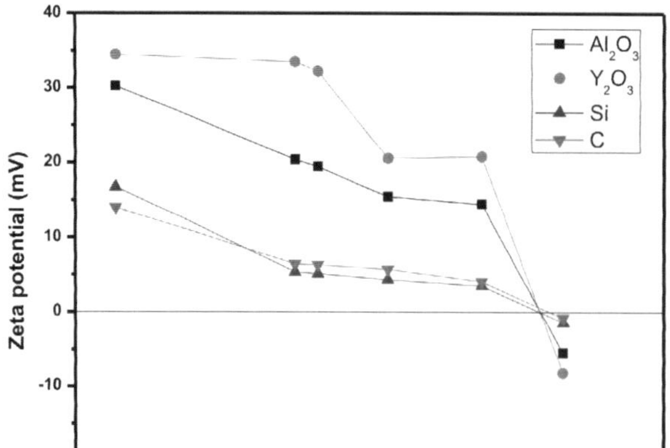

Figure 2. Zeta potential of the suspension containing amorphous polysilazane and sintering additives.

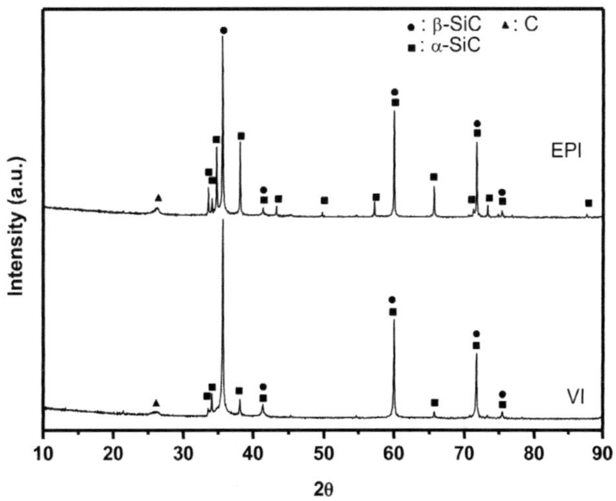

Figure 3. XRD patterns of sintered SiC_f/SiC composites prepared by different infiltration methods.

Figure 3 displays the XRD patterns of SiC_f/SiC composites prepared by vacuum infiltration and electrophoretic infiltration. The results showed the formation of α-SiC in the SiC_f/SiC composite infiltrated by EPI, while the structure of α-SiC along with β-SiC observed for VI samples. Moreover, the presence of excess C was observed both in EPI and VI, indicates that the amount of Si and C in the amorphous polysilazane was not stoichiometry to form SiC. The effect of the infiltration technic influences the crystal structure of SiC. Single-phase α-SiC formed in the SiC_f/SiC composites infiltrated by electrophoretic infiltration, while the presence of β-SiC detected in the vacuum infiltration. Compared to α-SiC, β-SiC is metastable at 1600-1900°C and might transform to α-SiC, then, the phase transformation from β- to α-SiC during sintering at high temperature is universal phenomena. Several factors enhance the change of β- to α-SiC, such as temperature, pressure, sintering additive, and sintering atmosphere [14]. Sintering atmosphere plays a vital role in the transformation of β- to α-SiC. For instance, the N_2 atmosphere inhibits the conversion from β- to α-SiC [15], while the Ar atmosphere increases the transformation. Since the Argon atmosphere used in this study, the β- to α-SiC transformation is inevitable, which is in agreement with the other report [16].

Moreover, the additives may enhance this β- to α-SiC transformation. Another possibility that enhances the transformation β- to α-SiC is the infiltration method itself, in particular for the EPI. In this study, amorphous polysilazane used as a precursor for the SiC instead of SiC powder. Since the amorphous polysilazane contain Si and C, the moving speed of these particles during EPI depends on the zeta potential. Although the zeta potential of those particles is similar, the ability to infiltrate SiC fabrics of those particles would be different during EPI. Therefore, the distribution of Si and C on the SiC fabric is not homogeneous, and lead to the formation of α-SiC without the presence of β-SiC. On the other hand, SiC_f/SiC composite that infiltrated by vacuum infiltration showed the presence of α-SiC and β-SiC, indicates that incomplete transformation of β- to α-SiC. Regardless of the formation of SiC, these results imply that amorphous polysilazane is an alternative as a precursor for SiC.

Table 1. Density, relative density and flexural strength of SiC_f/SiC composite using different infiltration method.

Infiltration Method	Density (g/cm^3)	Relative density (%)
Vacuum infiltration	2.68 ± 0.04	81.6
Electrophoretic infiltration	2.92 ± 0.05	90.9

Table 1 shows the density and relative density of SiC_f/SiC composite infiltrated by VI and EPI was 2.68 and 2.92 g/cm^3, respectively. The EPI process imparts a higher density compared to VI. However, it still lowers if compared with EPI using SiC powder instead of amorphous polysilazane.

IOP Conf. Series: Materials Science and Engineering **924** (2020) 012002 doi:10.1088/1757-899X/924/1/012002

Figure 4 presents the SEM images of SiC_f/SiC composite infiltrated by VI and EPI. Both of the infiltration methods revealed dense microstructure in the matrix area (Figure 4(a) and (c)). Nevertheless, the presence of pores was still observed in the fiber area. It shows that EPI results in smaller pores rather than VI so that it can produce a higher density. This is because the suspension of VI did not go through deep into the fiber area, resulting in many pores (Figure 4 (b)). Since the VI process depends on the properties of suspension, it seems the improvement of suspension is needed to infiltrate the SiC fabric deeply into the fiber area. Although the presence of pores observed in the SiC_f/SiC composite infiltrated by EPI showed the capability of infiltrating the SiC_f deeply (Figure 4 (d)) and showing better relative density than VI. However, the improvement is needed to achieve SiC_f/SiC composite near the theoretical density by EPI [17].

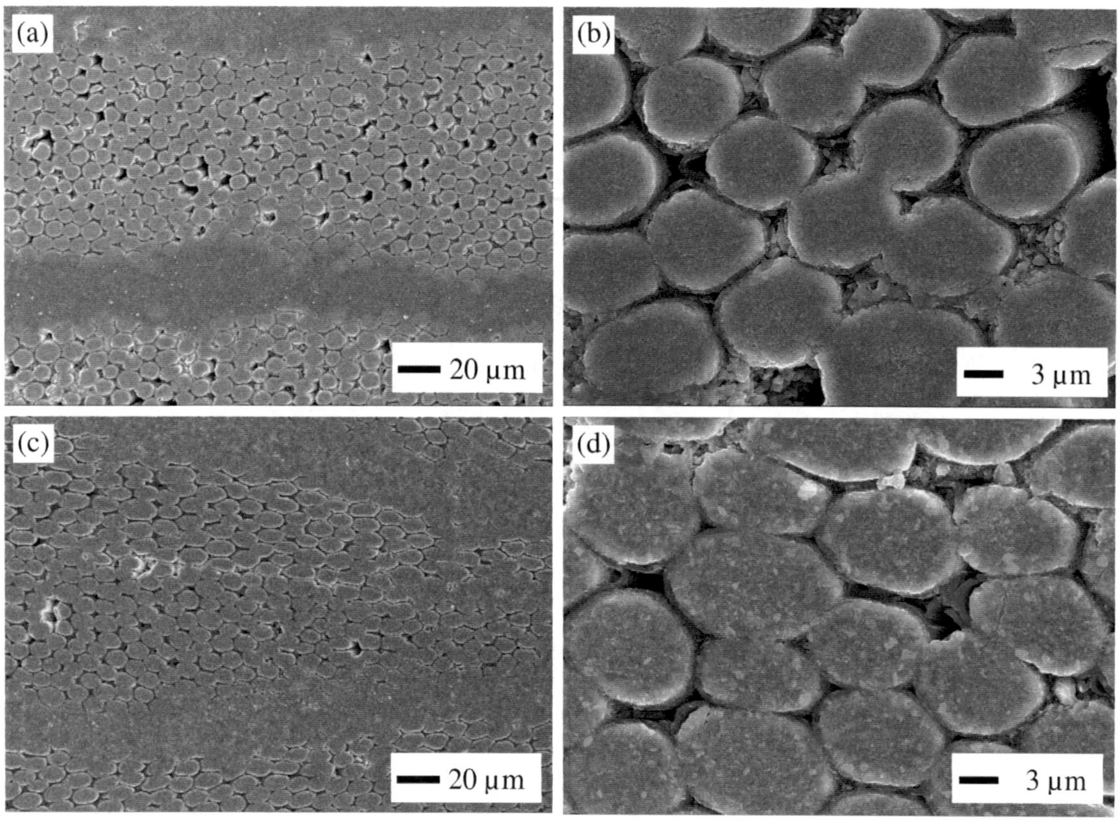

Figure 4. SEM images of sintered SiC_f/SiC composites at low magnification for (a) VI and (c) EPI, and at high magnification in fiber area for (b) VI and (d) EPI.

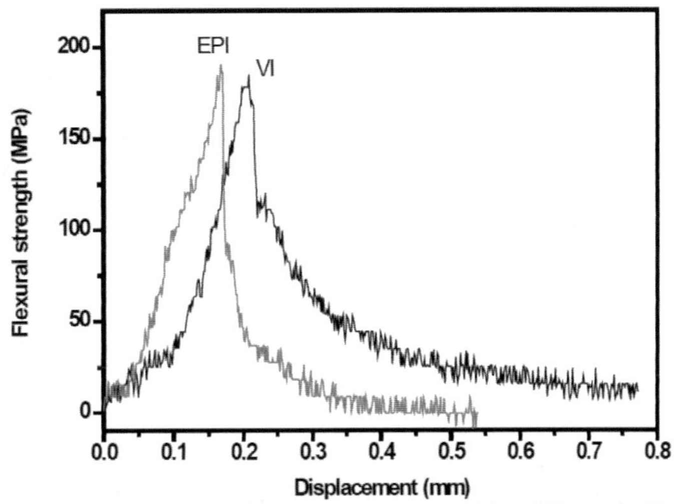

Figure 5. Flexural strength of SiC_f/SiC composites prepared by different infiltration methods.

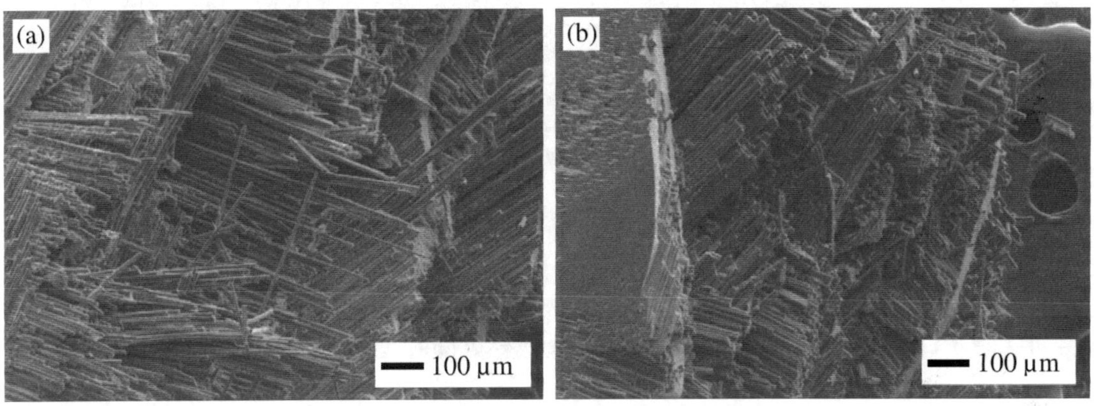

Figure 6. Fractured surface of SiC_f/SiC composites after bending test prepared by (a) vacuum infiltration and (b) electrophoretic infiltration.

Figure 5 shows the flexural strength of SiC_f/SiC composite prepared by different infiltration method. Although the density of SiC_f/SiC composite infiltrated by VI and EPI quite different, the flexural strength of these composites is almost similar. The average flexural strength for VI and EPI was 174 ± 5 and 200 ± 7 MPa, respectively. It should be noted that abrupt load after maximum flexural strength, i.e., typical ceramic, was not occurred in both samples, as shown in Figure 5. Some tail extension, which generally occurs in metal, observed in both methods. This tail extension indicates pseudo-ductility fracture behavior, i.e., nonlinear regions appear clearly after proportional limits and the load will go down gradually after the maximum load [18]. The elongation at VI is more significant than EPI, where it is related to the morphology of the two infiltration methods, as shown in Figure 6. Indeed, the primary mechanism of tail extension of both samples was delamination, however, the degree of delamination in VI (Figure 6 (a)) higher compared to EPI (Figure 6 (b)). Thus, we observed prolonged displacement in the VI compared to EPI, as shown in Figure 5. This can be explained due to the density of VI is lower than EPI, thus the delamination of woven fabric occurs easily. Contrary, dense matrix observed in SiC_f/SiC composites infiltrated by EPI, as shown in Figure 6 (b), with a little delamination. Based on the load-displacement graphics analyzed by a non-brittle fracture activity in the SiC fabric increases the toughness of SiC_f/SiC composites compared to other methods at the other reports [19].

4. Conclusion

SiC_f/SiC composites were successfully fabricated using an amorphous polysilazane as a precursor for SiC, and two infiltration methods were used to infiltrate SiC fabric: vacuum infiltration and electrophoretic infiltration. This study showed the possibility to used polysilazane for fabricating SiC_f/SiC composite with the slurry infiltration method. The dense structure in the matrix area was observed for VI and EPI samples, even though the pores exist in the fiber area. The average flexural strength for VI and EPI did not differ significantly. However, tail extension observed for both samples, showing in the toughness of this composite increase compared to a general monolithic SiC.

Acknowledgment
This research funded by a grant from the National Intensive Research System Innovation 2019 (INSINAS, No. 30/INS-1/PPK/E4/2019), Ministry of Research, Technology and Higher Education, Republic of Indonesia.

References

[1] Y. Katoh and L. L. Snead, "Silicon carbide and its composites for nuclear applications – Historical overview," *Journal of Nuclear Materials*, vol. 526, 2019.
[2] S. Yu, Z. Chen, Y. Wang, R. Luo, and S. Cui, "Effect of fabric structure on the permeability and regeneration ability of porous SiC_f/SiC composite prepared by CVI," *Ceramics International*, vol. 45, no. 9, pp. 11564-11570, 2019.
[3] L.Y. Wang, R.Y. Luo, and G.Y. Cui, "Effect of pyrolysis temperature on the mechanical evolution of SiC_f/SiC composites fabricated by PIP," *Ceramics International*, vol. 46, no. 2, pp. 1297-1306, 2020.
[4] S.P. Lee, H.K. Yoon, J.S. Park, Y. Katoh, A. Kohyama, D. H. Kim, and J.K. Lee, "Reaction sintering process of tyranno SA/SiC composites and their characterization," *Fusion Engineering and Design*, vol. 61–62, pp. 717-722, 2002.
[5] P. Yonathan, J.H. Lee, H.T. Kim, and D.H. Yoon, "Properties of SiC_f/SiC composites fabricated by slurry infiltration and hot pressing," *Mater. Sci. Technol.*, vol. 27, no. 1, pp. 257–263, 2011.
[6] M. Li, D. Yang, H. Wang, and X. Zhou, "Property evolvements in SiC_f/SiC composites fabricated by a combination of PIP and electrophoretic deposition at different pyrolysis temperatures," *Ceramics International*, vol. 45, pp. 15689–15695, 2019.
[7] P. Wang, F. Liu, H. Wang, H. Li, and Y. Gou, "A review of third-generation SiC fibers and SiC_f/SiC composites," *Journal of Materials Science & Technology*, vol. 35, no. 12, pp. 2743-2750, 2019.
[8] K. Shimoda, A. Kohyama, T. Hinoki, "High mechanical performance SiC/SiC composites by NITE process with tailoring of appropriate fabrication temperature to fiber volume fraction", *Compos. Sci. Technol.*, vol. 69, pp. 1623–1628, 2009.
[9] A. S. Sharma, P. Fitriani, and D. H. Yoon, "Microstructure-fracture behavior correlation of toughened SiC_f/SiC composites prepared by vacuum infiltration and hot pressing," *Ceramics International*, vol. 42, no. 7, pp. 8713-8723, 2016.
[10] K. Raju, H.W. Yu, J.Y. Park, and D.H. Yoon, "Fabrication of SiC_f/SiC composites by alternating current electrophoretic deposition (AC-EPD) and hot pressing", *J. Eur.Ceram. Soc*, vol. 35, pp. 503–511, 2015.
[11] D. H. Jeong, A. Septiadi, P. Fitriani, and D. H. Yoon, "Joining of SiC_f/SiC using polycarbosilane and polysilazane preceramic mixtures," *Ceramics Internasional*, vol. 44, no. 9, pp. 10443-10450, 2018.
[12] Y. Lee, J. H. Lee, D. G. Shin, A. Noviyanto, H. M. Lee, T. Nishimura, B. K. Jang, W. T. Kwon, Y. Kim, S. Kim, and Y. H. Han, "Phase transformation on spark plasma sintered dense polycarbosilane-derived SiC without additive," *Scripta Materialia*, vol. 143, pp. 188-190, 2018.
[13] P. Yonathan, J. H. Lee, D. H. Yoon, W. J. Kim, and J. Y. Park, "Improvement of SiC_f/SiC density by slurry infiltration and tape stacking," *Mater. Res. Bull.*, vol. 44, no. 11, pp. 2116–2122, 2009.

IOP Conf. Series: Materials Science and Engineering **924** (2020) 012002 doi:10.1088/1757-899X/924/1/012002

[14] C. Ragaru, M. Lancin, and C. Marhic, "β⇒α Phase transformation in SiC: Link between polytypism, dihedral angle and nucleation mechanism of α twins," *J. Eur. Ceram. Soc.*, 1999.
[15] C. M. Parish, T. Koyanagi, S. Kondo, and Y. Katoh, "Irradiation-induced β to α SiC transformation at low temperature," *Scientific Reports*, vol. 7, no. 1198, pp. 1-9, 2017.
[16] A. Noviyanto, D. H. Yoon, Y. H. Han, and T. Nishimura, "Effect of sintering atmosphere on the grain growth and hardness of SiC/polysilazane ceramic composites," *Advances in Applied Ceramics*, vol. 115, no. 5, pp. 272-275, 2016.
[17] K. Yoshida, "Application of Electrophoretic Deposition for Interfacial Control of High-Performance SiC Fiber-Reinforced SiC Matrix (SiC_f/SiC) Composites," *MAX Phases and Ultra-High Temperature Ceramics for Extreme Environments*, chapters 18, pp. 533-552, 2013.
[18] P. Fitriani, A. S. Sharma, S. Lee, and D. Yoon, "Formation of a Carbon Interphase Layer on SiC Fibers Using Electrophoretic Deposition and Infiltration Methods," *Journal of the Korean Ceramic Society*, vol. 52, no. 4, pp. 284-289, 2015.
[19] M. Li, D. Yang, H. Wang, and X. Zhou, "Fabrication of a 3D4d braided SiC_f/SiC composite via PIP process assisted with an EPD method," *Ceramics Internasional,* vol. 45, pp 11668–11676, 2019.

Densification of SiC from Amorphous Polysilazane with The Addition of β-SiC as a Filler Prepared by Hot Pressing Furnace

Galuh Sukmarani[1], Alfian Noviyanto[1,2,*], Retno Kusumaningrum[1], Alvin Muhammad Habieb[1], Fiqhi Fauzi[1], Wahyu Bambang Widayatno[3], Agus Sukarto Wismogroho[3], Muhamad Ikhlasul Amal[4], and Nurul Taufiqu Rochman[4]

[1] Nano Center Indonesia, Jl. PUSPIPTEK, South Tangerang, Banten 15314 Indonesia
[2] Department of Mechanical Engineering, Mercu Buana University, Jl. Meruya Selatan, Kebun Jeruk, Jakarta 11650, Indonesia
[3] Research Center for Physics, Indonesian Institute of Sciences, PUSPIPTEK, South Tangerang, Banten 15314, Indonesia
[4] Research Center for Metallurgy and Materials, Indonesian Institute of Sciences, PUSPIPTEK, South Tangerang, Banten 15314, Indonesia

*a.noviyanto@nano.or.id

Abstract. Dense silicon carbide (SiC) was successfully sintered from amorphous polysilazane (PSZ) using hot pressing at 1750°C for 1 hour under an applied pressure of 20 MPa in Ar atmosphere. The effect of β-SiC powder as a filler on the density, phase, microstructure and hardness were examined. Al_2O_3 and Y_2O_3 were used as sintering additives through the liquid phase sintering mechanism. The phase analysis showed the formation of SiC after sintering of amorphous PSZ. However, α-SiC was a dominant phase and the amount of α-SiC decreased with the addition of β-SiC powder. The relative density of sintered SiC was obtained in range 99.6 – 99.7% regardless of the addition of β-SiC powder. Sintered SiC from amorphous PSZ revealed significant shrinkage compared to sintered SiC from β-SiC powder, while the minimum shrinkage was achieved by the addition of 70% β-SiC into amorphous PSZ. However, the hardness of sintered SiC did not correlate with the addition of β-SiC powder, with highest hardness of 26.4 GPa, which is SiC from solely β-SiC powder. This result indicates that amorphous PSZ is an alternative precursor to fabricate dense SiC.

***Keywords*:** SiC, amorphous polysilazane, hot pressing, density, shrinkage.

1. Introduction

Silicon carbide (SiC) is a non-oxide ceramic material which has attracted enormous attention due to the excellent thermal, mechanical and chemical properties [1]. This material has been developed in many industrial application such as fusion reactor, semiconductor devices, optoelectronics, and heating elements in solar cell system [1–3]. However, a strong covalent bonding between Si and C leads to this material has a low diffusion which results in a great challenge to sintering SiC in high density [4]. Therefore, it requires a high temperature and pressure or using sintering additives. Dense SiC-based ceramics are commonly prepared by powder sintering, carbothermal reduction [5], chemical vapour deposition [6], and polymer-derived sintering [7]. Because of the remarkable potential for low-temperature processing of both non-oxide and oxide ceramics, the polymer-derived sintering has gathered a great interest in producing high-density SiC-based ceramics [8].

Polysilazane (PSZ) is one promising polymer precursor which results in S/C/N ceramics include SiC. The used PSZ as the polymer-derived precursor for SiC_xN_y has been reported by Obmann, *et al.* [9]. Moreover, Salameh *et al.* also reported the used PSZ to prepare SiAlCN ceramics [10]. Besides, effect milling and sintering atmosphere also have been studied by Noviyanto, *et al.* which showed that high density and purity SiC-based ceramic was obtained by milling and sintering in Ar atmosphere [11], [12]. However, due to large shrinkage of the PSZ, several infiltrations are needed to fabricate SiC_f/SiC composite using polymer impregnated and pyrolysis method [13].

Therefore, the aim of this study is to examine the effect of amorphous PSZ as a precursor to producing high-density SiC-based ceramic. Due to the importance of the SiC phase to improve its mechanical properties, β-SiC powders were used as a filler in various compositions to analyse its effect for SiC properties. The density, hardness, and shrinkage of the sintered samples from amorphous PSZ with/without the addition β-SiC were analysed to explain the influence of phase ratio and PSZ.

2. Experimental method

PSZ (KiON Ceraset Polysilazane 20, USA) and β-SiC powder (> 97.5 % purity, 4620KE, NanoAmor Inc., USA) were used in this study. PSZ was cured and cross-linked at 200°C for 90 minutes using a hot plate. The cross-linked powder was pyrolyzed in a nitrogen atmosphere at 1000°C for 2 h to produce amorphous PSZ powder. This amorphous powder was mixed with β-SiC powder and sintering additives (Al_2O_3 and Y_2O_3) using ball milling for 24 h. SiC ball was used to minimize the contamination during milling. Detail composition and sample name in this study are shown in Table 1. The mixed powder was sintered in argon atmosphere at 1750 °C for 1 h with pressure 20 MPa.

Table 1. Composition of the samples

Sample name	Amorphous PSZ (wt.%)	β-SiC (wt.%)	Sintering additives (wt.%)
100PSZ	90	0	10
30SiC70PSZ	63	27	10
50SiC50PSZ	45	45	10
70SiC30PSZ	27	63	10
100SiC	0	90	10

To evaluate the mechanical properties of the samples, Archimedes principle was used to measure the density while the hardness was measured by The Vickers indentation method using a hardness tester (MVK-H1, Mututoyo, Japan) with a 1 kg load and 10 s loading time. Scanning electron microscopy (SEM: S-4800, Hitachi at 15 kV and 10 μA) and high-resolution transmission electron microscopy (HR-TEM: Tecnai G2F20 S-twin, FEI operated at 200 kV) also were used to analysed the microstructure and particles of the sample. Meanwhile, X-ray diffraction (XRD: X'Pert-PRO MPD, PANalytical, The Netherlands) at room temperature using the Cu Kα line was used to identify phase generated during hot pressing.

3. Results and discussion

Table 2 shows the relative density of the sintered samples. The relative densities of 70SiC30PSZ, 50SiC50PSZ and 30SiC70PSZ were 99.7, 99.8 and 99.7%, respectively. The sintered sample of 50SiC50PSZ indicated the maximum density as high as 100SiC while the amorphous PSZ without the addition of β-SiC powder (100PSZ) showed the lowest density was 99.6%. The sintering condition in this study is quite mild, while the typical sintering condition for SiC using hot pressing furnace is temperature > 1800°C, pressure >25 MPa and sintering time > 1 h with the typical relative density 94 –

99.4% [14]. In this study, although the sintering temperature only 1750 °C with the applied pressure of 20 MPa and only for 1 h, the relative densities achieved in this study was 99.6 – 99.8%. The result reveals that amorphous PSZ may be used as a precursor for SiC, and the addition of β-SiC powder as a filler on amorphous PSZ did not decrease the density of the samples. Since the sintering additives were used in this study, the liquid phase sintering mechanism is convinced to occur during sintering and help the densification of these samples. The liquid can be formed from the melting of sintering additives in eutectic formation, which are Al_2O_3 and Y_2O_3 in this study. Nature high covalent bonding of SiC hampers the diffusion of Si and C, thus sintering SiC without the addition of additives requires very high temperature and applied pressure. For instance, fully dense SiC could only be achieved at 2500 °C with an applied pressure of 5 GPa [15].

Table 2. The density of sintered amorphous PSZ with β-SiC filler

Sample	Density (g/cm^3)	Relative density (%)
100SiC	3.205	99.8
70SiC30PSZ	3.202	99.7
50SiC50PSZ	3.204	99.8
30SiC70PSZ	3.201	99.7
100PSZ	3.197	99.6

Figure 1 shows the XRD pattern of the samples, which indicates the formation of SiC after sintering of amorphous PSZ. The phase analysis showed that there were two phases formed, namely α-SiC and β-SiC. No secondary phases observed in Figure 1, indicating that the high purity of SiC ceramic was obtained in this work. Although the elements in the amorphous PSZ were Si, C, N, and O [16], the formation of Si_3N_4 or SiO_2 or another Si-based material could be inhibited during sintering of these samples. Si_3N_4 is not a stable phase at high temperature and tend to decompose to another phase [12]. Therefore, the formation of SiC more likely from the decomposition of Si_3N_4, which reacts with C to form SiC. The similar condition also applied for the SiO_2. Consequently, single phase SiC was observed, as shown in Figure 1 (a), without the presence of secondary phases. Figure 1 (b) displays the variation of α- and β-SiC content in the samples. Indeed, the α-SiC phase was dominant in 100PSZ sintered sample, while SiC made from β-SiC powder revealed the β-SiC phase in the sintered sample. The formation was caused by high-temperature sintering about 1600 – 1900°C, which generally occurs the transformation of β- to α-SiC phase [17]. The transformation was also influenced by several factors such as sintering additive, sintering atmosphere, precursor, temperature and pressure [17, 18]. Moreover, the formation α-SiC on 100PSZ linked to the high number of entropy on amorphous PSZ, which caused the formation α phase instead β-SiC phase [19]. Furthermore, the phase analysis also reveals that the addition of β-SiC powder on amorphous PSZ leads to declining α-SiC phase on the sintered samples.

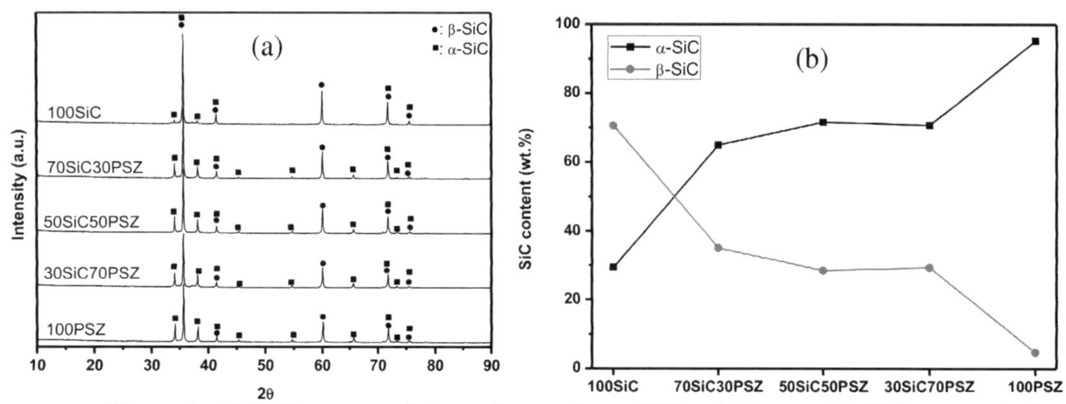

Figure 1. (a) XRD pattern of sintered amorphous PSZ and (b) α- and β-SiC content in the samples.

Figure 2 displays SEM images showing the microstructural of the samples after hot pressing sintering. The results demonstrated that 100SiC and 50SiC50PSZ had a tighter microstructure than 100PSZ, 70SiC30PSZ and 30SiC70PSZ. The result reveals that the addition of β-SiC powder produced denser the sintered samples, although it showed the large grain sizes, which is caused by PSZ as a precursor. The 100SiC indicated the finest microstructure, which is in contrast with 50SiC50PSZ that showed the largest grain sizes. Due to the phase difference in the samples where the high amount of β-SiC caused the smaller grain sizes. This results was consistent with previous work reported by Xu *et al* [17] whereby they explained that the α-SiC as a starting material would result in the larger grain grow rather than β-SiC. Moreover, the microstructure results of these samples also correspond to the hardness and density of the samples. A dense and fine microstructure leads to the sample had a high density and hardness [12].

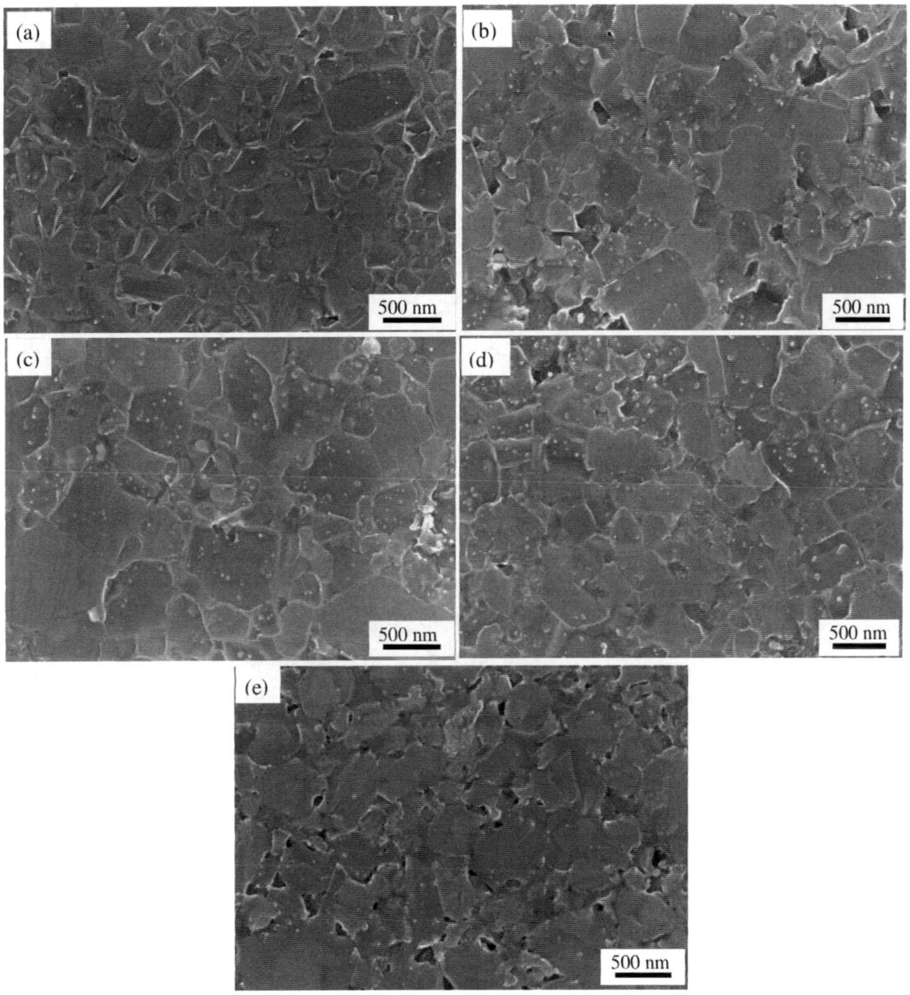

Figure 2. SEM images of sintered (a) 100SiC, (b) 70SiC30PSZ, (c) 50SiC50PSZ, (d) 30SiC70PSZ and (e) 100PSZ samples.

Figure 3 shows the shrinkage and hardness of the samples. The shrinkage was determined based on the sintered samples thickness, which was compared with sintered β-SiC thickness was 2.43 mm. The shrinkage of 70SiC30PSZ, 50SiC50PSZ, 30SiC70PSZ and 100PSZ were 4.9, 9.1, 10.3 and 13.6%, respectively. As a polymer precursor, shrinkage is one of a drawback which could not be prevented. Thus, to minimize the shrinkage, the β-SiC powder was added as a filler on the sintering of SiC from amorphous PSZ as a precursor. According to the results, it is clear that the rising portion of β-SiC powder produced a considerable decline of shrinkage in the sintered samples. The hardness of 70SiC30PSZ, 50SiC50PSZ and 30SiC70PSZ were 23.6, 22.4 and 25.3 GPa, respectively while 100SiC and 100PSZ were 26.4 and 24.5 GPa, respectively. Indeed, the highest hardness obtained by 100SiC due to fine grain size as explained in the previous paragraph, while the lowest hardness in the 50SiC50PSZ sample due to large grain, as shown in Figure 2 (c). As explained in the Hall-Petch equation, the fine grain size proportional with high hardness [20].

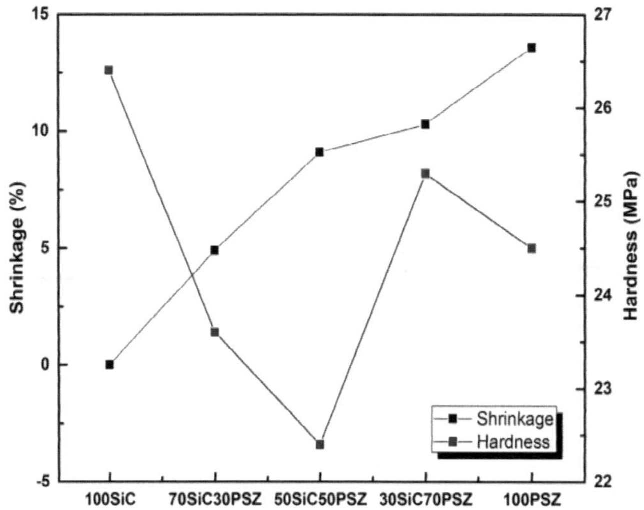

Figure 3. The shrinkage and hardness of the sintered amorphous PSZ with β-SiC as a filler in various composition.

4. Conclusion

Monolithic SiC was prepared by amorphous PSZ precursor and successfully sintered using hot pressing vacuum furnace. Certainly, the lowest shrinkage was obtained by the addition of 70% of β-SiC powder, while the highest shrinkage observes in the sintered SiC without the addition of β-SiC powder. It should be noted that the relative density of sintered SiC for all samples appeared very high, i.e., > 99%, which is near the theoretical density of SiC. Sintering SiC from amorphous PSZ causes the formation of α-SiC with minor β-SiC. This is due to the high number of entropies on amorphous PSZ, which caused the formation α phase instead of β-SiC phase. Finally, the addition of 50% of β-SiC powder into sintered SiC from amorphous PSZ generate large grain in the microstructure compared to sintered SiC without the addition of β-SiC powder. Thus, the lowest hardness was obtained in this sample. This study demonstrates that amorphous PSZ may be used as a precursor for SiC through the liquid phase sintering mechanism by the addition of sintering additives. The addition of β-SiC powder into amorphous PSZ succeed to minimize the shrinkage of amorphous PSZ without hampers the properties of sintered SiC.

Acknowledgment

This research was funded by a grant from National Intensive Research System Innovation 2019 (INSINAS, No. 30/INS-1/PPK/E4/2019), Ministry of Research, Technology and Higher Education, Republic of Indonesia and thank to Research Center for Physics, Indonesian Institute of Sciences for all the facilities which were given to conduct this research.

References

[1] M. Aizat, H. Mohd, M. Mustapha, and J. C. Kurnia, "Silicon carbide- from synthesis to application : a review," *MATEC Web Conf.*, vol. 131, pp. 1–6, 2017.

IOP Conf. Series: Materials Science and Engineering **924** (2020) 012003 doi:10.1088/1757-899X/924/1/012003

[2] H. Ou, Y. Ou, A. Argyraki, S. Schimmel, M. Kaiser, P. Wellmann, and K. Margareta, "Advances in wide bandgap SiC for optoelectronics," *Eur. Phys. J. B*, vol. 87, p. 58, 2014.

[3] M. Syväjärvi, Q. Ma, V. Jokubavicius, A. Galeckas, J. Sun, X. Liu, M. Jansson, P. Wellmann, M. Linnarsson, P. Runde, B. Andre, A. Thøgersen, S. Diplas, P. Almeida, O. Martin, D. Nilsen, A. Yu, and B. G. Svensson, "Solar Energy Materials & Solar Cells Cubic silicon carbide as a potential photovoltaic material," *Sol. Energy Mater. Sol. Cells*, vol. 145, pp. 104–108, 2015.

[4] C. Greskovich and J. H. Rosolowski, "Sintering of Covalent Solids," *J. Am. Ceram. Soc.*, vol. 59, no. l, pp. 336–343, 1976.

[5] J. Xiong and Q. Huang, "Crystallization behavior of polymer derived silicon carbide sintered through microwave heating technique," *J. Wuhan Univ. Technol. Sci. Ed.*, vol. 32, no. 6, pp. 1368–1373, 2017.

[6] H. Abderrazak, E. Selmane, and B. Hadj, "Silicon Carbide : Synthesis and Properties Silicon Carbide : Synthesis and Properties," *Prop. Appl. Silicon Carbide*, pp. 361–388, 2011.

[7] C. Hwang, Q. Yang, S. Xiang, V. Domnich, A. U. Khan, K. Y. Xie, K. J. Hemker, and R. A. Haber, "Journal of the European Ceramic Society Fabrication of dense B 4 C-preceramic polymer derived SiC composite," *J. Eur. Ceram. Soc.*, vol. 39, no. 4, pp. 718–725, 2018.

[8] Y. Li, L. Chen, L. Hong, K. Ran, Y. Zhan, and Q. Chen, "Fabrication of porous silicon carbide ceramics at low temperature using aluminum dihydrogen phosphate as binder," *J. Alloys Compd.*, vol. 785, pp. 838–845, 2019.

[9] R. Obmann, S. Schörpf, C. Gorsche, R. Liska, T. Fey, and T. Konegger, "Porous polysilazane-derived ceramic structures generated through photopolymerization-assisted solidi fi cation templating," *J. Eur. Ceram. Soc.*, vol. 39, no. 4, pp. 838–845, 2018.

[10] F. Salameh, C, Bernard, S, Gervais, C, Babonneau, A. Bruma, S. Malo, and P. Miele, "Aluminum-modified polysilazanes: synthesis, pyrolisis behaviour and microstructural evolution," *J. Eur. Ceram. Soc.*, vol. 39, no. 2–3, pp. 183–194, 2018.

[11] A. Noviyanto and T. Nishimura, "Effect of Milling on the Densification of SiC-Based Composites from Polysilazane," *Makara J.Technol*, vol. 20, no. 3, pp. 109–113, 2016.

[12] A. Noviyanto, D. Yoon, Y. Han, T. Nishimura, D. Yoon, and Y. Han, "Effect of sintering atmosphere on the grain growth and hardness of SiC / polysilazane ceramic composites," *Adv. Appl. Ceram.*, pp. 1–4, 2016.

[13] S. G. Lee, J. Fourcade, R. Latta, and A. A. Solomon, "Polymer impregnation and pyrolysis process development for improving thermal conductivity of SiCp / SiC – PIP matrix fabrication," *Fusion Eng. Des.*, vol. 83, pp. 713–719, 2008.

[14] S.-G. Lee, W.-H. Shim, J.-Y. Kim, Y.-W. Kim, and W. T. Kwon, "Effect of sintering-additive composition on fracture toughness of liquid-phase-sintered SiC ceramics," *J. Mater. Sci. Lett.*, vol. 20, no. 2, pp. 143–146, 2001.

[15] J. S. Nadeau, "Very high pressure hot pressing of silicon carbide," *Am. Ceram. Soc. Bull.*, vol. 52, no. 2, pp. 170–174, 1973.

[16] D. Galusek, S. Reschke, R. Riedel, W. Dreszler, Sajgalk, Z. Lenčéš, and J. Majling, "In-Situ Carbon Content Adjustment in Polysilazane Derived Amorphous SiCN Bulk Ceramics," *J. Eur. Ceram. Soc.*, vol. 19, Aug. 1999.

[17] H. Xu, T. Bhatia, S. A. Deshpande, N. P. Padture, A. L. Ortiz, and F. L. Cumbrera, "Microstructural Evolution in Liquid-Phase-Sintered SiC : Part I , Effect of Starting Powder," *J. Am. Ceram. Soc.*, vol. 84, pp. 1578–1584, 2001.

[18] C. Ragaru, M. Lancin, and C. Marhic, "A Phase Transformation in SiC : Link between Polytypism , Dihedral Angle and Nucleation Mechanism of Twins," *J. Eur. Ceram. Soc.*, vol. 19, pp. 2701–2709, 1999.

[19] K. A. Graeser, J. E. Patterson, J. A. Zeitler, and T. Rades, "The Role of Configurational Entropy in Amorphous Systems," *Pharmaceutics*, vol. 2, no. 2, pp. 224–244, May 2010.

[20] R. Vaßen, A. Kaiser, J. Förster, H. P. Buchkremer, and D. Stöver, "Densification of ultrafine SiC powders," *J. Mater. Sci.*, vol. 31, no. 14, pp. 3623–3637, 1996.

IOP Conf. Series: Materials Science and Engineering **924** (2020) 012004 doi:10.1088/1757-899X/924/1/012004

The Study of Soga Brown Colour Visual Digitization from Classical Batik of Yogyakarta

Almira Firmansyah*
Bina Nusantara University,
Jl. Kyai H. Syahdan No. 9, Kemanggisan, Kec. Palmerah, Jakarta 11480, Indonesia

**afirmansyah@binus.edu*

Abstract. Indonesian traditional textile has distinct signature textile culture on its natural dye, for example Classical Batik of Yogyakarta which is well known for its brown *soga* colour, a brownish colour obtained from the colourants combination of *jambal* bark (*Pelthophorum ferrugineum*), *tingi* bark (*Ceriops candolleana*), and *tegeran* wood (*Cudrania javanensis*). Along with those ingredients, Classical Batik of Yogyakarta also uses indigo plant (*Indigofera tinctoria*) as its blue colourant to help create darker brown colour, and *kapur tohor* (Calcium Oxide) for pre and post mordant treatment. The colour study output would be a recommendation as it would be a non-analog inventory documentation of Classical Batik of Yogyakarta natural dye. In order to make the colour visual data suitable for digital device, the research requires fabrics dyed with Classical Batik of Yogyakarta natural dye ingredients and technique as samples for colour identification study by spectrophotometry and by digitalization method to obtain RGB colour value. The natural dye colour digital format data is designed to facilitate data preservation and its acquisition by individuals whom have interest on textile natural dye.

Keywords : Natural Dye, Colour Identification, RGB Colour Value, *Soga* Brown, Classical Batik of Yogyakarta

1. Introduction

Colouration textiles using natural dye is unique within colour results and the creative process of making the colour, including the traditional Indonesian textiles products which have uniqueness in its textile culture. One of the traditional Indonesian textiles which has unique feature is the Classic Batik of Yogyakarta. In addition to a distinctive feature of the decoration, classical batik also has a unique colour system, which is a dark blue colour (*wulung*), sepia (*dragemsogan*), and the white colour that is easily obtained by naturally using the plant materials [1]. *Soga* dark brown colour is obtained from a combination of natural dyes *jambal* bark (Pelthophorum ferrugineum), the *tingi* bark (Ceriops candolleana) and tegeran wood (Cudrania javanensis). The dark blue colour is obtained from indigo natural dyes (Indigofera tinctoria). *Kapur tohor*, or the oxidize calcium, as a mordant early additive and fixation in the dyeing process.

The knowledge of the traditional textile dyeing tradition need to be preserved as it is part of the cultural richness of Indonesian textiles. Textiles genuine artefact storage as natural colours resources is not reliable for the textile object has a period of obsolescence that cannot be avoided, the physical fabric may be damaged due to the effects of weather and disruption of moths, especially within Indonesian natural conditions which gives unfavourable fabric preservation which as the result the archaeological remains of ancient Indonesia textile is difficult to be obtained [2]. The accuracy of the artefacts data identification can be questioned, because the artefacts colour object will gradually fade because of the results of time, hygiene, weather, and sunlight [3]. The means of information of traditional creativity needs to take advantage of current media and technology, such as mobile application, so that the distribution of information is accessible to the modern society. In digital technology, large amounts of information are compressed into a smaller size and then it will be restored to its original size when data information is required [4]. Information digitalization allows the information transfer that can be done through a variety of media with a clear program, both from traditional media such as print newspapers and magazines to contemporary electronic media like computers. Data compression of colour and colouration into a digital format which to be applied to mobile applications is expected to be an alternative media inventory, as well as facilitating access to information for individuals who have an interest in natural textile dyeing.

2. Experimental method

2.1. Materials

Natural dyes commonly extracted from plants and animals, and the colour of Classical Batik of Yogyakarta is obtained from plants products. The plant tissue contained colour pigment, which produced the colour. Javanese ancestor since hundred years ago use natural dyes due to the natural colour, which looks elegant and durable [5].

The sample object for the colour test identification research will be made from colour dyes and colouration techniques of Classical Batik of Yogyakarta. Method of colouring fabrics in the making of the sample is divided into two kinds, namely colour dyeing with batik technique and colour dyeing without batik technique, with dyeing variations for three, six, and nine times. The raw material used for the sample material colouring are indigo colour dye from *Indigofera tinctoria* plant to produce a blue colour, the combination of colour dye from the *jambal* bark, *tingi* bark, and tegeran wood to produce *soga* brown colour, and front-end *mordant* in a form of *tohor* calcium. Total of the samples are 30 pieces natural dyed fabric.

Figure 1. An examples of *jambal* bark (a), *tingi* bark (b) and *tegeran* wood (c).

2.2. Methods

Colour identification is done by using a spectrophotometer to obtain the value of the length of the colour wave, and the process of digitizing the samples to obtain discrete values at each pixel colour digital image with a sample size of 64 pixels x 64 pixels. The results of the identification test is to be standardized by converting it into RGB (Red, Green, Blue) notation, adapted to the needs of digital media designing that uses the RGB colour system [6].

Table 1. Sample A1-A2, D1-D2 and J1-J2 for colour test identification using Classical Batik of Yogyakarta colouration technique.

No.	Sample Code	Fabric Sample	Technique	Colouring Process		Mordant	
				Indigo	*Soga*	Pre	Post
1	A1		Batik	3 times	0	*Kapur tohor*	*Kapur tohor*
	A2		Without batik	3 times	0	*Kapur tohor*	*Kapur tohor*
2	D1		Batik	0	3 times	*Kapur tohor*	*Kapur tohor*
	D2		Without batik	0	3 times	*Kapur tohor*	*Kapur tohor*
3	J1		Batik	6 times	3 times	*Kapur tohor*	*Kapur tohor*
	J2		Without batik	6 times	3 times	*Kapur tohor*	*Kapur tohor*

2.2.1. Spectrophotometer. Spectrophotometry is the use of instruments to determine the concentration of colored species. The instrument used for spectrophotometry is spectrophotometer, which measures the amount of light absorbed or the intensity of color at a given wavelength [7].

Table 2. Spectrophotometry Result for Sample A1-A2, D1-D2 and J1-J2.

Sample Code	WL (nm)	RFL	K/S
A1 – A2	420 nm – 460 nm	15.474 – 29.070	2.309 – 0.865
	700 nm	19.951	1.606 – 0.285
D1 – D2, J1 – J2	700 nm	7.739 – 47.881	5.499 – 0.285

Based on the results of the spectrophotometry scanning, the result of wavelength obtained is adrift in the value range of 400 nm to 700 nm. In the sample using only indigo dye, the maximum colour value with the highest reflective rate is found in the wavelength of 420 nm, 430 nm, 440 nm,

IOP Conf. Series: Materials Science and Engineering **924** (2020) 012004 doi:10.1088/1757-899X/924/1/012004

450 nm, 460 nm, and 700 nm. As within the sample, it was only using *soga* brown dye and samples with double dye obtain the maximum colour value of 700 nm.

2.2.2. Colour digitization. Every digital-capture system is capable of transforming an object into a digital image. Nonetheless the digital image has archival value as a digital representation. Digital image-archives should facilitate scientific documentation and study of works of the object, therefore the objective is to have the capability to estimate quantitative information about the object, not creates a visually equivalent representation for a defined output device [8]. Raw data of the captured image has advantage that the high bit depth allows to correct for exposure level, white balance, sharpness, etc., without significant loss in image quality where digital image supposed to be as exact match as the original captured object [9].

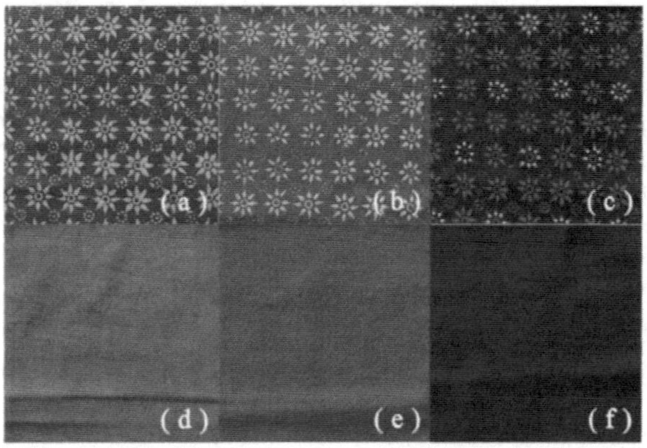

Figure 2. Samples of digitized natural dyed fabric sample image (64 pixel x 64 pixel): sample A1 (a), D1 (b), J1 (c), A2 (d), D2 (e) and J2 (f)

Table 3. An example of RGB colour results conclusion on sample A1 and sample A2

Sample	RGB Value (Red, Green, Blue)			
	AVG	MED	MIN	MAX
A1 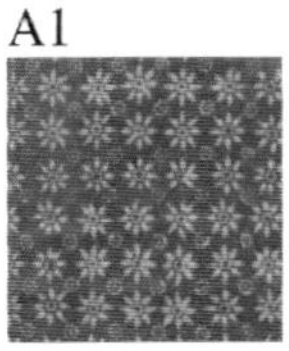	**93.7, 119.13, 135.35** (93.7, 119.13, 135.35) 	**86, 114, 131** (86, 114, 131) 	**7, 51, 80** (7, 51, 80) 	**238, 224, 217** (238, 224, 217) 
A2 	**82, 127, 141** (82, 127, 141) 	**82, 128, 142** (82, 128, 142) 	**46, 78, 93** (46, 78, 93) 	**110, 156, 168** (110, 156, 168) 

IOP Conf. Series: Materials Science and Engineering **924** (2020) 012004 doi:10.1088/1757-899X/924/1/012004

Where as based on the digitization of colour, it is obtained 120 colour results along with its RGB notation which become the recommendation for digital colour library of natural dyes Classical Batik of Yogyakarta. The colour results classified into several groups of keyword for folksonomy purpose of featured menu recommendations, which each keyword group works as electronic tag for the colour data and content. The classified keywords are based on the name of the colour group, the colourant ingredient group, the mordant material group, the name of the traditional textile group, and the region group.

3. Results identification and color standardization

RGB colour model is considered as the base colour space for various applications due to no transformations required displaying information on the screen [10]. Therefore the application of RGB colour model includes computer graphics, image processing, analysis and storage. Colour data, which will be better to be displayed on digital media as content is taken from the colour of the identification results in the process of digitizing the image, because the identification of the resulting colour display quite varied colour range and is considered to represent the original colour of the fabric sample. The conversion of RGB colour data as the result of the conversion colour wavelength value from the identification test using the spectrophotometer are considered less representative of the original colour because there are significant differences between the colours in the RGB colour sample object with wavelength conversion results.

In addition to significant differences in colour, profiling colour identification results by spectrophotometry has limitations on the resulting colour variations based on wavelength value conversion in RGB colour notation. Where as based on the digitization of color, it is obtained 120 color results along with its RGB notation which become the recommendation for digital color library of natural dyes Classical Batik of Yogyakarta.

4. Conclusion

This study is a response to the needs of non-analog inventory documentation for the result of Indonesia natural dyes, particularly the natural dyes used in colouring Classical Batik of Yogyakarta. The study of the colour identification test is done in two ways, namely by spectrophotometry and by digitizing the image. The result of colour identification is then converted into RGB notation as a standard colour notation, which can be applied as digital content. The design of the digital content can be recommended as material for interactive digital color library features.

Acknowledgment

The author thanks the late Mr. Hendri Suprapto from Batik Bixa, Yogyakarta, who provided insight and expertise assistance with *soga* brown natural dye technique for Classical Batik of Yogyakarta that greatly assisted the research.

References

[1] S. D. Kartika, *Archipelago culture: Study of the concept of the mandala and the concept of the triloka / buana on tree life on classical batik,* Bandung: Rekayasa Sains, 2007. *(Indonesian)*
[2] Hasanudin, *Batik pesisiran: Track the influence of the santri trade ethos on batik decoration,* Bandung: PT Kiblat Buku Utama, 2001. *(Indonesian)*
[3] Milner, *The ashford book of dyeing,* Wellington: Brifget Williams Book Limited, 1992.
[4] J. Hartley, *Communication, cultural, & media studies: Konsep kunci* (Kartika Wijayanti, terj.), Yogyakarta: Jalasutra, 2010.
[5] Widyatmono, Nugraha & Ayu, Arik 2014 *Eco-friendly of natural color technique of traditional javanese batik* 10.13140/2.1.3935.2645.
[6] Suyoto W, Dasuki L A, and Prasetya R G 2012 The Algorithm to Detect Color Gradation on Silk *J. IJCSI Intl. Journal.Comp. Sci.* **9**

[7] Nouraldein Mohammed Hamad, Mosab 2018 *Colorimetery & spectrophotometry* 10.13140/RG.2.2.35282.45764.

[8] Berns, Roy 2001 The Science of Digitizing Paintings for Color-Accurate Image Archives: A Review *Journal of Imaging Science and Technology* **45**

[9] Kurecic, Maja & Darko, Agić & Mandić, Lidija 2007 Color management implementation in digital photography *Journal of Information and Organizational Sciences* **31**

[10] Ibraheem, Noor & Hasan, Mokhtar & Khan, Rafiqul Zaman & Mishra, Pramod 2012 Understanding Color Models: A Review *ARPN Journal of Science and Technology* **2**

IOP Conf. Series: Materials Science and Engineering **924** (2020) 012005 doi:10.1088/1757-899X/924/1/012005

Hedonic Test of Kombucha Coffee

Raden Nana Kusdiana*, Vinsensius Ferdi, Indra Kusumawardhana, Farah Levyta
Bina Nusantara University, Jl. K.H. Syahdan No. 9, Palmerah, Jakarta 11480, Indonesia

rkusdiana@binus.edu

Abstract. Kombucha is a fermented drink that has a health benefit. Originally, kombucha made from a combination of sweetened tea that mixed with a symbiotic culture of bacteria and yeast to ferment. Moreover, the making of kombucha starts to vary through an infusion method such as herbs infusion, flowers or fruits, and vegetable mixture, and even to substitute with the coffee ground. Indonesia is the fifth largest consumer of coffee in the world. Therefore, this article aims to study an upcycling of leftover coffee grounds to produce kombucha. This form of fermentation method with inoculum Kombucha tea, which then known as Kombucha coffee (KC). The empiric study was conducted using a quantitative method, where a secondary data was supported with the primary result of questionnaires to eighty panelists. The findings of this research suggested that the aroma and taste of KC widely accept by the respondents. Additionally, KC's aroma, taste, and texture proved more hedonic value than tea kombucha. Although the study has a limitation such as time and the number of respondents. But, the result of the study stimulates a further discussion to investigate the significance of KC towards business and economic benefits.

***Keywords*:** Kombucha, coffee, fermentation, hedonic test

1. Introduction

Consumption of fermented foods and drinks increases to 149% in 2018. This shows developments in the consumption of kombucha, a popular fermented beverage for health-conscious people [1]. Kombucha was originally only made with tea ingredients. But along the way, it was known that any liquid can be the basic ingredient of kombucha as long as it contains sugar. Coffee has a distinctive taste and delicious aroma that differ from other product. Indonesia is the 5th largest coffee consuming country in the world since 282.000 kilogram coffee was consumed throughout Indonesia just in the period of year of 2017 - 2018 [2]. Hence, one of the best ways to make coffee grounds into a high-value product is to turn it into kombucha. Through this upcycling concept, coffee grounds would get a chance to be consumed within the wider community. Consequently, coffee could be fermented with the kombucha culture inoculum. Kombucha culture is not pure culture but is a kind of symbiotic culture between bacteria and yeast. The combination of bacteria and yeast is hereinafter referred to as SCOBY (Symbiotic Culture of Bactery and Yeast) consisting of several bacteria and yeast. The most important bacteria are Acetobacter xylinum and some yeasts that will carry out the fermentation and oxidation processes. The culture will convert sugar into alcohol and produce important substances, including glucuronic acid, acetic acid, lactic acid, vitamins, amino acids and antibiotic substances [3].

The result of the coffee liquid fermentation is called KC, which contains acid content, catechins, alcohol and caffeine. Thus, KC can also be drunk as a healthy drink that follow it predecessor such as KT [4]. Furthermore, based on the empiric study, kombucha based on robusta coffee has the ability to inhibit the growth of Salmonella typhi bacteria [5]. Additionally, other research found that 15% sugar concentration in KC had the highest or most preferred organoleptic value [6].

Therefore, this study aims to determine people's acceptance of KC as a substitute for KT and the level of people's liking between the two such as color, aroma, and taste. The result of the study would benefit to introduce KC to wider market and enhance its economic value.

State of the Art: Based on the research "Chemical and Microbiological Characteristics of Kombucha from Various High Phenol Leaves During Fermentation" conducted by Dwi Wistiana and Elok Zubaidah in 2015, one of the organoleptic analyzes conducted showed that the color of kombucha based on coffee leaves had the most preferred level high compared to kombucha made from tea leaves, bay leaves, soursop leaves, guava leaves, and betel leaves. This is similar to the study of the authors who used coffee grounds as the basis for kombucha [7].

Based on the research "The Effect of Soursop Leaves Kombucha Fermentation Process on Vitamin C Levels" conducted by Irham Falahuddin, Ike Apriani, and Nurfadilah in 2017, vitamin C levels in kombucha are increasing along with the length of fermentation carried out [8].

Based on the research "Antibacterial Effects of Robusta Coffee Fermented with Kombucha on Salmonella typhi" conducted by Romanna Julia Duma Simanjuntak in 2017, kombucha based on robusta coffee has the ability to inhibit the growth of Salmonella typhi bacteria [9].

Based on the research "Effect of Variation Concentration of Sucrose Sugar and Fermentation Time on Kombucha Coffee Making" conducted by Marni Otace Wulan Napitupulu, Setyohadi, and Linda Masniary Lubis in 2015, the more sugar concentrations in kombucha coffee making, the value of the organoleptic taste test will increase as soon as also the longer the fermentation, the value of taste organoleptic test will increase [10].

Based on the research "Antioxidant Activity, Total Phenolic Levels, and Caffeine Levels in Fermented Coffee Kombu Fermentation in Various Sugar Concentrations" conducted by Lusiawati Dewi, Susanti Pudji Hastuti, and Agustian Lucky Silana in 2014, 15% sugar concentration in coffee kombucha had the highest or most preferred organoleptic value [11].

Organoleptic Test and pH Changes of Kombucha Palm Sugar Beverage from Different Types of Coffee which are affected by Fermentation Time. This research from the Journal of Pharmacy and Science draws the conclusion that coffee-based kombucha experiences an increase in pH and a decrease in organoleptic value of taste, color, and aroma as long as fermentation [12].

Study on the Use of Toraja Coffee as Kombu Fermentation Media. This study draws the conclusion that coffee solution can be used as a medium for making kombucha. The percentage of dilution of coffee to water also affects the organoleptic value of taste, color, and aroma. Dilution of 8.5 grams of fresh coffee (not coffee grounds) per 250ml of water is most preferred [13].

Effects of Medium Variation and Fermentation Time on the Antioxidant and Antimicrobial Properties of Kombucha. This research from the Malaysian Journal of Fundamental and Applied Sciences draws the conclusion that the antioxidant content in kombucha made from tea has decreased when entering the 7th day of fermentation [14].

Enhancement of the Functional Properties of Coffee Through Fermentation by "Tea Fungus" (Kombucha). This research from the Journal of Food Processing and Preservation draws the conclusion that there is chemical evidence of increased levels of antioxidants and amylase enzymes in coffee fermented into kombucha [15].

IOP Conf. Series: Materials Science and Engineering **924** (2020) 012005 doi:10.1088/1757-899X/924/1/012005

Consumer Acceptability of a Kombucha Coffee Prototype with Traditional Coffee Characteristics. This previous study draws the conclusion that aroma is the most preferred characteristic with the highest value in the sensory evaluation of coffee-based kombucha [4].

2. Experimental method

This study uses experimental research methods to obtain the effect of a particular treatment on others under controlled conditions [16]. In this case the treatment given is substitution of basic ingredients kombucha using coffee grounds and testing the level of preference for kombucha made from tea.

Sensory evaluation is a scientific method of product evaluation using the five senses consisting of the mouth as a taste appraiser, the nose as an aroma appraiser, and the eyes as a color and appearance evaluator. This study uses affective testing in which the type of research is in the form of preference or hedonic tests for panelists who do not need to be trained [17].

In this study the population used was active students at Bina Nusantara University. Population is defined as the area of generalization of objects or subjects that have certain characteristics and qualities to be studied and conclusions are determined [16]. The sampling technique used is simple random sampling where the sampling of members of the population is carried out randomly without regard to strata in the population [16]. The number of untrained panelists used as a sample of this study is based on the recommendations of the Committee on Sensory Evaluation of the Institute of Food Technologist, which is 80 people for preference tests [18].

The authors use the scale measurement sensory rating of six hedonic scale to determine the level of panelist preference for kombucha products. The scale is composed of numerical numbers 1 to 6 that indicate the response information (1) Dislike, (2) Neutral, (3) Rather like, (4) Like, (5) Really like, and (6) Very like. The panelists' personal evaluation of a number of those determined in the study sample was collected by means of the questionnaires. The numerical data collected is then processed using statistical analysis instruments so that differences in the value of the final product can be obtained [18]. Statistical Descriptive Analysis is a way of describing or describing data that has been collected as it is without making generally accepted conclusions or generalizations. This analysis is used to describe the sample data, not to make conclusions that apply to the entire population [16]. Additionally, Paired sample T test is conducted. This is a comparative analysis between two samples that have correlations. Comparative analysis (difference) means the analysis used to find out the difference between two variables (data). Paired sample T test decision making is based on the probability value (sig) of the calculation with SPSS. If the probability value (sig) is greater than the significant level (α) then the null hypothesis (Ho) is accepted. Conversely, if the probability value (sig) is smaller than the significant level (α) then the null hypothesis is rejected [19].

3. Results and discussion

3.1. Descriptive Frequency Analysis

Table 1 display the information as follow. In the color frequency of sample A, as many as 1.3% (1 panelist) chose dislike, 16.3% (13 panelists) chose neutral, 18.8% (15 panelists) chose rather like, 42.5% (34 panelists) chose like, 10% (8 panelists) chose really like, and 11.3% (9 panelists) voted very like. So that the biggest frequency is 42.5% showing likeness to the color of kombucha with the basic ingredients of tea. Thus, in the frequency of aroma sample A, as many as 13.8% (11 panelists) chose dislike, 28.8% (23 panelists) chose neutral, 22.5% (18 panelists) chose rather like , 21.3% (17 panelists) chose like, 10% (8 panelists) chose really like, and 3.8% (3 panelists) voted very like. So that the biggest frequency is 28.8% showing neutral to the flavor of kombucha with the basic ingredients of tea. Hence, in the sample taste frequency A, 30% (24 panelists) chose dislike, 20% (16 panelists) chose neutral, 20% (16 panelists) chose rather like, 18.8% (15 panelists) chose like, 8.8% (7 panelists) chose really like, and 2.5% (2 panelists) voted very like. So that the biggest frequency is 30% showing dislike of the taste of kombucha with the basic ingredients of tea.

Table 1. Kombucha uses tea (Sample A).

Kombucha uses tea (Sample A)												
	Frequency			Percent			Valid Percent			Cumulative Percent		
	Color	Flavor	Taste	Color	Flavor	Taste	Color	Flavor	Taste	Color	Flavor	Taste
Dislike	1	11	24	1.3	13.8	30.0	1.3	13.8	30.0	1.3	13.8	30.0
Neutral	13	23	16	16.3	28.8	20.0	16.3	28.8	20.0	17.5	42.5	50.0
Rather Like	15	18	16	18.8	22.5	20.0	18.8	22.5	20.0	36.3	65.0	70.0
Like	34	17	15	42.5	21.3	18.8	42.5	21.3	18.8	78.8	86.3	88.8
Really Like	8	8	7	10.0	10.0	8.8	10.0	10.0	8.8	88.8	96.3	97.5
Very Like	9	3	2	11.3	3.8	2.5	11.3	3.8	2.5	100	100	100
Total	80	80	80	100	100	100	100	100	100			

Table 2. Kombucha uses coffee grounds (Sample B).

Kombucha uses coffee grounds (Sample B)												
	Frequency			Percent			Valid Percent			Cumulative Percent		
	Color	Flavor	Taste	Color	Flavor	Taste	Color	Flavor	Taste	Color	Flavor	Taste
Dislike	6	6	10	7.5	7.5	12.5	7.5	7.5	12.5	7.5	7.5	12.5
Neutral	10	12	10	12.5	15.0	12.5	12.5	15.0	12.5	20.0	22.5	25.0
Rather Like	17	12	15	21.3	15.0	18.8	21.3	15.0	18.8	41.3	37.5	43.8
Like	25	23	25	31.3	28.8	31.3	31.3	28.8	31.3	72.5	66.3	75.0
Really Like	15	19	12	18.8	23.8	15.0	18.8	23.8	15.0	91.3	90.0	90.0
Very Like	7	8	8	8.8	10.0	10.0	8.8	10.0	10.0	100	100	100
Total	80	80	80	100	100	100	100	100	100			

Table 2 display the information as below. In the color frequency of sample B, as much as 7.5% (6 panelists) chose dislike, 12.5% (10 panelists) chose neutral, 21.3% (17 panelists) chose rather like, 31.3% (25 panelists) chose like, 18.8% (15 panelists) chose really like, and 8.8% (7 panelists) chose very like. So that the biggest frequency is 31.3% showing likeness to the color of kombucha with coffee grounds. Moreover, in the scent frequency of sample B, 7.5% (6 panelists) chose dislike, 15% (12 panelists) chose neutral, 15% (12 panelists) chose rather like, 28.8% (23 panelists) chose like, 23.8% (19 panelists) chose really like, and 10% (8 panelists) voted very like. So that the biggest frequency is 28.8% showing likeness to the flavor of kombucha with coffee grounds. Furthermore, in the sample taste frequency B, 12.5% (10 panelists) chose dislike, 12.5% (10 panelists) chose neutral,

18.8% (15 panelists) chose rather like, 31.3% (25 panelists) chose like, 15% (12 people) chose really like, and 10% (8 panelists) voted very like. So that the biggest frequency is 31.3% showing the taste of kombucha with coffee grounds.

3.2 Statistical Descriptive Analysis

Table 3. Statistical Descriptive Analysis.

Statistical Descriptive Analysis					
	N	Minimum	Maximum	Mean	Std. Deviation
Color A	80	1	6	3.78	1.211
Color B	80	1	6	3.68	1.357
Valid N (listwise)	80				
Flavor A	80	1	6	2.96	1.345
Flavor B	80	1	6	3.76	1.425
Valid N (listwise)	80				
Taste A	80	1	6	2.64	1.434
Taste B	80	1	6	3.54	1.475
Valid N (listwise)	80				

The table above on statistical descriptive analysis prove the information as below. In term of descriptive color statistics between tea-based kombucha (Color A) and coffee grounds-based kombucha (Color B), the average value of color A was 3.78 and color B was 3.68. This tells the data that panelists prefer the color of KT with an average difference of 0.1. Both have a minimum value of 1 (dislike) and a maximum of 6 (very, very like). On the descriptive statistics of flavor from tea-based kombucha (Flavor A) with kombucha made from coffee grounds (Flavor B), the average flavor value of A was 2.96 and the flavor of B was 3.76. It shows that the community prefer the flavor of KC with an average difference of 0.8. Both have a minimum value of 1 (dislike) and a maximum of 6 (very like). Moreover, the descriptive statistics of taste from tea-based kombucha (Taste A) against kombucha made from coffee grounds (Taste B) shows the average value of taste A was 2.64 and taste B was 3.54. These prove that community preferred the taste of KC with an average difference of 0.9. Both have a minimum value of 1 (dislike) and a maximum of 6 (very like).

Table 4. Paired Sample T Test.

Paired Samples Test									
		Paired Differences					t	df	*Sig. (2-tailed)*
		Mean	*Std. Deviation*	*Std. Error Mean*	*95% Confidence Interval of the Difference*				
					Lower	*Upper*			
Pair 1	Color A – Color B	.100	1.506	.168	-.235	.435	.594	79	.554
Pair 2	Flavor A – Flavor B	-.800	1.504	.168	-1.135	-.465	-4.756	79	.000
Pair 3	Taste A – TasteB	-.900	1.620	.181	-1.260	-.540	-4.971	79	.000
Pair 4	Total A – Total B	- 1.813	3.865	.432	-2.673	-.952	-4.194	79	.000

Table 4 above present the result of the calculation of the paired sample T test whose decision making is based on the probability value (sig). In the color indicator, the probability value (sig) is greater than the significant level (α), so the null hypothesis is accepted (0.554 (sig)> α). While in the flavor indicator, the probability value (sig) is smaller than the significant level (α), so the null hypothesis is rejected (0,000 (sig) <α). Thus, the taste indicator, the probability value (sig) is smaller than the significant level (α), so the null hypothesis is rejected (0,000 (sig) <α).

Table 5. Results of Kombucha Tea Assessment.

Results of Kombucha Tea and Coffee Grounds Assessment						
	Kombucha Tea Assessment			**Kombucha Coffee Grounds Assessment**		
	Color	**Flavor**	**Taste**	**Color**	**Flavor**	**Taste**
Dislike	1	11	24	6	6	10
Neutral	13	23	16	10	12	10
Rather Like	15	18	16	17	12	15
Like	34	17	15	25	23	25
Really Like	8	8	7	15	19	12
Very Like	9	3	2	7	8	8
Total	80	80	80	80	80	80

On the result of Kombucha Tea Assessment, the table shows the data as follow. The color of KT, 1 panelist chose dislike, 13 panelists chose neutral, 15 panelists chose rather like it, 34 panelists chose to like, 8 panelists chose really like, and 9 panelists chose very like. Therefore, panellists like KT's colour. Moreover, towards the flavor of kombucha using tea, 11 panelists chose dislike, 23 panelists chose neutral, 18 panelists chose rather like it, 17 panelists chose to like, 8 panelists chose really like, and 3 panelists chose very like. So that the scale most chosen is neutral to the scent of KT. Thus, in term of the taste of kombucha using tea, 24 panelists chose dislike, 16 panelists chose neutral, 16 panelists chose rather like it, 15 panelists chose like, 7 panelists chose raelly like, and 2 panelists chose very like. So, the panellist do not like the taste of KT.

On the other hand, the result of Kombucha using coffee grounds Assessment display the information as follow. The color of kombucha using coffee grounds presents 6 panelists chose dislike, 10 panelists chose neutral, 17 panelists chose rather like it, 25 panelists chose like, 15 panelists chose really like, and 7 panelists chose very like. So, the community is like the color of KC. On the test of the flavor of kombucha using coffee grounds, 6 panelists chose dislike, 12 panelists chose neutral, 12 panelists chose rather like it, 23 panelists chose like, 19 panelists chose really like, and 8 panelists chose very like. Therefore, the community accept the flavor of KC. Finally, the taste of kombucha using coffee grounds prove that 10 panelists chose dislike, 10 panelists chose neutral, 15 panelists chose rather like it, 25 panelists chose like, 12 panelists chose really like, and 8 panelists chose very like. This means the community like the taste of KC.

4. Conclusion

Based on the results of the paired sample T test, it can be concluded that the color of kombucha made from coffee grounds is not acceptable to the community. But, the flavor and the taste of kombucha made from coffee grounds were acceptable.

Additionally, based on the results of the statistical descriptive analysis test, the level of preference on color indicators for tea-based kombucha with an average value of 3.78 is more preferable than coffee-based kombucha with an average value of 3.68. Whist, the level of preference on aroma indicators for kombucha based coffee grounds with an average value of 3.76 is more liking by the community rather than kombucha made from tea (average value of 2.96). Furthermore, the level of preference towards flavor indicators, for kombucha based on coffee grounds with an average value of 3.54 prevails kombucha made from tea with an average value of 2.64.

Therefore, based on data, whereas the colour of Kombucha based on coffee grounds is darker, the community prefer the color of Kombucha based on tea, which is light brown. But on the Flavor and Taste indicator, panelists prefer Kombucha based on coffee grounds than tea based Kombucha. This is because the community likes the flavor and taste of coffee-based Kombucha that is stronger and delicious than tea-based Kombucha.

References

[1] L. Saxe, "*Fermented foods are up 149% - as long as they're unfamiliar*", Forbes.com, February 6, 2019. [Online]. Available: https://www.forbes.com/sites/lizzysaxe/2019/02/06/fermented-foods-are-up-149-percent-as-long-as-theyre-unfamiliar [accessed June 22, 2019]
[2] International Coffee Organization, "*World Coffee Consumption*", 2019. [Online]. Available: http://www.ico.org/prices/new-consumption-table.pdf [accessed June 22, 2019]
[3] World Health Organization, "*2018 Health SDG Profile: Indonesia*", 2018. [Online]. Available: http://www.searo.who.int/entity/health_situation_trends/cp_ino.pdf [accessed June 22, 2019]
[4] Lin. L M "*Consumer Acceptability of a Kombucha Coffee Prototype with Traditional Coffee Characteristics.*" Family and Consumer Science, 2017, p 81. [Abstract]. Available: ProQuest https://pqdtopen.proquest.com/ [accessed November 13, 2019]
[5] R. J. D. Simanjuntak, *Effect of Variation Concentration of Sucrose Sugar and Fermentation Time on Kombucha Coffee Making*, Bandar Lampung: Medical Education, Faculty of Medical, University of Lampung, 2017
[6] L. Dewi, S. P. Hastuti and A. L. Silana, *Antioxidant Activity, Total Phenolic Levels, and Caffeine Levels in Fermented Robusta Coffee Combo in Various Sugar Concentrations,* 2015
[7] D. Wistiana, and E. Zubaidah, "Journal of Food and Agro-Industry", *Chemical and Microbiological Characteristics of Kombucha from Various High Phenol Leaves During Fermentation*, Vol. 3(4), pp. 1446-1457, 2015.
[8] I. Falahuddin, I. Apriani, and Nurfadilah, "Biota Journal", *Effect of Soursop Leaf Kombucha Fermentation Process on Vitamin C Levels*, Vol. 3(2), pp. 90-95, 2017
[9] R. J. D. Simanjuntak, *Effect of Variation Concentration of Sucrose Sugar and Fermentation Time on Kombucha Coffee Making*. Bandar Lampung: Medical Education, Faculty of Medical, University of Lampung, 2017
[10] M. O. W. Napitupulu, Setyohadi, and L. M. Lubis, "Journal of Food and Agriculture Engineering", *Effect of Variation Concentration of Sucrose Sugar and Fermentation Time on Kombucha Coffee Making*, Vol. 3(3), pp. 316-322, 2015
[11] L. Dewi, S. P. Hastuti, and A. L. Silana, "National Microbiology Seminar Paper ", *Antioxidant Activity, Total Phenolic Levels, and Caffeine Levels in Fermentation of Robusta Coffee Combo in Various Sugar Concentrations*, 2014
[12] K. A. P Lestari, Surahmaida, R. Darmawan, and L. Sadiyah,"Journal of Pharmacy and Science", *Organoleptic test and change in pH of kombucha palm coffee drinks from various types of coffee that are affected by the fermentation time*, Vol. 4(1), pp. 15-18, 2019
[13] E. M. Marampa, Study on the Use of Toraja Coffee as Kombu Fermentation Media. Salatiga: Universitas Kristen Satya Wacana, 2019
[14] B. M. Vohra, S. Fazry, F. Sairi, and O. Babul-Airianah, "Malaysian Journal of Fundamental and Applied Sciences", *Effects of medium variation and fermentation time on the antioxidant and antimicrobial properties of kombucha*, pp. 298-302, 2019

IOP Conf. Series: Materials Science and Engineering **924** (2020) 012005 doi:10.1088/1757-899X/924/1/012005

[15] M. I. Watawana, N. Jayawardena, and V. Y. Waisundara, "Journal of Food Processing and Preservation", *Enhancement of the functional properties of coffee through fermentation by "tea fungus"(kombucha)*, Vol. 39, pp. 2596-2603, 2015
[16] Sugiyono, *Quantitative, Qualitative, and R&D Research Methods*, Bandung: Alfabeta, 2017
[17] I. S. Surono, P. Waspodo and B. S. Abbas, *New Product Research and Development,* Jakarta: Binus Media & Publishing, 2015
[18] D. R. Adawiyah, B. Nurtama, D. Herawati, D. Indrasti and E. Syamsir, *Sensory Evaluation. Publication of Limited Circulation,* 2015
[19] F. J. Gravetter and L. B. Forzano *Research Methods for the Behavioral Sciences,* Canada: Nelson Education, 2015

IOP Conf. Series: Materials Science and Engineering **924** (2020) 012006 doi:10.1088/1757-899X/924/1/012006

The Effect of Mechanochemical on The Formation of Calcium Titanate ($CaTiO_3$) Prepared by High Energy Milling

Fiqhi Fauzi[1], Alvin Muhammad Habieb[1], Alfian Noviyanto[1,2,*], Retno Kusumaningrum[1], Galuh Sukmarani[1], Edela Uswah Dien Muhammad[3], Valleta Jovanka Widodo[3], Diva Calista Amalia[3], Didik Aryanto[4], and Nurul Taufiqu Rochman[5]

[1]Nano Center Indonesia, Jl. PUSPIPTEK, South Tangerang, Banten 15314 Indonesia
[2]Department of Mechanical Engineering, Mercu Buana University, Jl. Meruya Selatan, Kebun Jeruk, Jakarta 11650, Indonesia
[3]Department of Metallurgy and Materials Engineering, Faculty of Engineering, University of Indonesia, West Java, Depok 16424, Indonesia
[4]Research Center for Physics, Indonesian Institute of Sciences, PUSPIPTEK, South Tangerang, Banten 15314, Indonesia
[5]Research Center for Metallurgy and Materials, Indonesian Institute of Sciences, PUSPIPTEK, South Tangerang, Banten 15314, Indonesia

*a.noviyanto@nano.or.id

Abstract. Single-phase calcium titanate ($CaTiO_3$) was successfully synthesized by mechanical milling and the solid-state reaction of $CaCO_3$ and TiO_2. The speed of high energy ball milling was 700 rpm with the ball and jar were made from stainless steel. The milling time and ball to powder ratio was 10 h and 50 h, respectively. After milling for 10 h, the mixed powder of $CaCO_3$ and TiO_2 experienced heavy milling, which indicated by the average particle size before and after milling was > 1 μm and 85.56 ± 35.62 nm, respectively. Furthermore, the XRD pattern of milled powder revealed the disappearance of $CaCO_3$ peaks and a considerable reduction of TiO_2 peaks after milling for 10 h. Moreover, the presence of $CaTiO_3$ peaks in the milled powder was noticeably detected in the XRD pattern, showing the mechanical alloying of $CaCO_3$ and TiO_2 was occurred. The milled powder was calcined at 800, 900 and 1000°C for 2 h. The results showed the formation of a single phase of $CaTiO_3$ after calcination at any temperatures. However, the samples indicated the presence of Fe_2O_3, which from the milling media. The presence of impurities after milling is inevitable due to friction between ball and jar. Further study is needed to obtain the optimum condition of mechanical milling to minimize the contamination.

Keywords: *Calcium titanate; mechanical milling; raw materials; calcination; impurities*

1. Introduction

Calcium titanate, $CaTiO_3$, is one of a group of metal compounds with a perovskite structure, has long been known as a dielectric ceramic with a high constant dielectric [1]. Hence, $CaTiO_3$ has important applications in microwave communication systems owing to ferroelectric properties [2]. Recently,

$CaTiO_3$ is considered as an orthopedic implant material due to the excellent biocompatibility properties [3–6].

Conventional technique to obtain $CaTiO_3$ is a solid state reaction between $CaCO_3$ or CaO and TiO_2 at a high temperature of 1350°C [7]. The heating conditions play an important role in producing crystallized $CaTiO_3$ powder [8]. However, this process is challenging to produce fine microstructure of $CaTiO_3$ due to grain growth during heating at high temperature. There are several methods to synthesize $CaTiO_3$, namely through the wet chemical method, for example, a colloidal sol-gel [9], alkoxide [10], polymeric precursors [11], spray pyrolysis [12], and hydrothermal [1]. A typical characteristic of $CaTiO_3$ that synthesize by wet chemistry is fine particle size, i.e., nanoscale range, and homogeneous in size. In addition to wet chemistry method, mechanical milling is also used to synthesize $CaTiO_3$ from $CaCO_3$/CaO and TiO_2. Mechanical milling of these powder has been effective to induce the reaction faster than a conventional solid-state reaction, called mechanochemical activation.

The mechanochemical process is the combination of the mechanical and chemical processes that used a grinding mill with a high intensity such as planetary mill, high energy milling, and oscillatig mills. This process causes defects are being formed. Moreover, the mechanochemical process did not use many organic solvents needed in industrial chemical processes, making them environmental-friendly [13]. High-energy ball mills have been used in producing ultra-fine powder in the sub-micron to the nanometer range. In addition, this process causes severe and intense mechanical action on solid surfaces, which causes the physical and chemical changes in the area near the surface where solids come in contact with mechanical forces, therefore have a great impact on the phase composition, morphology, particle size, and surface area [14]. Currently, this method is used to synthesize inorganic materials because it shows several advantages, such as a reduction in sintering temperature [15–17]. Based on this background, this work aims to synthesize $CaTiO_3$ through mechanochemical method. The effect of milling parameter and their relationship with the calcination temperature to produce a single-phase $CaTiO_3$ was investigated thoroughly in this study.

2. Experimental method

$CaCO_3$ calcite (Kanto Chemical Co. Inc., p.a. 99,5%) and TiO_2 rutile (Kanto Chemical Co. Inc., p.a. 99,0%) were used as raw materials in this study. The powder was mixed with a mass weight based on stoichiometric calculations of $CaTiO_3$. The mixed powder was milled for 10 h in the high energy ball milling (HEM Shaker Mill, Fisika Laboratoria, Indonesia) that has speed of 700 rpm with the ball to powder ratio was 50. Stainless steel mill jar and ball were used in this study. In order to increase the effectiveness of milling and homogeneous mixture of the powders, ethanol was used as an immersion medium during milling. Subsequently, the milled powder was calcined at 800, 900 and 1000°C for 2 h with heating rate 5°C/minute in the air atmosphere.

The morphology of milled and calcined powder was observed using scanning electron microscopy (SEM, FEI Quanta 650, USA). The average particle size was estimated by measuring the diameter of 100 particles from the SEM images and analyzed statistically. The crystalline phase after milling and calcination was examined by X-ray diffraction (XRD, Shimadzu Maxima-X 7000, Japan) with CuK-α radiation (λ=1.5418).

3. Results and discussion

Figure 1 displays SEM images of the microstructure of a mixture of $CaCO_3$ and TiO_2 samples before and after milling. It is obvious that, the mixture of $CaCO_3$ and TiO_2 before milling distinguish two particles that differ in size, as shown in Figure 1 (a). The big and fine particle in Figure 1 (a) were $CaCO_3$ and TiO_2, respectively. After being milled for 1 hour (Figure 1 (b)), the particle shape begins to collapse and the size becomes smaller with the same contour. No evidence the presence of big particle, which indicates that $CaCO_3$ powder is already pulverized after milling for 1 h. Moreover, it is found that the particle shape is irregular, which is more likely from the $CaCO_3$ powder. Further milling time until 10 h, revealing the existence of slab shaped particle, as shown in Figure 1 (c). This means that the milling process is related to the agglomerate process to produce a small particle size and a wider surface [18]. Indeed, the particle size of the milled powder decreased with increasing milling time. The average particle size for 1 and 10 h milling was 185 ± 57 and 85 ± 35 nm, respectively. This is confirming the advantage of high energy milling to obtain the nanoparticle powder only in 10 h of milling time.

Figure 1. SEM of the $CaCO_3$-TiO_2 powder mixtures (a) before milling and after milling for (b) 1 h and (c) 10 h.

Figure 2 shows the XRD pattern of a mixture of $CaCO_3$ and TiO_2 before and after milling for 10 h. The diffraction peaks of the sample before milling had high intensity and highly crystalline phase of $CaCO_3$ and TiO_2, which reveals the composition of initial powder that consists of $CaCO_3$ and TiO_2 powder. However, a dramatic change occurred after milling for 10 h, where $CaCO_3$ peaks disappear. The disappearance of $CaCO_3$ peaks implied that the crystalline phase of $CaCO_3$ become amorphous, thus could not detected in the XRD. Contrary, although TiO_2 powder experienced the same heavy milling with $CaCO_3$, small peaks of TiO_2 still exist after 10 h of milling. This difference is more likely due to the difference in hardness of $CaCO_3$ and TiO_2, where the hardness of $CaCO_3$ and TiO_2 was 3.0 and 6.5-7.0 GPa, respectively [19]. This reduction in peak energy implies the formation of amorphous material from the milling material. This is consistent with research conducted by several researchers when high energy milling was carried out on oxide materials [8]. One notable finding in this study is the formation of $CaTiO_3$ phase after milling for 10 h, as shown in Figure 2. The formation of a new phase during mechanical milling is called mechanical alloying [20]. Milling causes a decrease in particle size, hence the diffusion paths to be shortened. Moreover, the particle that went through heavy milling has a lot of energy in their surface. Therefore, the kinetic energy of the system increases so that it could provides energy for the reappearance of new crystallization [21]. However, the formation of $CaTiO_3$ induced by mechanical milling depends on the milling condition, i.e., time, speed of milling and ball to powder ratio. For instance, a single phase of $CaTiO_3$ formed after mechanical milling for 15 h with the ball to powder ratio was 20 and speed of milling was 300 rpm

[22]. Furthermore, another study showed the $CaTiO_3$ was obtained after milling for 70 h with the ball to powder ratio was 70 [23]. In this present study, the milling condition is moderately mild, where the milling time only for 10 h and the ball to powder ratio is 50, yet the $CaTiO_3$ was already observed. However, after milling for 10 h, Fe peak identified, which it might be the contamination from the milling media and ball mill that made from stainless steel. Impurities during mechanical milling are inevitable, in particular for the high ball to the powder ratio, which increases the wear between ball and mill jar or ball and ball. Therefore, the optimum condition during mechanical milling needs to improve for minimizing the contamination.

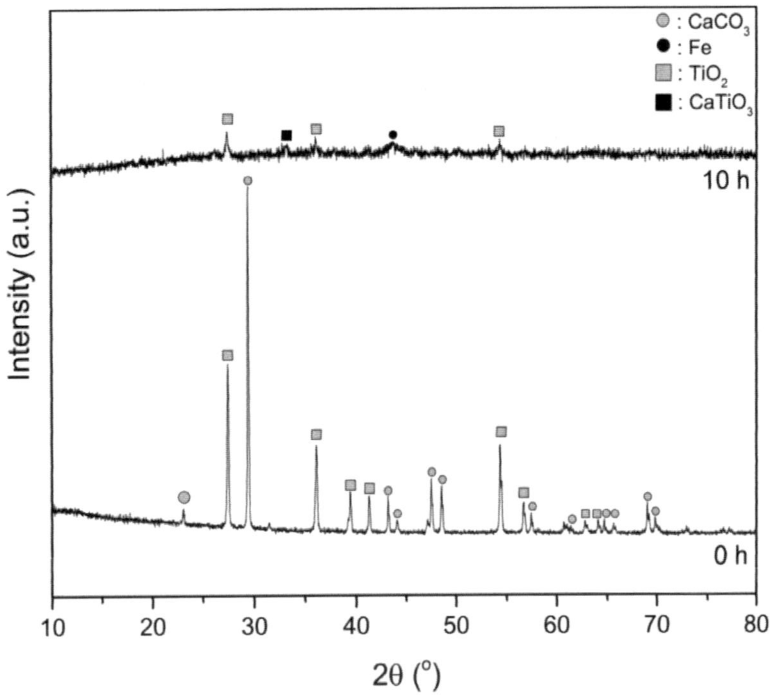

Figure 2. X-ray diffractometer pattern of $CaCO_3$-TiO_2 mixture before milling (0 h) and milling for 10 h (10 h).

Figure 3 shows XRD patterns of the milled powder after calcination at 800, 900, and 1000°C, for 2 hours in an air atmosphere. Despite the presence of Fe_2O_3 as impurities from the milling media, $CaTiO_3$ phase was formed at all calcination temperatures. Moreover, at the lowest calcination temperature, i.e., 800°C, no excess of $CaCO_3$ or TiO_2 were found, indicating that all of those phases react completely and formed $CaTiO_3$ phase, as shown in Figure 3. This is due to heavy milling that decreases the particle size of those powder, as shown in Figure 1 (c). Moreover, smaller particle size has a large surface area, thus the reaction between two particles take places without a significant effort. It should be noted that the complete reaction between $CaCO_3$ and TiO_2 to form $CaTiO_3$ occurred at 1350°C [7], while in this study at 800°C. This formation temperature of $CaTiO_3$ in this study is relatively low compared to other studies. For instance, Vukotic *et al.* [17] reported the formation of $CaTiO_3$ at 850°C, while Evans *et al.* [23] reported single phase $CaTiO_3$ were obtained at 920°C. Moreover, Berbenni and Marini [16] reported the synthesis of $CaTiO_3$ from the mechanically milled powder of $CaCO_3$ and TiO_2 at 750 – 850°C indicates the formation of $CaTiO_3$ along with CaO and TiO_2 at any temperature. Furthermore, as shown in Figure 3, the crystallinity of $CaTiO_3$ peaks increased with increasing temperature.

Figure 4 shows the SEM image of milled powder after calcination at 1000°C for 2 h. Although a high temperature was used, i.e., 1000°C, the particle size of $CaTiO_3$ showed 288 ± 128 nm,

IOP Conf. Series: Materials Science and Engineering **924** (2020) 012006 doi:10.1088/1757-899X/924/1/012006

which is fine particle size. This results was consistent with previous report by Cavalcante *et al* [11] whereby they explained that the increase in temperature promotes an increase in the average crystallite size or particle size. This relates to aggregate production and nuclei formation. Sreckovic [24] reported mechanical activation of powders during grinding improves their reactivity and further, the calcination process accelerates the formation of a new phase by a solid-state reaction. Moreover, almost homogenous particle size was obtained in this study with the unperfect spherical shape of $CaTiO_3$.

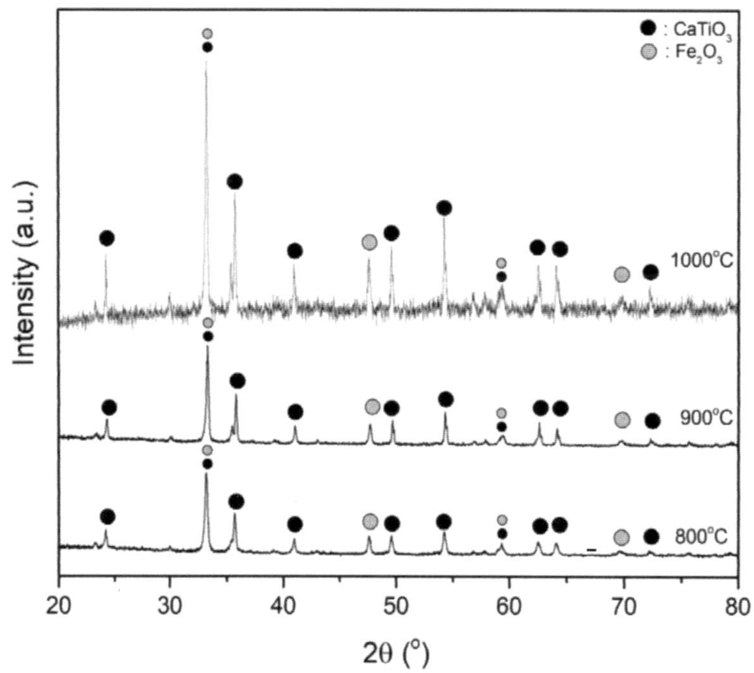

Figure 3. X-ray diffractometer pattern of $CaTiO_3$ with furnace temperature variations: 800°C, 900°C and 1000°C.

Figure 4. SEM of the $CaCO_3$-TiO_2 powder mixtures mechanically activated 10 h and furnace 1000°C.

4. Conclusion

Single-phase calcium titanate ($CaTiO_3$) was successfully obtained by high energy milling of raw powder followed by calcination. XRD pattern of milled powder revealed the powder experience heavy milling, which indicated by the disappearance of $CaCO_3$ peaks and a significant decrease of TiO_2 peaks. The loss of crystallinity of milled powder implied amorphous powder was obtained. Moreover, $CaTiO_3$ peak was observed in the milling time of 10 h, appearing that mechanical alloying between $CaCO_3$ and TiO_2 occurred during milling. However, Fe peak was also found in the milled powder, which the contamination from the milling media that is inevitable. Calcination of the milled powder led to the formation of $CaTiO_3$ along with the Fe_2O_3 as impurities regardless of the calcination temperature. Therefore, it can be concluded that single phase $CaTiO_3$ can be obtained at a relatively low temperature induced by mechanical milling. However, further study needed to minimize the contamination by optimizing the milling parameters.

Acknowledgment

Support from Laboratory of High Resistance Materials, Research Center for Physics, Indonesian Institute of Sciences, especially to Dr. Agus Sukarto Wismugroho and Dr. Wahyu Bambang Widayatno, we gratefully acknowledged.

References

[1] A. Maddu, L. Permatasari, and A. Arif, "Structural and dielectric properties of $CaTiO_3$ synthesized utilizing Duck's eggshell as a calcium source," *J. Ceram. Process. Res.*, vol. 18, no. 2, pp. 146–150, 2017.
[2] Y. Hu, Y. Zhang, H. Liu, and D. Zhou, "Microwave dielectric properties of PTFE/$CaTiO_3$ polymer ceramic composites," *Ceram. Int.*, vol. 37, no. 5, pp. 1609–1613, 2011.
[3] D. Mondal, L. Nguyen, I. H. Oh, and B. T. Lee, "Microstructure and biocompatibility of composite biomaterials fabricated from titanium and tricalcium phosphate by spark plasma sintering," *J. Biomed. Mater. Res. - Part A*, vol. 101 A, no. 5, pp. 1489–1501, 2013.
[4] N. T. B. Linh, D. Mondal, and B. T. Lee, "In vitro study of catio3-hydroxyapatite composites for bone tissue engineering," *ASAIO J.*, vol. 60, no. 6, pp. 722–729, 2014.
[5] K. Manjunath and C. G. Thimmanna, "Studies on Synthesis, Characterization and Applications of Nano CaTiO3 Powder," *Curr. Nanomater.*, vol. 1, no. 2, pp. 145–155, 2016.
[6] J. E. Arce *et al.*, "Calcium phosphate-calcium titanate composite coatings for orthopedic applications," *Ceram. Int.*, vol. 42, no. 8, pp. 10322–10331, 2016.
[7] G. Branković, V. Vukotić, Z. Branković, and J. A. Varela, "Investigation on possibility of mechanochemical synthesis of $CaTiO_3$ from different precursors," *J. Eur. Ceram. Soc.*, vol. 27, no. 2–3, pp. 729–732, 2007.
[8] S. Palaniandy and N. H. Jamil, "Influence of milling conditions on the mechanochemical synthesis of $CaTiO_3$ nanoparticles," *J. Alloys Compd.*, vol. 476, no. 1–2, pp. 894–902, 2009.
[9] F. R. Cesconeto, M. Borlaf, M. I. Nieto, A. P. N. de Oliveira, and R. Moreno, "Synthesis of $CaTiO_3$ and $CaTiO_3/TiO_2$ nanoparticulate compounds through Ca^{2+}/TiO_2 colloidal sols: Structural and photocatalytic characterization," *Ceram. Int.*, vol. 44, no. 1, pp. 301–309, 2018.
[10] A. P. Rodríguez *et al.*, "In vitro efficacy of $CaCO_3$ content in $CaTiO_3$- $CaCO_3$ composites for bone growth," *J. Hard Tissue Biol.*, vol. 27, no. 3, pp. 250–256, 2018.
[11] L. S. Cavalcante *et al.*, "Synthesis, structural refinement and optical behavior of $CaTiO_3$ powders: A comparative study of processing in different furnaces," *Chem. Eng. J.*, vol. 143, no. 1–3, pp. 299–307, 2008.
[12] S. Lanfredi, F. Storti, L. P. M. Simões, E. Djurado, and M. A. L. Nobre, "Synthesis and structural characterization of calcium titanate by spray pyrolysis method," *Mater. Lett.*, vol. 201, pp. 148–151, 2017.
[13] P. Kaur, R. Arora, S. K. Shahi, and J. S. Shahi, "Green Synthesis of Nano-sized Calcium Titanate $CaTiO_3$ Using Solid State Mechano-Chemical Solventless Method and its

IOP Conf. Series: Materials Science and Engineering **924** (2020) 012006 doi:10.1088/1757-899X/924/1/012006

Characterization," *Int. J. Res. Eng. Appl. Manag.*, vol. 4, no. 01, pp. 106–110, 2018.

[14] D. Stoyanova *et al.*, "Mechanical milling of hydrothermally obtained $CaTiO_3$ powders—morphology and photocatalytic activity," *Nano-Structures and Nano-Objects*, vol. 18, p. 100301, 2019.

[15] M. Jean and V. Nachbaur, "Determination of milling parameters to obtain mechanosynthesized $ZnFe_2O_4$," *J. Alloys Compd.*, vol. 454, no. 1–2, pp. 432–436, 2008.

[16] V. Berbenni and A. Marini, "Mechanical activation of calcium titanate formation from $CaCO_3$-TiO_2 mixtures," *J. Mater. Sci.*, vol. 39, no. 16–17, pp. 5279–5282, 2004.

[17] V. M. Vukotić, T. Srećković, Z. Marinković, G. Brankovic, M. Cilense, and D. Arandjelović, "Mechanochemical Synthesis of $CaTiO_3$ from $CaCO_3$ - TiO_2 Mixture," *Mater. Sci. Forum*, vol. 453–454, pp. 429–434, 2004.

[18] J. W. P. Boke, "Calcium Carbonate Particle Size Effects on Titanium Dioxide Light Scattering in Coatings," no. June, San Luis Obispo, 2013.

[19] C. Suryanarayana, "Mechanical alloying and milling," *Mech. Alloy. Milling*, vol. 46, pp. 1–472, 2004.

[20] C. C. Koch, "Synthesis of nanostructured materials by mechanical milling: Problems and opportunities," *Nanostructured Mater.*, vol. 9, no. 1–8, pp. 13–22, 1997.

[21] S. Sahoo, U. Dash, S. K. S. Parashar, and S. M. Ali, "Frequency and temperature dependent electrical characteristics of $CaTiO_3$ nano-ceramic prepared by high-energy ball milling," *J. Adv. Ceram.*, vol. 2, no. 3, pp. 291–300, 2013.

[22] S. Manafi and M. Jafarian, "Synthesis of perovskite $CaTiO_3$ nanopowders with different morphologies by mechanical alloying without heat treatment," *Int. J. Phys. Sci.*, vol. 8, no. 23, pp. 1277–1283, 2013.

[23] I. R. Evans, J. A. K. Howard, T. Sreckovic, and M. M. Ristic, "Variable temperature in situ X-ray diffraction study of mechanically activated synthesis of calcium titanate, $CaTiO_3$," *Mater. Res. Bull.*, vol. 38, no. 7, pp. 1203–1213, 2003.

[24] T. Sreckovic, "Sintering of Mechanically Activated Powders Tatjana Sreckovic," *Adv. Sci. Technol.*, vol. 45, pp. 619–628, 2006.

Effects of Pouring Temperature Variations on Fatigue Properties of Al-Cu Alloy After Remelting

Hizba Muhammad Sadida, and Priyo Tri Iswanto*
Department of Mechanical and Industrial Engineering, Faculty of Engineering, Universitas Gadjah Mada (UGM)
Jl. Grafika No.2, Yogyakarta, 55281, Indonesia

*priyotri@ugm.ac.id

Abstract. The aim of this research is to observe the casting temperature variations effects on fatigue properties of Al 2024 with different types of casting temperature. Three different pouring temperatures were carried out 688, 738, 788 °C, and the cast mold was preheated and kept constant at 220°C. The speciment was heat-treated by T6 process in which this method was heat treated in electrical furnace at 500 °C for 2 hours continued by fast quenching into water. Finally, the sample cast was treated by artificial aging method with the temperature at 190 °C. The results show that at 688°C pouring temperature, it has the highest tensile test results than to other variations which is 138.37 MPa, and for the fatigue strength at more than $2x10^6$ cycles is 62.3 MPa. It shows that fatigue life of aluminum alloy 2024 could be improved by T6 heat treatment and artificial aging in which it indicates that the number of the fatigue strength depends on the number of the maximum tensile strength of the material because the increase in the maximum tensile strength will increase the fatigue strength.

***Keywords*:** Al 2024, T6 heat treatment, fatigue properties, pouring temperatures

1. Introduction

Al 2024 is a well-knwon heat treatable alloy widely applied in aerospace industries and automotive because of its good damage tolerance and low density [1]. This material also can be produced by different casting methods, such as., semi-solid casting, and direct-chill casting. However, casting defects such as microcracks, porosities, hot tearing and macrosegregation have damaging effects on mechanical properties. Therefore, optimizing and detecting are the notable parameters which is able to eliminate defects that would be advantage for the product [2].

Remelting process could reduce the toughness of aluminum alloy caused by porosity due to an increase in hydrogen gas when the metal transforms from solid to liquid then to solid. This identifies that the ability of a material to deform plastically and absorb energy before and after damage is reduced because the material undergoes a process of remelting [3]. The process of remelting in cast aluminum also affects the cracking characteristics of aluminum alloy cracks, for instance, there is a decrease in hardness and fatigue (N) cycles. Fatigue testing results indicate that the remelting process can reduce fatigue life [4]. A518,A713, and A206 are utilized to assist effect on hot tearing and also on the grain size on three different pouring temperatures namely 700, 750, and 780 °C. As a result, it has a significant effect which is would be very useful for other materials [5]. The pouring temperature is a critical parameter for predicting crack susceptibility. The results show a good correlation between grain size, pouring temperature and crack susceptibility, so it is proven that hot tearing significantly decreases with decreasing pouring temperature[6]. T6 heat treatment and aging temperature will improve the number of failure cycles in which casting metal defects in the form of cracks during the formation of metals such hot tears are one of the sources of crack initiation and could affect the strength of fatigue[7]. In this research, The heat treatment solution included heating the sample in an electric furnace to reach a temperature of 535 °C, for 4 hours and then quickly cooling with room temperature water. It has a purpose to avoid structure diffusion, and to clarify the material structure is still in the α phase. Finally, the process of artificial aging in which the speciments were put and rehetead into the furnace at 150 °C for 2 hours until reaching room temperature in the furnace to release heat [8].

The purpose of this work is to analyze the effects of pouring temperature variations on fatigue strength of Al-Cu alloy after remelting.

2. Experimental Methods

2.1 Materials

The base material was aluminum commercial bar wrought alloy 2024. The Table 1 shows chemical composition Al-Cu after remelting process with using Shimadzu EDX-8000. Tensile testing has been carried out by Akhyar [8].

Table 1. Chemical Composition wt/ %.

Al	Cu	Mg	Mn	Fe	Si	Zn	Ca	Ta	Cr
94.347	3.119	1.406	0.556	0.166	0.150	0.131	0.054	0.025	0.011

Table 2. Tensile Properties of Al-Cu Pouring Temperature Variations.

Casting Temperature (°C)	Ultimate Tesile Strength (MPa)	Elongation (%)
688	138.37	0.84
738	83.96	0.88
788	103.47	1.12

2.2 Casting Process

The base material aluminum alloy with a length of 1,500 mm and diameter of 50.8 mm was cut and put into electrical furnace. Casting method used gravity cast. Then, Al 2024 alloy was melted at 688, 738, and 788 °C and poured into carbon steel EMS/17330 with a constant temperature 220 °C for all type casting conditions and the temperature was measured using a K-type thermocouple on the furnace.

2.3 Heat Treatment T6

Heat treatment T6 is a heat treatment process in the certain temperature that was done with various pouring temperature, and the temperatures were controlled (+ 2 °C) electrical resistance furnace. Firstly, the temperature is heated (T_0) until a temperature of 500 °C and held on the temperature for 2 hours, on this stage, to ascertain the β phase will be solute on the α phase (solutionized). Secondly, the sample is soaked in water (T1=25 + 1 °C) to avert diffusion of the structure, in addition to clarify the structure of the material is still in the α phase. A supersaturated solid solution in the α phase is reheated at a (α+ β) temperature area (T2). Based on previous research [8], T2 was set at 190 °C for artificial aging (T6) for 5 hours and then water quenched in water at room temperature. As a follow methods act as an aging method through temperature release until the samples reach room temperature. The solution treatment leads to three phenomena in the material; firstly, eutectic-Cu fragmentation and spheroidization, secondly, dissolution of precipitate particles, and lastly, chemical homogenization [9].

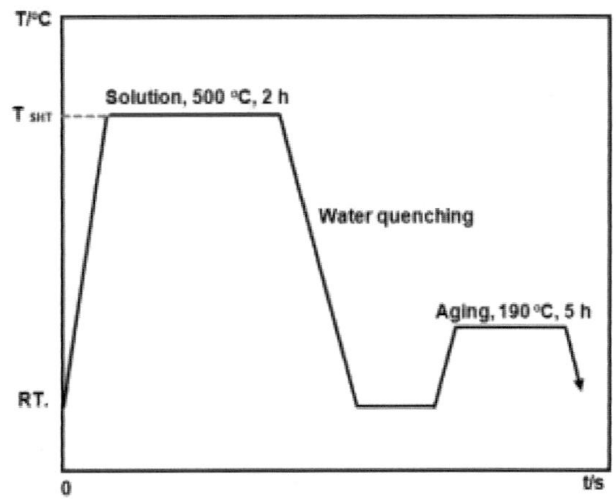

Figure 1. T6 Heat treatment scheme.

2.4 Mechanical Properties Test

Fatigue testing of rotary bending is to determine the fatigue of test specimens using the Shimadzu ONO'S type H-6 bending fatigue tester. The shape and size of the test specimens carried out following the JIS Z2274 standard shown in Figure 2. Rotary bending fatigue testing is performed at the ratio R = -1. The maximum cycle is determined at $2x10^6$ under the condition of the non-fractured test specimen. Loading test used a voltage amplitude of 30 to 60% of the maximum tensile strength with an increase of 5% [10]. Hardness testing is to determine effect of pouring temperature after remelting with pouring temperature variations 688, 738, and 788°C. Hardness testing used Brinell method and device

2.5 Microstructure

Microscopic analyzing was carried using an Olympus optical microscope and Optilab device. The samples were set up using standard metallographic procedures be composed of grinding, polishing and etching using Keller's reagent made of 5 ml HNO3 (95% concentration), 190 ml H2O, 3 ml HCl, and 2 ml HF.

3. Results and Discussion

3.1 Hardness Testing

Hardness testing in this study was carried out by the Brinell method. Tests were carried out using a load of 613 kg with an indenter ball-shaped with a diameter of 2.5 mm. Observations were made after doing the indentation. The former indentation was observed using a microscope with a scale1 mm = 38 strips.

Figure 2 shows the highest hardness value which was obtained at a pouring temperature of 738 ° C with a Brinell hardness of 146.675 kgf / mm², whereas the lowest hardness test was 105.324 kgf / mm² with a temperature of 788 °C, and a temperature variation of 688 ° C showing a value of 110.886kgf / mm². Based on these data, the decrease in hardness value occurs along with the increase in pouring temperature, due to the larger the grain size and the higher the pouring temperature which could cause easy dislocation.

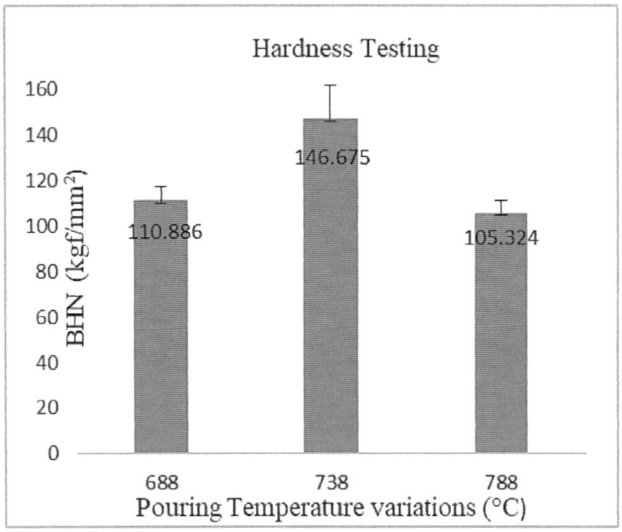

Figure 2. Effect of Pouring Temperature on Hardness.

3.2 Fatigue Properties

Fatigue rotary bending test with aluminum alloy 2024 used a percentage of loading ranging from 40 to 65% of the maximum tensile strength. Tensile testing has been carried out by Akhyar [8].The highest fatigue life is at the pouring temperature variation 688°C (the test specimen did not break when it reached $2x10^6$ cycles) was 55.3 MPa at 40% of the maximum tensile strength. At the pouring temperature variation 738°C obtained the highest voltage amplitude of 37.8 MPa at 45% of the maximum tensile strength which is the smallest amplitude compared to the other pouring temperatures, whereas at the casting temperature variation 788°C the largest amplitude percentage obtained at 40% of the maximum tensile strength with 41 .4 MPa working tensile stress. However, the lowest cycle is at a temperature variation of 688 where at a tensile stress of 89.9 MPa or 65% of the maximum tensile stress only obtained 103,100 cycles, whereas other variations at a percentage of 65% of the maximum tensile stress reach above 2,000,000 cycles with all specimens broken condition.

As illustrated in Figure 3 that each variation has a pattern that tends to be the same as the results of the tensile test results where each pouring temperature increases will decrease except at the pouring temperature 788°C which rises slightly. When seen from the percentage of voltage amplitude to the highest variation of the pouring temperature is obtained at a temperature variation of 688°C that is equal to 40% of the maximum tensile strength. Then it can be concluded on the testing of fatigue rotary bending that the fatigue strength depends on the maximum tensile strength value of the material because the percentage of the amplitude of stress that is relatively equal in other words an increase in maximum tensile strength will increase fatigue. fatigue strength in cast aluminum alloys is affected by defects originating from the casting process such as cracks that are formed due to the greatest defect area and cracks growth propagation that follows the grooves in the smaller defect concentration area [10-12].

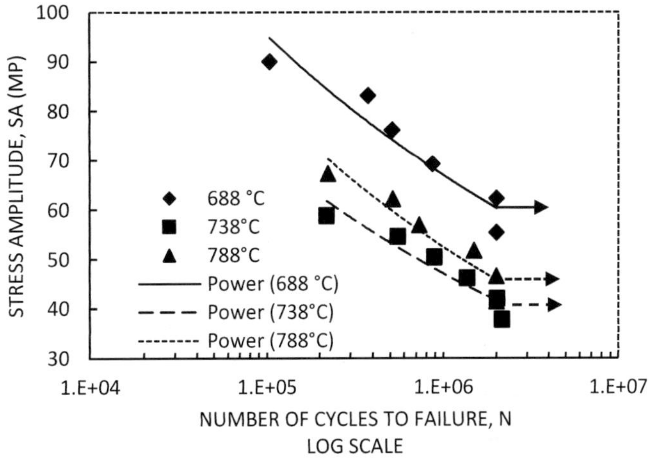

Figure 3. S-N Curves (R=-1) of Al-Cu, Effect Pouring Temperature.

3.3 Microstructure

Observation of the structure with an optical microscope was carried out on the surface of the specimen. Microstructure observation results show that the influence of an increase in pouring temperature for casting in the casting process affects the microstructure that results from re-casting.

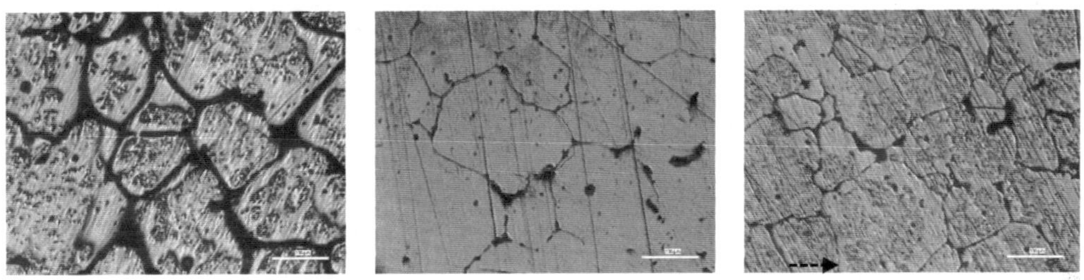

(a) (b) (c)

Figure 4. Microstructure of Al-Cu after remelting by different pouring temperature (a) 688°C (b) 738°C (c) 788°C.

Figure 4 shows the pouring temperature 688°C having a relatively smaller grain size among other pouring temperature variations, while the 788°C pouring temperature variation has a relatively larger grain size among other variations of the pouring temperature. Grain boundaries are marked with a red line in Fig. with a higher temperature, then a decrease in temperature requires a longer or (Δt) longer time. Aluminum cast which has a longer pouring time temperature (Δt), the greater it will be to release the stress due to the freezing of liquid metal. this results in larger grain sizes.The higher the pouring temperature, the more porosity will appear, and it will be easy to be contaminated by hydrogen gas which causes the trapping of hydrogen gas in the liquid metal. the presence of porosity defects in the specimen can reduce the mechanical properties of a material.

4. Conclusion

Effect of pouring temperature of Al-Cu of fatigue properties after remelting has been analyzed. result can be concluded :The highest hardness value was obtained at a pouring temperature of 738 °C 146,675 kgf/mm². The highest of fatigue life is at 688°C pouring temperature with $2x10^6$ cycles at working tensile stress 62.3 MPa. Fatigue strength depends on the number of the maximum tensile strength of the material because the increase in the maximum tensile strength will increase the fatigue strength.

Acknowledgements
The authors gratefully acknowledge the financial support of Hibah RTA 2019 Universitas Gadjah Mada.

References

[1] M. H. Ghoncheh, S. G. Shabestari, A. Asgari, and M. Karimzadeh 2018 Nonmechanical criteria proposed for prediction of hot tearing sensitivity in 2024 aluminum alloy *Trans. Nonferrous Met. Soc. China* **28** 5 848–857

[2] M. H. Ghoncheh and S. G. Shabestari, 2015 Effect of Cooling Rate on the Dendrite Coherency Point During Solidification of Al2024 Alloy **46**

[3] Ghasem Eisaabadi *et al.* 2014 Materials Science & Engineering A The effect of remelting on the melt and casting quality in Al – 7 % Si – Mg castings **605** 203–209

[4] A. Mohamed, Y. El-Madhoun, and M. N. Bassim 2005 The effect of tempering on low cycle fatigue behavior of Al-2024 *J. Mater. Process. Technol.* **162–163** 362–366

[5] A. K. Birru, D. B. Karunakar, and M. M. Mahapatra 2012 A study on hot tearing susceptibility of Al-Cu, Al-Mg, and Al-Zn alloys *Trans. Indian Inst. Met.* **65** 1 97–105

[6] M. R. Nasresfahani and B. Niroumand 2014 A new criterion for prediction of hot tearing susceptibility of cast alloys *Metall. Mater. Trans. A Phys. Metall. Mater. Sci.* **45** 9 3699–3702

[7] P. T. Iswanto, H. Akhyar, V. Malau, and R. Wirawan 2016 Effect of T6 Heat Treatment and Artificial Aging on Mechanics and Fatigue Properties of A356 . 0 Aluminum Alloy Produced by 350 rpm Centrifugal Casting In: IEEE Procceeding of 6th International Annual Engineering Seminar (InAES) 1-3 August Yogyakarta *Indonesia*

[8] H. Akhyar, P. T. Iswanto, and V. Malau 2018 Non treatment, t4 and t6 on tensile strength of al-5.9cu-1.9mg alloy investigated by variation of casting temperature *Mater.Sci. Forum* **929** 56–62

[9] M. J. Roy and D. M. Maijer 2014 Materials Science & Engineering A Response of A356 to warm rotary forming and subsequent T6 heat treatment *Mater. Sci. Eng. A* **611** 223–233

[10] T. Ludian and L. Wagner 2007 Effect of age-hardening conditions on high-cycle fatigue performance of mechanically surface treated Al 2024 *Mater. Sci. Eng. A* **468–470** 210–213

[11] Y. Chen, C. Liu, J. Zhou, and F. Wang 2019 Effect of alternate corrosion factors on multiaxial low-cycle fatigue life of 2024-T4 aluminum alloy *J. Alloys Compd.* **772** 1–14

[12] T. Soysal and S. Kou 2019 Effect of filler metals on solidification cracking susceptibility of Al alloys 2024 and 6061 *J. Mater. Process. Technol.* **266** 421–428

IOP Conf. Series: Materials Science and Engineering **924** (2020) 012008 doi:10.1088/1757-899X/924/1/012008

Distribution of Hardness, Surface Roughness and Wettability of AISI 316L Induced by Shot Peening with Different Duration and Shooting Distance

Priyo Tri Iswanto[1*], Akhyar[2], Arif Faqihudin[1], and Hizba Muhammad Sadida[1]
[1]Department of Mechanical and Industrial Engineering, Universitas Gadjah Mada (UGM) Jl. Grafika No.2, Yogyakarta,55281, Indonesia
[2]Department of Mechanical and Industrial Engineering, Syiah Kuala University, Jl. Syech Abd. Rauf., No.7, Darussalam, Banda Aceh, 23111, Yogyakarta, Indonesia.

*priyotri@ugm.ac.id

Abstract. A research is conducted to analyze the effect of shot peening on distribution of hardness, surface roughness, and wettability of AISI 316 L. The shot peening process was performed on the surface of a 316L stainless steel sample with a shooting duration of 2, 4, 15, and 30 minutes, using a 0.4 mm diameter steel ball with a hardness of 40-50 HRC, the pressure of compressor held constant at 8 bar with nozzle diameter 5 mm, and the distance between the nozzle and the surface sample is 6 cm and 12 cm for each shooting duration. The results show that the shot peening can change the surface roughness. The closer the distance and the longer the duration of shot peening will reduce the surface roughness. Surface hardness increase with increasing of shooting duration and hardness distribution rate declines as the farther the distance from the surface of the shot peening.

***Keywords*:** Shot peening duration, shot peening distance, 316L stainless steel, surface roughness, hardness distribution, wettability.

1. Introduction

As one of the branches of science in the field of materials, biomaterial plays an important role in health problems. Currently, biomaterials continue to be developed as one of the healing methods in bone implant problems. Some materials that qualify as implant materials include austenitic stainless steel, titanium alloys, and cobalt alloys [1]. The material that is widely used for implanted materials is titanium and stainless steel, because it has high strength, good ductility, and can be received by the body [2]. Each year more than 2500 cases of implant failure occur in Canada [3]. Orthopedic bone plate implant fail in their application till 5.8% [4]. The most common cause of orthopedic bone plate implant failure is due to material fatigue and corrosion [5].

SS 316L is widely used as an implant material for making artificial joints, bone plates, stents, prostheses, as this material is corrosion resistant, has good mechanical properties and is relatively cheaper than other metal biomaterials [6]. There are several techniques for improving the physical and mechanical properties of materials by engineering the properties of the steel and its alloys. Austenitic stainless steel cannot use heat treatment so cold treatment is used, ie shot peening. The shot peening process is a process that is performed by firing high-speed steel balls on the surface of the work piece in order to provide a compressive residual stress on the surface of a component that can improve the material properties of the dynamic load [7]. Surface treatment will be very effective in improving fatigue and corrosion properties [8].

Stainless steel 316L material has been widely observed for biomedical applications, preparation for feeding, structural and submarine [9]. Research on the surface characterization of the material with severe shot peening then performed plasma nitriding at 400°C for 4 hours was performed on 316L stainless steel [10]. The effect of severe shot peening on microstructures and mechanical properties on the 31L sample was studied. The result is that the structure morphology on the surface tends to narrow down to nano size, forming the compression residual stress[7]. Variations of ultrasonic shot peening parameters such as ball material and diameter, sonotrode amplitude and coverage were studied in AISI 316L [11]. The effect of the relationship between surface hardening and roughness due to ultrasonic damping of the padding was investigated in AISI 316L [12]. Local corrosion, such as pitting and crevice corrosion were evaluated on austenitic AISI 316L stainless steel [13]. Surface hardness enhancement and surface compressive residual stress after shot peening delay crack formation and suppress crack growth rate [14]. Roughness of surface layer and droplet angel are significantly affect bio-adhesion process of orthopedic implant. Roughness of surface layer and small droplet angel can increase adsorption of protein which initiates the bone improvement in human bone [15].

The aim of this experiment is focused on the effect of shot peening treatment with the shooting distance of 6 and 12 cm, 2, 4, 15, and 30 minutes duration at each distance against surface roughness, hardness distribution and wettability on 316L stainless steel material.

2. Experimental method

2.1. Materials

The material used in this study was SS 316L which has chemical composition presented in Table 1.

Table 1. Chemical composition (%wt).

C	Mn	Si	Cr	Ni	Mo	S	P	Fe
0.03	1.08	0.97	16.78	10.87	1.89	0.003	0.04	Balance

2.2. Sample Preparation

The 3 mm SS 316L samples were cut into 37.5 x 20 mm sizes of 8 samples. The plate is then smoothed with sandpaper that has been set in the sander orbital sander or gradually polishing machine with mesh size 100-5000 and polished with auto sol. The shot peening process is performed on the shot peening tool shown in Figure 1. The tool consists of components such as a 3HP 120L 10bar 380V compressor, a nozzle diameter of a 5 mm, a steel ball with a 0.4 mm diameter and hardness 40-50 HRC, rubber hose, container box for steel ball tank on top of the table, sample clamp, and container box as the place to put sample that will be given shot peening treatment. The samples were given treatment with a distance of 6 cm and 12 cm, and the duration at each distance was 2, 4, 15, and 30 minutes. Pressure on the compressor was maintained at 8 bar during the shot peening process.

Figure 1. Shot peening apparatus.

Surface roughness testing was performed with Surfcom 120A ver 2.2 with stylus profilometer type with cut off setting of 0.8 mm, length 8 mm, and meas-mag 2000. Samples that have been given shot peening treatment, cleaned first with alcohol to remove impurities which may affect the accuracy of roughness data retrieval.

Surface hardness testing conducted to determine the hardness distribution on a cross-section sample of shot peening process. The test is done by using micro Vickers tool with Buchler micro met 2100 series. The hardness test begins by cutting the material transversely then indented with 25 grams of the load for 10 seconds to get a small indentation at a narrow distance. The indentation was performed 10 times starting with a distance of 15 μm from the surface of the shot peening result and continued to the next indentation in the material with a distance of 15 μm each. Therefore, the total distance measured in inner hardness is 150 μm. Then the indentation diagonal is measured and calculated to obtain the hardness distribution value on the piece of material.

The wettability test is carried out by using a wettability test instrument consisting of a holder for the material site, measuring cups for the liquid site, and dropper drops. Prior to testing, the surface of the sample is first cleaned with alcohol. This test is done by measuring the contact angle between the droplet and the sample surface. Testing of wettability is done by dripping water as much as 4 points on the surface of the sample, which is then taken 3 nearest data and taken the average of the three angles of contact. After dropping on the surface of the sample, the droplet is then shot with the camera at 5 seconds after the water drip. The droplet image is then processed with image software, resulting in an average contact angle from each sample.

3. Results and discussion

3.1. Surface roughness

The surface roughness test result is the mean surface roughness value (Ra) obtained from the test on each sample. The shot peening treatment has an effect on the surface roughness of SS 316L as shown in Figure 2. The shot peening treatment makes the surface roughness of the material increasing compared to the surface roughness of raw material or non-treatment (NT). This provides information that the surface of the material changes to be rougher. The shot peening process provides a scar or indentation resulting from the impact of steel balls shot at high speed and pressure on the surface of the material. Thus, on the surface of the material will form an indentation pit which will make the surface of the material becomes rougher.

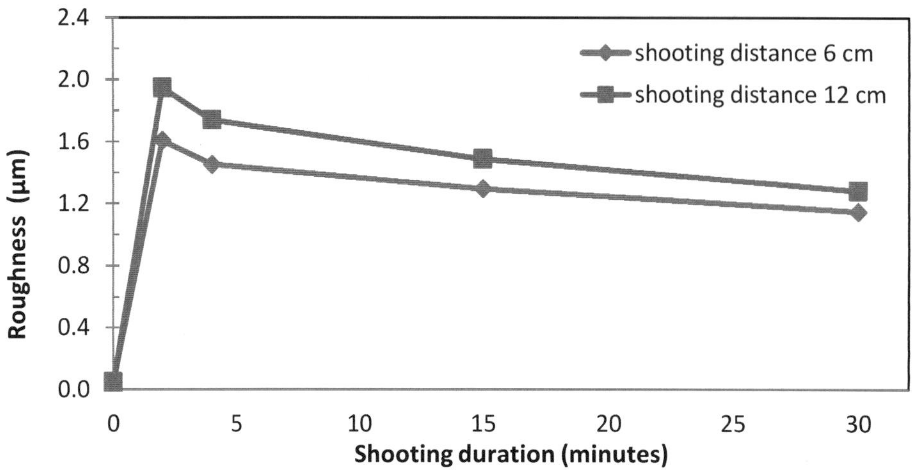

Figure 2. Surface Roughness of Shot Peened SS316 L.

The duration of the shot peening process also gives effect to the surface roughness of the material. The subjects given a shot treatment for 4 minutes had a lower surface roughness than the material with shot peening treatment for 2 minutes. This decline in roughness values occurs with the longer duration of shot peening. The shot peening treatment causes an irreversible plastic deformation. From these data, there was an increase in roughness at 2-minute duration for both shooting distance. This increase occurs because of the formation of the pit at several locations on the surface of the sample. Unlike the case when the duration of 4, 15, and 30 minutes, the value of roughness tends to fall. This is because the entire surface of the test sample begins to be covered by a new pit due to the impact of steel balls over and over again. Thus, the height of the mound to the pit is reduced and the depth of the pit is no longer affected by repeated collisions. The longer the duration of shot peening, the more the number of steel balls are hit and hit the surface of the material, so that the surface area that has not been hit by steel balls will be shot by the impact of steel balls as well as causing the surface of the material to become flattered and smoother.

The different of shooting distance in the shot peening process also gives different roughness values. This provides information that the sample shot at a distance of 6 cm has a smoother surface or a low roughness than the sample shot at a distance of 12 cm. This indicates that the further the shooting distance is used, the higher the resulting surface roughness. The increase in roughness is due to the small collision intensity of the ball so as not to strike all sample surfaces. In contrast to the closer the shooting distance, the intensity of the steel ball shooting becomes larger and shooting another part of the surface.

3.2. Hardness Distribution

Hardness distribution is resulted from hardness testing at cross-section. This test is carried out using a small load of 10 gf with a duration of 10 seconds indentation to obtain a small indentation and distance as close as possible between indents. To obtain cross-sectional hardness test data, indented 10 times with each distance 15 µm. The hardness distribution in cross-section of the sample which has been subjected to shot peening. Figure 3 shows that the further distance from the surface of the hardness value decreases. Indentation point with distance 15 µm from surface treatment of shot peening, has high hardness value. While, the indent point with a distance of 150 µm from the surface of the shot peening treatment results, has a low hardness. A decrease in the value of hardness occurs as the distance from the surface of the shot peening results. This indicates that the hardening effect of the shot peening sample occurs only on the surface of the material up to a certain depth. This decrease occurs because the effect of the dislocation density cannot reach into the larger material.

Figure 3(a) shows cross-sectional hardness distribution data on shot peening material with 6 cm range. While the hardness cross-section data distribution on shot peening material with 12 cm shot distance is shown in Figure 3(b). Both shot distance, it showed that the cross-sectional distribution of hardness are identical or resemble each other, as the distance from the shot peening surface further decreases the hardness. However, the two graphs have the difference that the slope of the graph line in Figure 3(a) is more sloping than the slope of the graph in Figure 3(b). This occurs because the difference in the distribution of indented hardness values at each point provides a smaller difference for the shot peening material with a distance of 6 cm. With close proximity, then the intensity of the collision will be more and more so that the dislocation density is formed greater. In contrast to the peak shot material with a distance of 12 cm, the intensity of the shot impact is smaller so as to provide a greater difference in the distribution of hardness values at each indentation point. This shows that shot peening by using a smaller firing range will have a deeper effect of hardness.

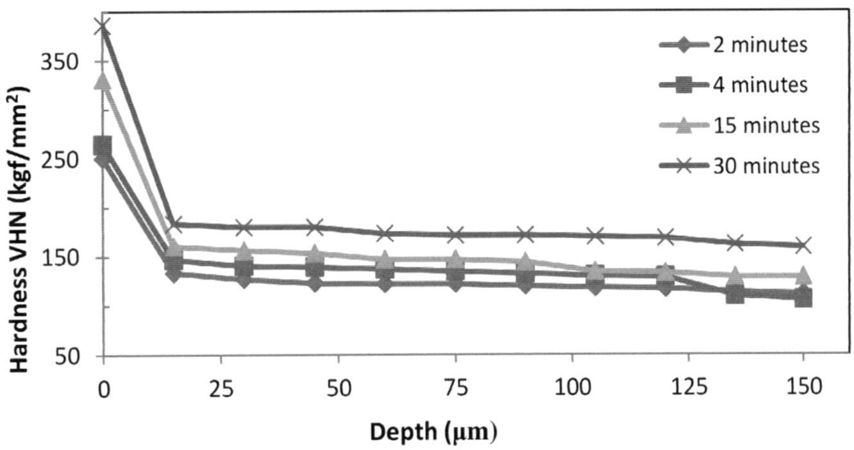

(a)

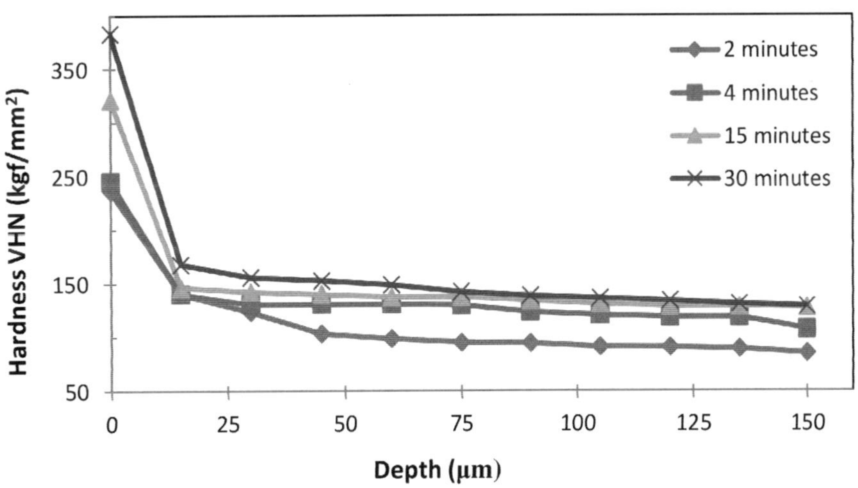

(b)

Figure 3. Hardness distribution on shot peened material with a shooting distance 6 cm (a) and a shooting distance 12 cm (b).

3.3. Wettability

Shot peening affects the surface properties of SS 316L, so it can change the wetting properties of the implantable material. The wettability test is done by dripping water onto the surface of NT samples and shot peening results on 316L stainless steel material. Water droplets above the surface of the sample are drawn so that they can be analyzed and determined by the wetting properties of the sample. The result of the wettability test to distance and duration of shot peening can be seen in Figure 4.

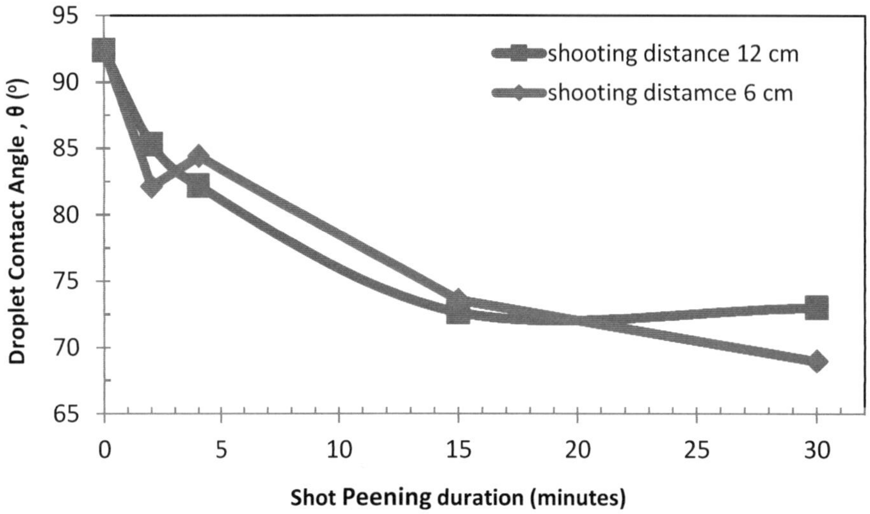

Figure 4. Droplet Contact Angle of Shot Peening SS 316 L.

It was found that the shot peening process decrease the droplet contact angle from the original 92.42° for the NT as the control sample, then decrease to 82.13° for the sample with a 6 cm shot peening distance and dropped to 85.29° for the sample with distance shoot 12 cm. Droplet contact angle after shot peening process with duration of 2 minutes in both shooting distance decrease angle which is inversely proportional to the increase of it roughness. The shot peening reduces the droplet contact angle from ± 92° (control sample) to a droplet angle of approximately 68-85°. From the data, we can know that NT 316L stainless steel classified as having hydrophobic properties ($\theta > 90°$). Meanwhile, the sample shot peening results differ from the wetting character of the NT where the roughness value is different. Shot peening's material has droplet contact angles ranging from 68-85° ($\theta < 90°$) so they have hydrophilic properties.

4. Conclusion

The effect of shot peening treatment with the shooting distance of 6, and 12 cm, then 2, 4, 15, and 30 minutes duration at each distance against surface roughness, hardness distribution and wettability on 316L stainless steel material was observed. The conclusion of this experiment is increased of surface roughness by shot peening process on AISI 316L. Moreover, the longer the shooting duration and the closer the shooting distance will reduce the roughness of the surface of the material, so it obtains a low roughness value or a smoother surface. Shot peening on AISI 316L gives a deepening effect of hardness. Surface hardness significantly increase with increasing of shooting duration and hardness distribution rate declines as the farther the distance from the surface of the shot peening. Shot peening on AISI 316L causes a decrease in droplet contact angle and causes a change in the properties of hydrophobic to hydrophilic.

Acknowledgment

This work is supported by the Research Fund of Ministry of Research, Technology, and Education, Republic of Indonesia (Hibah PDUPT- Universitas Gadjah Mada) and The Departement of Mechanical and Industrial Engineering (Hibah DTMI - Universitas Gadjah Mada).

References

[1] R. Rokicki, T. Hryniewicz, and K. Rokosz 2008 Modifying Metallic Implants with Magnetoelectro polishing *Med Device Ind* **30** 102-111

[2] S.M. Perren, R. Mathys, and O. Pohler 2000 Priciples of Fracture Management: Implants and Materials In Fracture Fixation New York:AO Publishing

[3] Holt R. T., and Wallace W., 1980, Failure analysis of Some Orthopedic Implant, National Research Council of Canada (Ontario) Div. of Mechanical Engineering, Canada

[4] Ogbemudia A.o., Umebese P. F. A., 2006, Implant failure in ostesynthesis of fracture of long bones, JMBR; A Peer-review Journal of Biomedical Sciences, Vol. 5 No. 2, pp. 75-78

[5] Davis, J.R., 2003, Handbook of Materials for Medical Devices, ASM International, pp: 51-135.

[6] A. A. Ahmed, M. Mhaede, M. Basha, M. Wollmann, and L. Wagner 2015 Surface & Coatings Technology The effect of shot peening parameters and hydroxyapatite coating on surface properties and corrosion behavior of medical grade AISI 316L stainless steel *Surf. Coat. Technol* **280** 347–358

[7] D. Cahyadi 2005 Mechanical Static Characteristics of UNS G10450 Steel Experiencing Shot Dizzy Process in Indonesian *Traksi Indonesia* **3** 2

[8] Nishida S., N. Hattori, S.W. Wang, J. Y. Ye, 1999, Surface Treatment V, WIT press, pp. 161-169

[9] S. Bagherifard, S. Slawik, I. Fernández-Pariente, C. Pauly, F. Mücklich, and M. Guagliano, 2016 Nanoscale surface modification of AISI 316L stainless steel by severe shot peening *Mater. Des.*, **102** 68–72

[10] M. Jayalakshmi, P. Huilgol, B. R. Bhat, and K. U. Bhat 2016 Microstructural characterization of low temperature plasma-nitrided 316L stainless steel surface with prior severe shot peening *Mater. Des* **108** 448–454

[11] J. Marteau, M. Bigerelle, P. E. Mazeran, and S. Bouvier 2015 Relation between roughness and processing conditions of AISI 316L stainless steel treated by ultrasonic shot peening *Tribol. Int.*, **82** 319–329

[12] J. Marteau and M. Bigerelle 2015 Relation between surface hardening and roughness induced by ultrasonic shot peening *Tribol. Int* **83**105–113

[13] T. Liptâkovâ, M. Lovišek, A. Alaskari, and B. Hadzima 2016 Surface state effect of welded stainless steels on corrosion behavior *Acta Metall. Slovaca* **22** 1 44–51

[14] Iswanto P.T., S. Nishida, N.Hattori, 2006, Influence of Mean Stress on the StressControlled Fatigue of Structural Stainless Steels, Japan Soc. Str. Fract. Mats. Inter. J. pp. 1-7

[15] B. Arifvianto, M. Mahardika, P. Dewo, P. T. Iswanto, and U. A. Salim, Effect of surface mechanical attrition treatment (SMAT) on microhardness, surface roughness and wettability of AISI 316L, *Mater. Chem. Phys.*, vol. 125 (2011), no. 3, pp. 418–426.

IOP Conf. Series: Materials Science and Engineering **924** (2020) 012009 doi:10.1088/1757-899X/924/1/012009

Study of Effectiveness of Integrated Public Space

T. Herry Rachmatsyah* **and Chicilia**
Universitas Prof. Dr. Moestopo (Beragama)
Jl. Hang Lekir I No. 8, Jakarta Pusat 10270, Indonesia

*herry.rachmatsyah@dsn.moestopo.ac.id

Abstract. Child Friendly Integrated Public Space is one of the important policies/programs of the Special Capital Region of Jakarta Provincial Government to realize Jakarta as a Child Decent City. It is a place and/or open space that combines the activities of citizens by implementing the 10 Main Program for the Family Empowerment and Welfare with the Child Decent City Program. To find out the effectiveness of the Kenanga Child Friendly Integrated Public Space in Cideng Subdistrict, indicators for fulfillment of functions were used as contained in Article 4 of the Governor Regulation of the Special Capital Region of Jakarta Province Number 123 Year 2017 concerning Management and Requirements for Facilities and Infrastructure of Child Friendly Integrated Public Spaces. Data were collected through observations, document reviews, and in-depth interviews with informants consisting of the Secretary of the Family Empowerment and Welfare Activator Team, consultants, managers, and user communities. The collected data were analyzed using quantitative and qualitative descriptive analysis techniques. The results of the study concluded that the Kenanga Child Friendly Integrated Public Space has been effective in all indicators of fulfillment of functions, but still requires assistance from the government in terms of maintenance of infrastructure and facilities as well as the development of a management information system that is supported by a complete set of hardware, software, databases, as well as computer networks and data communications.

Keywords **:** Effectiveness, Child Friendly Integrated Public Space, Child Decent City

1. Introduction

Indonesia has ratified the Convention on the Rights of the Child through the Presidential Decree Number 36 of 1990 concerning Ratification of the Convention on the Rights of the Child and has signed the World Fit for Children Declaration. Therefore, Indonesia is obliged to improve and fulfill the rights of children. The next step taken was to amend the 1945 Constitution and to issue the of the Minister of Women Empowerment and Child Protection Regulation Number 2 of 2009 concerning Child Decent Regencies/Cities Policy. Jakarta then became one of the cities that had to be developed into Child Decent City based on the Minister of Women Empowerment and Child Protection Decree Number 56 of 2010 concerning Appointment and Stipulation of Provinces that Develop Child Decent Regencies/Cities. One of the things done by the Special Capital Region of Jakarta Provincial Government to accelerate the achievement of Child Decent City is to build child friendly integrated public spaces.

According to Garau, public space is a place where people perform a number of functions that are shared with all: moving from place to place; enjoying public recreational and cultural opportunities; visiting other parts of the city or other private locations; shopping; meeting others or simply strolling [1]. Public space is an important element for a city. Public space can help build a sense of togetherness, civic identity, and culture. Public space also facilitates social capital, economic development, and community revitalization. Having access to public space does not only improve the quality of life but is also the first step towards community empowerment and greater access to institutional and political space. The active and sustainable use of public space will lead to a well-maintained, healthy and safe urban environment, making a city an attractive place to live and work. Public spaces can consist of open spaces (for example roads, walkways, parks and gardens) and protected spaces where all important civic functions can be carried out in all weather conditions (for example schools, libraries and museums). In 1999 UNICEF developed a children's rights-based program called Child Friendly Spaces to support and protect children. In Indonesia, this program was then implemented into Child Friendly Integrated Public Spaces. Referring to the Governor Regulation of the Special Capital Region of Jakarta Province Number 123 Year 2017 concerning Management and Requirements for Facilities and Infrastructure of Child Friendly Integrated Public Spaces, child friendly integrated public space is a place and/or open space that combines the activities of citizens by implementing the 10 Main Program for the Family Empowerment and Welfare with the Child Decent City Program [2]. What is meant by a child is a person who is not yet 18 years old, including a fetus who is still in the womb. Research on child friendly integrated public spaces has actually been conducted by other researchers.

Research conducted by Hernowo and Navastara concluded that the Bahari Child Friendly Integrated Public Space in South Jakarta had a variety of characteristics that must be present in a public space. It was complete, safe, had many trees, was comfortable, had many activities, and easily accessible [3]. Research conducted by Purnama on the South Gandaria Child Friendly Integrated Public Space in South Jakarta concluded that the public space was able to attract the interest of residents to come there because of its diverse facilities [4]. Research conducted by Herlina and Nadiroh on the Utama Child Friendly Integrated Public Space in West Jakarta concluded that the public space was not maximal and optimal in fulfilling children's rights [5]. Research conducted by Prakoso and Dewi on 10 Child Friendly Integrated Public Spaces spread across various regions in Jakarta concluded that children had a sense of attachment and positive assessment of the child friendly integrated public spaces [6]. Research conducted by Rahmiati and Prihastomo on the Kambang Iwak Park Child Friendly Integrated Public Space in Palembang, South Sumatra concluded that the facilities and elements in the public space were not specifically specific to children and might be harmful to them. The public space was actually a pleasant and comfortable public space, but safety and health factors had not been considered in its design [7].Research conducted by Lestari and Prima on Traffic Park Child Friendly Integrated Public Space and Janggan Park Child Friendly Integrated Public Space in Denpasar, Bali concluded that kindergarten schools utilized both parks for learning activities outside the classroom as a learning model [8]. Research conducted by Nurhasbi on Kalijodo Child Friendly Integrated Public Space in West Jakarta concluded that the performance of the public space had not been satisfactory in terms of safety, health, security, comfort, attractiveness, accessibility, and facilities [9]. Research conducted by Mustaram on Nusa Indah Child Friendly Integrated Public Space in West Jakarta concluded that in addition to having a positive impact, the existence of the public space also had a negative impact in the form of disorder, waste and reduced security [10]. Research conducted by Utami, Mugnisjah, dan Munandar on Sungai Bambu Child Friendly Integrated Public Space in North Jakarta, Kenanga Child Friendly Integrated Public Space in Central Jakarta, Bahari Child Friendly Integrated Public Space in South Jakarta, Kembangan Child Friendly Integrated Public Space in West Jakarta Barat, and Cililitan Child Friendly Integrated Public Space in East Jakarta concluded that these public spaces as active spaces, ecological spaces, individual spaces, and cultural spaces were not yet optimal in supporting child development [11]. Research conducted by Faedlulloh, Prasetyanti, and Indrawati on Sungai Bambu Child Friendly Integrated Public Space dan Sunter Jaya Berseri Child Friendly Integrated Public Space in North Jakarta concluded that the public spaces were not merely a means of gathering and playing for children, but also accommodating social interaction that could increase the potential for deliberative democracy [12]. Research conducted by Simatupang

IOP Conf. Series: Materials Science and Engineering **924** (2020) 012009 doi:10.1088/1757-899X/924/1/012009

on the Kalijodo Child Friendly Integrated Public Space in West Jakarta concluded that the public space was the result of social processes that were influenced and shaped by the interrelation between structuring forces and various actors [13].

All of these studies generally examined the design aspects of a public space by relating it to the characteristics that must be possessed by a child friendly integrated public space. The success of a child friendly integrated public space is not solely in the hands of the town planner, urban designer, or architect; but it relies also on people adopting, using and managing the space. Effectiveness is one measure to assess this success. Effectiveness is also used as one of the criteria for evaluating policies/programs. According to Dunn, effectiveness refers to the achievement of a valued outcome. Effectiveness is measured in terms of units of products or services or their monetary value [14]. So far there has been no research on child friendly integrated public spaces that highlights aspects of public space as a policy/program that after implementation must be evaluated for effectiveness using indicators contained in the policy/program. This then became the background of this research. The problem raised in this study was "How was the effectiveness of the Kenanga Child Friendly Integrated Public Space in Cideng Subdistrict, Central Jakarta?" Thus, the purpose of this study was to describe the effectiveness of the Kenanga Child Friendly Integrated Public Space in Cideng Subdistrict, Central Jakarta.

2. Method

This research was descriptive in accordance with the purpose of the study which was to describe the phenomenon that was the object of research. The Kenanga Child Friendly Integrated Public Space was chosen as a research location because it was the first generation of child friendly integrated public space which was a pilot project in Jakarta (inaugurated on May 30, 2015) [15]. Data was collected through observations, document reviews, and in-depth interviews with 7 informants consisting of 1 Secretary of the Family Empowerment and Welfare Activator Team, 1 consultant, 2 managers, and 3 user communities. The collected data were analyzed using quantitative and qualitative descriptive analysis techniques. For the purposes of quantitative analysis, scoring was used to measure effectiveness. Informants were asked to provide value using ineffective, less effective, effective, and very effective criteria. Not effective given a perception value of 1, less effective given a perception value of 2, effectively given a perception value of 3, and very effective given a perception value of 4. The indicators assessed were the fulfillment of functions as contained in Article 4 of the Governor Regulation of the Special Capital Region of Jakarta Province Number 123 Year 2017 concerning Management and Requirements for Facilities and Infrastructure of Child Friendly Integrated Public Spaces: public open parks; a place for community interaction of all ages; partnership infrastructure and facilities between the regional government and the community in fulfilling children's rights; part of infrastructure and facilities for Child Decent City; green open space and groundwater absorption site; infrastructure and facilities for citizens' social activities including the development of knowledge and skills of the Family Empowerment and Welfare cadres; efforts to increase family income, family consultation and information center; green, organized, beautiful, and comfortable family yard; and management information system. In-depth interviews were conducted with the results of assessments given by the informants. Qualitative analysis was used further to understand and explain the results of quantitative analysis.

3. Results and discussion

To find out the effectiveness of the Kenanga Child Friendly Integrated Public Space, measurements were made on indicators of fulfillment of functions as contained in Article 4 of the Governor Regulation of the Special Capital Region of Jakarta Province Number 123 Year 2017 concerning Management and Requirements for Facilities and Infrastructure of Child Friendly Integrated Public Spaces. Table 1 shows that the public space can be considered effective overall. All indicators that were assessed had effective and very effective values where the effective perception value (80.95%) was more than the very effective perception value (19.06%). However, specifically for the part of infrastructure and facilities for Child Decent City indicator and the management information system indicator, none of the informants rated it very effectively. Each indicator assessment result will be discussed sequentially.

Table 1. Fulfillment of Child Friendly Integrated Public Space Functions.

Number	Functions	Level of Effectiveness							
		Ineffective	%	Less Effective	%	Effective	%	Very Effective	%
1	Public open parks	0	0	0	0	5	71,43	2	28,57
2	A place for community interaction of all ages	0	0	0	0	5	71,43	2	28,57
3	Partnership infrastructure and facilities between the regional government and the community in fulfilling children's rights	0	0	0	0	6	57,14	1	42,86
4	Part of infrastructure and facilities for Child Decent City	0	0	0	0	7	100	0	0
5	Green open space and groundwater absorption site	0	0	0	0	4	57,14	3	42,86
6	Infrastructure and facilities for citizens' social activities including the development of knowledge and skills of the Family Empowerment and Welfare cadres	0	0	0	0	6	57,14	1	42,86
7	Efforts to increase family income; family consultation and information center	0	0	0	0	6	57,14	1	42,86
8	Green, organized, beautiful, and comfortable family yard	0	0	0	0	5	71,43	2	28,57
9	Management information system	0	0	0	0	7	100	0	0
	Total	0	0	0	0	51	80.95	12	19,06

3.1. Public open parks

The Kenanga Child Friendly Integrated Public Space was designed as a public open park where all people can visit it for free. It is located on Jl. Makian No. 1, Cideng Subdistrict, Gambir District, Central Jakarta with an area of 2,583m^2. It was built on the right location in a densely populated residential area (consisting of 10 RW and 113 RT) with relatively high poverty rates. The existence of the public space is very much needed because in these locations there are not enough infrastructure and facilities available, especially the playground needed for children's growth and development as well as other infrastructure and facilities expected by the local residents. It has a building (indoor) with an area of 144m^2, in which there is a library, multipurpose room, lactation room or nursing room, and a manager's room. For outdoor, various game facilities such as swings and sports arenas such as futsal court, basketball court, and badminton court are provided in accordance with the needs of residents during social mapping prior to the construction of the public space. It is also equipped with an interactive park (for playing and relaxing), a nutritional park (for farming), and supporting facilities for the park (parking area, park bench, manager post, lighting, etc.). Various facilities provided allow children to be able to play happily because these facilities have been designed to be friendly and safe for children.

3.2. A place for community interaction of all ages

The Kenanga Child Friendly Integrated Public Space can be utilized by all groups of residents starting from the age of zero, toddlers, to the elderly. All can come visit and take advantage of the facilities in the public space. Around 80 children visit it to play and learn and use various facilities every day. Even on holidays, the number of children who visit can reach 150 people. It does not carry out activities only for children, the management also organizes various activities that involve the active role of the Family Empowerment and Welfare cadres, mothers, teenagers, and the elderly such as training, games, learning forums, socialization and a forum for meetings between local neighborhood association and subdistrict or district parties. It has also become a place for residents to exercise and gather to communicate with and interact with fellow citizens.

3.3. Partnership infrastructure and facilities between the regional government and the community in fulfilling children's rights

The development of the Kenanga Child Friendly Integrated Public Space was carried out by the Special Capital Region of Jakarta Provincial Government by involving the private sector, PT Pembangunan Jaya in the form of Corporate Social Responsibility funding for the construction and provision of facilities. The development process begins with a social mapping of the public space site conditions carried out by PT Pembangunan Jaya together with the Community, Women, and Family Planning Empowerment Agency and the Family Empowerment and Welfare Activator Team by involving the participation of residents and community leaders using focus group discussion to compile the design and activities of the public space to be built. This was done so that the public space that was built, fits the needs of children and local residents.

3.4. Part of infrastructure and facilities for Child Decent City

Child friendly integrated public space development is an implementation of the Minister of Women Empowerment and Child Protection Regulation Number 11 of 2011 concerning Child Decent Regency/City Development Policy and the Minister of Women Empowerment and Child Protection Regulation Number 12 of 2011 concerning Child Decent Regency/City Indicators. The Kenanga Child Friendly Integrated Public Space has been equipped with a variety of adequate infrastructure consisting of multipurpose rooms, libraries, sports fields (basketball, futsal, and badminton), interactive parks (for playing and relaxing), nutrition parks (for planting), children's playgrounds (in the form of children's play facilities), nutritional ponds (for raising fish), pedestrian paths (for jogging, walking, and cycling), and supporting facilities for parks (parking areas, park benches, management posts, lighting, etc.). Various facilities provided in the public space allow children to play happily. Children feel happy because they get a place to gather with their peers to play futsal, swing, and other games. The existence of infrastructure and facilities provided in the public space is not only felt by children who live around its location, even children who live far from it, such as children who live in Duri Pulo Subdistrict. They came to visit to use the various game facilities available in the public space. For two to three months after the inauguration of the Kenanga Child Friendly Integrated Public Space, many parties supported the public space, but entered into the Phase II (2016) and Phase III (2017) development, support from the government diminished because the Kenanga Child Friendly Integrated Public Space was considered to be independent and the government concentrated more on other child friendly integrated public spaces which were built in the Phase II and Phase III. In Phase I (2015), budgeting and maintenance were the responsibility of the Office of Empowerment, Child Protection, and Population Control, Central Jakarta Administrative City while in 2016 the budgeting was handed over to Cideng Subdistrict but its maintenance was still the responsibility of the Office of Empowerment, Child Protection, and Population Control, Central Jakarta Administrative City. In fact, if there is a damage and then reported by the managers to the Office of Empowerment, Child Protection and Population Control Central Jakarta Administration City, the report has never been followed up. This causes the managers together with residents to initiate self-help to repair and/or replace damaged infrastructure and facilities. The managers hope that the public space will continue to get assistance at least in terms of maintenance of infrastructure and facilities.

3.5. Green open space and groundwater absorption site

The land of the Kenanga Child Friendly Integrated Public Space that can be built is limited to a maximum of $1500m^2$, while the rest is used as a green open space which also functions as a groundwater absorption site. In the public space, 15 biopore holes have been built, which were the Special Capital Region of Jakarta Province Water Service program. Through these biopore holes, rain

IOP Conf. Series: Materials Science and Engineering **924** (2020) 012009 doi:10.1088/1757-899X/924/1/012009

water can be collected so that it can get back into the ground so that the availability of groundwater for locations around the public space is never dry and always clean.

3.6. Infrastructure and facilities for citizens' social activities including the development of knowledge and skills of the Family Empowerment and Welfare cadres

The Kenanga Child Friendly Integrated Public Space has carried out various activities that not only include the activities for children but also has carried out activities for the Family Empowerment and Welfare cadres, mothers, teenagers, and the elderly. The existence of a multipurpose room can be utilized by community members as a place for training and socialization to develop knowledge and skills. The activities that have been carried out include training in the use and management of waste in collaboration with Rumah Pelangi and traditional dance training for children. It has also carried out other activities such as the elderly medical examination and free dental examination in collaboration with Obor Foundation, the Faculty of Medicine Universitas Atmajaya, and Prodia Clinical Laboratory; Healthy with Milo; as well as futsal tournaments for children and adolescents. In addition to the multipurpose room, there is also a library that is very useful for children and adolescents to get reading resources so that they can add knowledge and insight.

3.7. Efforts to increase family income; family consultation and information center

The Kenanga Child Friendly Integrated Public Space has conducted various training to produce something that can be sold, for example training in knitting in collaboration with Crafts Community, training in making bread and cakes in cooperation with PT Bogasari, training in growing vegetables and fruit, and training in processing vegetables and fruit after harvest. The products of the citizens' production can also be exhibited and sold at the Family Empowerment and Welfare Mart and bazaar held at the public space. The efforts made are not only in the form of increased incomes but also in reducing expenses by implementing a low-cost market where goods are sold at subsidized prices that are lower than market prices so that citizens can meet their daily needs with lower expenses. It also seeks to become a center for information and family consultation by holding various socialization activities such as socialization of clean and healthy behavior in cooperation with PT Unilever, socialization of parenting, socialization of the dangers of narcotics and juvenile delinquency, and socialization of prevention of violence against women and children.

3.8. Green, organized, beautiful, and comfortable family yard

The Kenanga Child Friendly Integrated Public Space was designed as a family yard that is overgrown with various trees and plants to add to the environmental greenish, order, beauty, and comfort for local residents who live in a densely populated area.

3.9. Management information system

The Family Empowerment and Welfare has a database containing citizen data that can be used to determine the potential for community participation if the managers want to carry out an activity in the Kenanga Child Friendly Integrated Public Space. The public space is also a place to socialize the results of the coordination meeting between local neighborhood association and subdistrict or district parties. The benefits of the public space are not only for citizens but also for the government. Various forms of citizen data collection placed at the public space can be filled out by residents where the data can later be used in the formulation, implementation, monitoring, and evaluation of policies/programs by the government. However, the public space does not yet have a management information system that is supported by a complete set of hardware, software, databases, as well as computer networks and data communications that can produce information that is useful for the managers in making decisions and performing managerial functions.

IOP Conf. Series: Materials Science and Engineering **924** (2020) 012009 doi:10.1088/1757-899X/924/1/012009

4. Conclusion

Child friendly integrated public space is a place and/or open space that combines the activities of citizens by implementing the 10 Main Program for the Family Empowerment and Welfare with the Child Decent City Program. One of the pilot and longest operating child friendly public spaces in Jakarta is the Kenanga Child Friendly Integrated Public Space. The success of the public space depends very much on the people who adopt, use, and manage the public space. Effectiveness is one measure to assess this success. To find out the effectiveness of the Kenanga Child Friendly Integrated Public Space, measurements were made on indicators of fulfillment of functions as contained in Article 4 of the Governor Regulation of the Special Capital Region of Jakarta Province Number 123 Year 2017 concerning Management and Requirements for Facilities and Infrastructure of Child Friendly Integrated Public Spaces: public open parks; a place for community interaction of all ages; partnership infrastructure and facilities between the regional government and the community in fulfilling children's rights; part of infrastructure and facilities for Child Decent City; green open space and groundwater absorption site; infrastructure and facilities for citizens' social activities including the development of knowledge and skills of the Family Empowerment and Welfare cadres; efforts to increase family income, family consultation and information center; green, organized, beautiful, and comfortable family yard; and management information system. The results shows that the Kenanga Child Friendly Integrated Public Space can be considered effective in fulfilling these functions. However, there are two things that need attention. First, it still requires assistance from the government in terms of maintaining infrastructure and facilities. Secondly, it is necessary to develop a management information system that is supported by a complete set of hardware, software, databases, as well as computer networks and data communications that can produce information that is useful for the managers of the public space in making decisions and performing managerial functions.

References

[1] Pietro Garau, *Global Public Space Toolkit: From Global Principles to Local Policies and Practice*, Nairobi: United Nations Human Settlements Programme (UN-Habitat), 2015

[2] Governor Regulation of the Special Capital Region of Jakarta Province Number 123 Year 2017 concerning Management and Requirements for Facilities and Infrastructure of Child Friendly Integrated Public Spaces

[3] Endi Hernowo and Ardy Maulidy Navastara, "Characteristics of the Bahari Child Friendly Integrated Public Space in Cilandak District, South Jakarta", Jurnal Teknik ITS Vol. 6 No. 2, pp. 567-570, June 2017

[4] Muhammad Sega Purnama, "Application of the Concept of Child Friendly Integrated Public Spaces in Settlements Case Study of the South Gandaria Child Friendly Integrated Public Space", Faktor Exacta Vol. 10 No. 4, pp. 401-405, December 2017

[5] Netti Herlina and Nadiroh, "Strategic Role of Child Friendly Integrated Public Spaces in the Context of Fulfilling Children's Rights to the Environment", Jurnal Pendidikan Usia Dini Vol. 12 No. 1, pp. 104-117, April 2018

[6] Susinety Prakoso and Julia Dewi, "Children's Sense of Attachment to Child Friendly Integrated Public Spaces", NALARs Vol.17 No. 1, pp. 1-10, January 2018

[7] Desti Rahmiati and Bondan Prihastomo, "Identification of the Implementation of Child Friendly Integrated Public Spaces Concept in the Kambang Iwak Park Palembang", Vitruvian Vol. 8 No. 1, pp. 29-42, October 2018

[8] Putu Indah Lestari and Elizabeth Prima, "The Role of Child Friendly Integrated Public Spaces for Early Childhood", Jurnal Obsesi Vol. 4 Issue 1, pp. 471-481, December 2019

[9] Muhammad Nurhasbi, "Performance Evaluation of the Kalijodo Child Friendly Integrated Public Space West Jakarta", Jurnal Mahasiswa Jurusan Arsitektur Vol. 7 No. 1, pp. 1-12, March 2019

[10] Agnatasya Listianti Mustaram, "Analysis of the Existence Of Child Friendly Integrated Public Space in the Housing Complex Location Object of Study: Nusa Indah Child Friendly Integrated

Public Space, West Jakarta", Jurnal Muara Sains, Teknologi, Kedokteran, dan Ilmu Kesehatan Vol. 2 No. 2, pp. 492-501, October 2018

[11] Putri Kharisma Utami, Wahju Qamara Mugnisjah, and Aris Munandar, "Benefit-Based Urban Community Participation In Shaping Children Friendly Public Park", Jurnal Lanskap Indonesia Vol. 8 No. 2, pp. 28-38, October 2016

[12] Dodi Faedlulloh, Retnayu Prasetyanti, and Indrawati, "Initiating Deliberative Democracy-Based Public Space: A Study on Dynamic Management of Child Friendly Integrated Public Space in North Jakarta", Spirit Publik Vol. 12 No. 2, pp. 43-60, October 2017

[13] Sahala Simatupang, "Social Process in the Production of Public Space 'Kalijodo Child Friendly Integrated Public Space' in Jakarta", Jurnal Arsitektur Vol. 5 No. 1, pp. 43-55, August 2017

[14] William N. Dunn, *Public Policy Analysis: An Integrated Approach Sixth Edition*, New York: Routledge, 2018

[15] Riena Juliannisa Surayuda. "Community Center and Collective Memory Contestation: Case Study of the Kenanga Child Friendly Integrated Public Space in Cideng, Central Jakarta", Masyarakat Vol. 21 No. 2, pp. 233-261, August 2016

IOP Conf. Series: Materials Science and Engineering **924** (2020) 012010 doi:10.1088/1757-899X/924/1/012010

Effect of Nano Hydroxyapatite in Toothpaste on Controlling Oral Microbial Viability

Aulia Rifada[1*], Bannan Muthi'atul Af'idah[1], Wafa Aufia[1], Agnia Vibriani[1], Maria Maghdalena[1], Kurniawan Eko Saputro, Dwi Wahyu Nugroho[1], M. Arief Iskandar[1], Arief Cahyanto[2], Alfian Noviyanto[1,3,‡], and Nurul Taufiqu Rochman[4]

[1]Nano Center Indonesia, Jl. PUSPIPTEK South Tangerang, Banten 15314 Indonesia

[2]Departement of Dental Materials Science and Technology, Faculty of Dentistry, Padjadjaran University

[3]Department of Mechanical Engineering, Mercu Buana University, Jl. Meruya Selatan, Kebun Jeruk, West Jakarta 11650, Indonesia

[4]Research Center for Metallurgy and Materials, Indonesian Institute of Sciences, PUSPIPTEK Tangerang Selatan, Banten 15314, Indonesia

*auliarifada@nano.or.id;
‡a.noviyanto@nano.or.id

Abstract. The aim of this study is to examine the effect of nano hydroxyapatite in the toothpaste and its effect on the pH and microbial activity. Nano-hydroxyapatite (nHAp) is considered one of the materials that have high biocompatibility for biomimetic material due to its chemical and morphological similarity with dental apatite. Additionally, it has been documented to possess antibacterial potentials. The present study was conducted to identify the relationship between oral microenvironment pH change and its role in the Streptococcus mutans viability, a common pathogen in the oral cavity. Change in pH is closely related to number of Streptococcus mutans as the main cariogenic organism and acid-producing bacteria. The study was carried out using 0.25%, 0.7%, and 1.5% concentration of HAp in toothpaste formulation and commercial toothpaste as control. Our studies showed that the most significant pH fall was observed in 1.5% and 0.7% nHAp. However, in 60 minutes, all nHAp groups were able to restore pH into neutral, especially in the 0.7% nHAp, which reached a pH of 7. As a comparison, the commercial toothpaste only returned to 6 within 1 hour. Experimental method of this study is In vitro oral microenvironment pH and microbiological analysis on teeth fragment. The detailed data about microstructure and antibacterial activity will be presented.

Keywords **:** Nano-hydroxyapatite, Streptococcus mutans, Dental, Toothpaste Formulation.

1. Introduction

Dental caries is the most common oral biofilm-associated disease caused by a complex interaction between cariogenic bacteria, hosts (saliva and teeth), and carbohydrate consumption [1]. The important factor that determines the onset of dental caries is the presence of Streptococcus mutans as the major cariogenic bacteria. S. mutans is acidogenic bacteria which has the ability to ferment carbohydrate into acid. By the time, the acid will accumulate then diffuse through the pores of enamel and dentin tubules. The acid condition could dissolve the mineral content both in dentin and enamel, thus demineralization process occur. For a prolonged time, the cavity is formed in the teeth [2]. Actually, the caries process consists of the dynamic continuous cycle between demineralization and remineralization. Remineralization is the natural repair process where calcium and phosphate ions deposited back into enamel or dentin. Remineralization can occur in nearly neutral conditions (> 5.5), while demineralization happens in an acid environment (< 5.5). However, in caries cases, demineralization is dominated by the remineralization process, thus leading to tooth decay [1-3]

Rajesh (2015) in his study had described the critical condition for demineralization of teeth is in acidic pH (< 7), in which a solution is saturated with a particular mineral [4]. Naturally, saliva plays a major role in recovering pH from the acid condition into neutral. Saliva also serves as a calcium and phosphate ions to repair ion loss in the enamel. Moreover, the ability of saliva is dependent on the quantity of secretion, systemic disease, and saliva flow rates [5]. Therefore, to help saliva roles in enhancing the remineralization process, fluoride is often used. However, it is important to consider that inappropriate fluoride application could cause enamel fluorosis.

Nano-hydroxyapatite (nHAp) is considered one of the materials that have high biocompatibility for bone tissue engineering and dental implant. Due to its chemical and morphological similarity with dental apatite, nHAp is widely used as biomimetic material to repair enamel or dentin defects. Hydroxyapatite could induce enamel remineralization through formation of a new layer of HAp, thus promote calcium and phosphate deposition [6]. Recently, some denitrifies has contain nHAp in their formulation, like Sangi Co. (Japan). However, to our knowledge, there is no information about hydroxyapatite effect in helping salivary pH recovery to support enamel remineralization process. The aim of this study is to examine the effect of nano hydroxyapatite (nHAP) in the toothpaste by concentration of nHAP 0.25, 0.70, and 1.5 % to the salivary pH and microbial activity.

2. Experimental method

2.1. Sample preparation

Bovine molar teeth were purchased from the slaughterhouse. The teeth were cleaned from soft tissue and debris, then the crown portion was sectioned to separate from the root, using handpiece and diamond disc. After that, the crown portion was cut into 4 × 4 mm pieces with ± 1.5 mm thickness. Teeth fragment were covered with acid-resistant nail varnish, except for enamel surface, then resulting 16 mm^2 open areas. Each enamel surface of the specimen was polished using sandpapers to remove the disturbance area. Then, teeth fragment was stored in 40°C prior to use. The fragment teeth were treated by the toothpaste that contains nHAp of 0.25, 0.7 and 1.5%. The samples were named 0.25nHAp, 0.7nHAp and 1.5nHAp, respectively. For comparison purpose, the fragment teeth were also treated by a commercial toothpaste (CT), distilled water (K-) and ampicillin (K+).

2.2. In vitro oral microenvironment pH and microbiological analysis

Teeth fragment was put on 96-well plate, then coated with artificial saliva for 90 minutes at 37°C. After removing unbound saliva, 200 µL of Streptococcus mutans ATCC 25175 (2×10^7 CFU/mL) in Brain Heart Infusion (BHI, Oxoid Limited, Hampshire, UK) broth supplemented with 1% sucrose (Himedia Laboratories, Mumbai, India) was added. Biofilms were allowed to growth for 24 hours at 37°C, anaerobically (10% CO_2, 80% N_2, 10% H_2).

After 24 hours of incubation, bacteria suspension was discharged, then the biofilm was gently washed using Phosphate Buffer Saline (PBS). The dental plaque formed in the teeth was treated with 200 µL of slurry toothpaste, correspond with the groups. Then, treated-dental plaque was incubated for 1 minutes at rotary shaker (85 rpm). Subsequently, teeth fragment was moved into 1,5 mL microtube filled with BHI broth + 1% sucrose. pH measurement was done using pH strip test, every 10 minutes in 1 hour. Furthermore, the total of Streptococcus mutans bacteria in dental plaque also counts using total plate count method at the end of pH measurement.

2.3. Remineralization capability

Fragment teeth were immersed in 20 mL demineralizing solution (2 mmol/L Ca^{2+}, 2 mmol/L PO_4^{3-}, 0.075 mol/L acetate) for 4 days at 37°C, to produce artificial carious lesions. After 4 days, remineralization procedure using artificial saliva and toothpaste sample was done for 14 days. A three minutes treatment with toothpaste was given twice daily (08.00 AM and 15.00 PM) during remineralizing step. Furthermore, fragment teeth were cleaned using toothbrush under distilled water, then stored again in artificial saliva at 37°C. Artificial saliva was changed every 48 hours. Finally, on the last day, fragment teeth were washed with deionized water, dried in room temperature, then observed the enamel surface under Scanning Electron Microscope (FEI Quanta 650) with an Oxford INCA/ENERGY-350 microanalysis system. Furthermore, quantification of hole area in the enamel surface was done using image processing with software ImageJ.

3. Results and discussion

Figure 1 shows the pH of dental plaque pH for 60 minutes. All treatment groups were experienced a pH decrease after 20 minutes. Significant pH fall was observed in toothpaste with the addition of 0.25, 0.7, and 1.5 nHAp. However, in 60 minutes, all nHAp samples were able to return the pH into neutral condition, showing the pH of 7. In particular of 0.7nHAp, the teeth that treated with this condition able to restore the pH to 6.7 in 50 minutes, which is the highest pH among all samples. On the other hand, CT samples only returned the pH to 6 within 1 hour, which is lower compared to nHAp samples. Indeed, the K- samples has a trend that continues to decline until reach pH 5 after 1 hour, while K+ tends to be stable in the range of 7 to 6.5. Carbohydrate fermentation by oral bacteria will result an organic acid as the final product via the glycolytic pathway [7-9]. Acidic conditions in tooth enamel can cause demineralization, because hydroxyapatite as the main component of enamel has a high solubility in acid solutions. Therefore, maintaining oral microenvironment pH in the neutral condition is an important method to prevent enamel demineralization [10,11]. According to the results are shown in Figure 1, nHAp is capable to to restore oral microenvironment pH back to neutral in order to promote enamel remineralization.

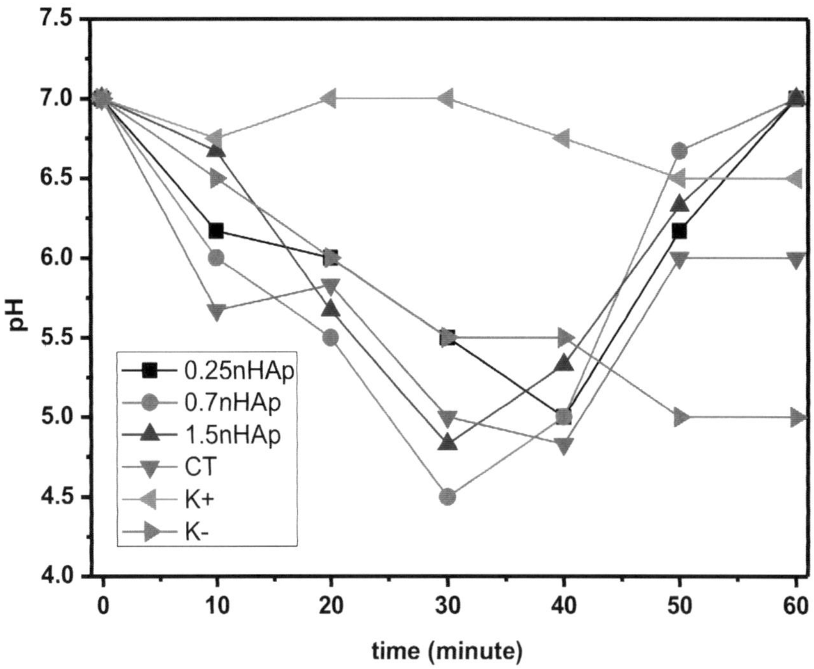

Figure 1. pH of dental plaque pH for 60 minutes.

Table 1. Relationship between pH value and number of Streptococcus mutans after 60 minutes.

	0.25nHAp	**0.7nHAp**	**1.5nHAp**	**CT**	**K+**	**K-**
	7	7	7	6	6.5	5
Number of bacteria (CFU/mL)	2.46×10^6	6.67×10^4	5.56×10^5	6.67×10^4	1.26×10^7	3.50×10^8

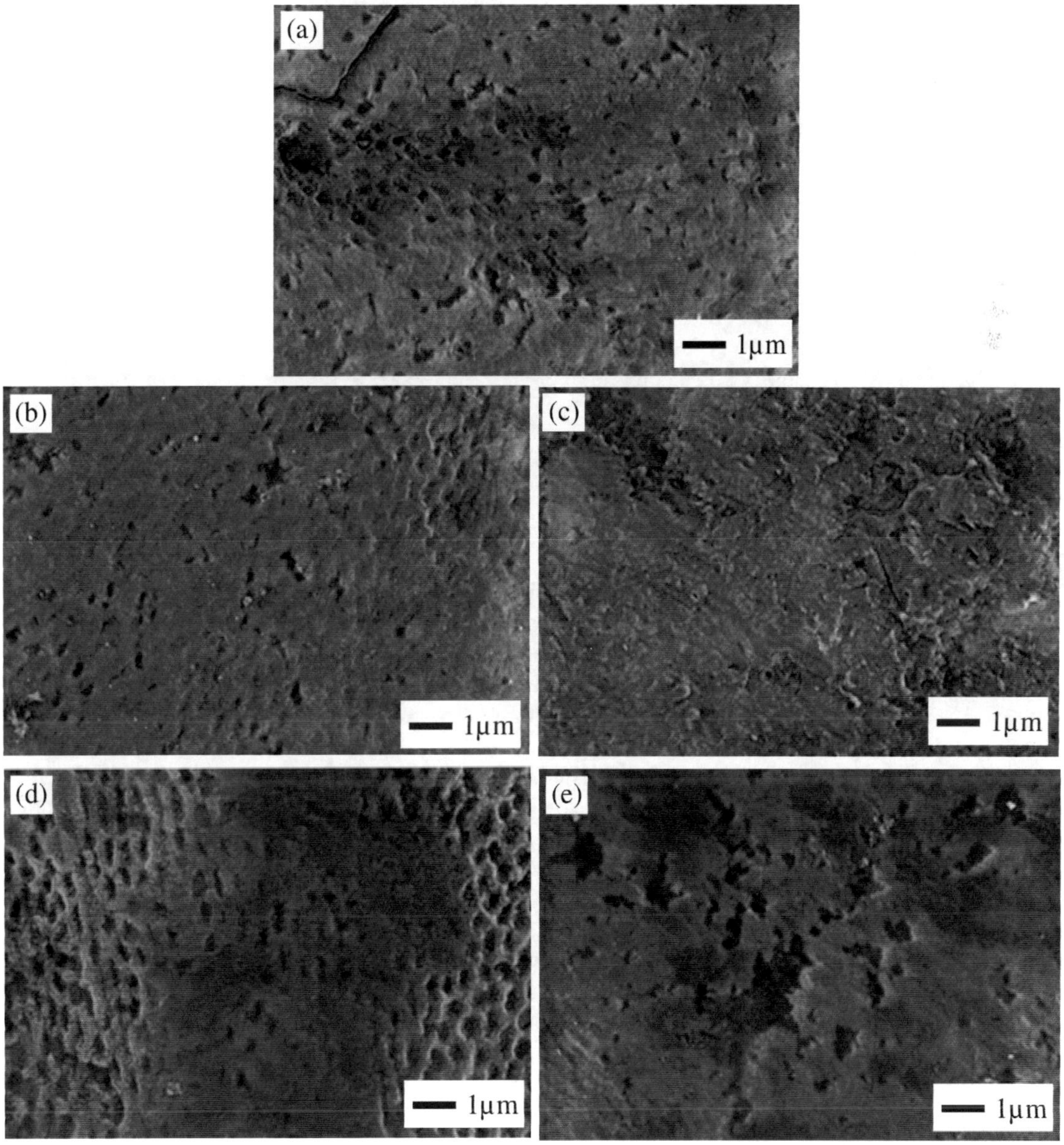

Figure 2. SEM images of the enamel surface after 14 days treatment (a) 0.25nHAp, (b) 0.7nHAp, (c) 1.5nHAp, (d) CT and (e) K-.

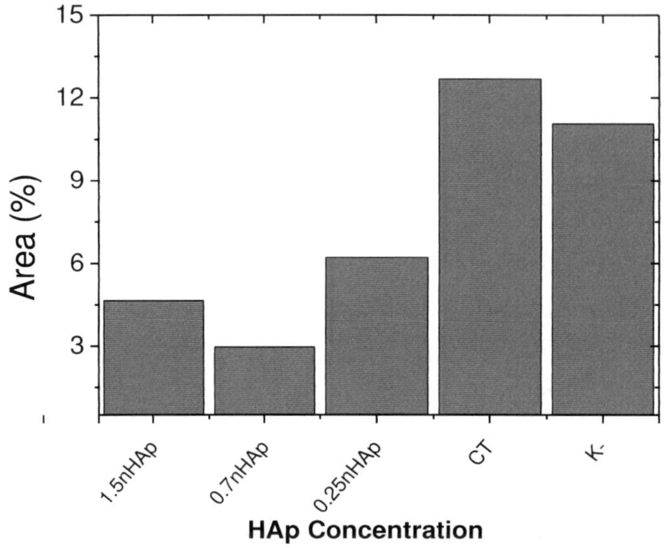

Figure 3. Percentage of total holes area from SEM image in Figure 2 (Processing image was done using ImageJ).

Change in pH is closely related to number of Streptococcus mutans as the main cariogenic organism and acid-producing bacteria. Table 1 shows the number of Streptococcus mutans and pH after 60 minutes. It should be noted that the amount of Streptococcus mutans in the nHAp samples was lower compared to K+ and K- samples. Although the final pH of nHAp samples after 60 minutes was similar, the total Streptococcus mutans is difference. As shown in Table 1, the total Streptococcus mutans for 0.25nHAp, 0.7nHAp and 1.5nHAp was 2.46×10^6, 6.67×10^4 and 5.56×10^5 CFU/mL, respectively. This difference is more likely corresponding to pH after 50 minutes, as shown in Figure 1. 0.7nHAp has the lowest Streptococcus mutans since its has highest pH after 50 minutes, while 0.25nHAp showed the lowest pH among the nHAp samples, therefore, the amount of Streptococcus mutans is the highest compared to other nHAp samples. The total number of Streptococcus mutans in the 0.7nHAp is equal with the CT samples, despite the fact that they have difference pH number. In contrast, K+ and K- samples revealed the highest amount of bacteria Streptococcus mutans, even though in the K+ treated by ampicillin.

The observed pH dynamics graph consisted of several stages, as shown in Figure 1. The first stage, which is in the 0 to 10 minutes, is the stage of adaptation or resting phase of Streptococcus mutans, so that there is a slightly change in pH. Furthermore, in the 10 - 40 minutes, Streptococcus mutans is in the active stage that metabolizes sucrose, which was indicated by the pH begins to fall. In the K+ samples, there was a decrease in pH, however, the decrease of pH is lower compared to the toothpaste with nHAp samples. The third stage is the recovery stage where the pH starts return to normal, which is observed in 40 - 60 minutes, especially in the nHAp toothpaste group. This result is in agreement with the other reports [12].

Generally, the recovery stage is covered by the role of saliva as buffer capacity [5]. However, in this research, the buffer capacity is performed by bacterial medium and nHAp, which bound in the enamel surface. Under acidic conditions ($pH \leq 5.5$) hydroxyapatite begins to dissolutions and release its ions, following the Equation (1) [13-15]:

$$Ca_{10}(PO_4)_6(OH)_2 \rightarrow 10Ca^{2+} + 6PO_4^{3-} + 2OH^- \quad (1)$$

Calcium ions reacts with water molecules to form $Ca(OH)_2$ which increases the pH. Moreover, phosphate ions also can help a pH buffer [8, 16]. The buffer capacity is not observed in K- and K+. In the K- and K+, the pH was observed decreasing gradually within 1 hour. This is because the acid production and growth of remaining bacteria in K+ and K-, which was shown in Table 1.

Figure 2 shows the SEM images of the enamel surface after 14 days treatment using nHAp, CT and K-. Indeed, the enamel surface of K- samples (Figure 2 (e)) showed rough surface, which might be from the demineralization since the pH of this samples is very low and had the large number of Streptococcus mutans. Surprisingly, the CT samples (Figure 2 (d) showed almost no difference compared to K- samples (Figure 2 (e)) in term of porosity and roughness. It seems that CT samples could not inhibit enamel demineralization activity after 14 days, showing very rough surface of enamel, as shown in Figure 2 (d). In fact, the pH and the number of bacteria of CT samples is comparable with nHAp samples after 60 minutes, however, for long treatment time, CT samples could not maintain the optimum condition for remineralization. In contrast, the number of porosities of the demineralized area decreased along with the increasing of nHAp content (Figure 2 (a) – (c)), showing a better appearance compared to K- and CT samples after 14 days treatment.

Anti-bacterial activity of nHAP is supported by the ability to recovery salivary pH to prevent the growth of *S. mutans* and inhibit adhesion of bacteria on the tooth surface [17-20]. Hydroxyapatite in nanosize is also considered as a remineralizing agent due to its morphology and chemical features more similar to biological apatite [21,22]. The application of nHAp as toothpaste may produce the outermost layer that serves as a barrier and protect the enamel against acid. nHAp will penetrate through enamel lesion and serve as a template to attract Ca^{2+} and PO_4^{3-} from surrounding environment. As the nHAp concentration increased, the rate and amount of Ca^{2+} and PO_4^{3-} deposition will increase [23,24]. This can be observed in 1.5nHAp, where the surface had minimal pores area, as shown in Figure 2 (c). Quantitatively, the holes area in 1.5nHAp and 0.7nHAp is not significantly different (Figure 3).

4. Conclusion

The effect of nHAp in the toothpaste on the pH and microbial activity was successfully investigated in this study. The results showed that the addition of nHAp in the toothpaste improve the ability to recover the pH in the first of 60 minutes. The pH of all nHAp samples after 60 minutes were 7, while the CT samples only 6. Moreover, the pH is correlated with the number of *Streptococcus mutans* bacteria, which showed the number of the bacteria is lower in the teeth that treated by nHAp compared to K+ and K- samples. The number of bacteria in the 0.25 nHAp, 0.7 nHAp and 1.5 nHAp was 2.46×10^6, 6.67×10^4 and 5.56×10^5 CFU/mL, respectively. Moreover, the treatment of the enamel surface for 14 days revealed that the nHAp prevent the damage of the enamel compared to the CT or K- samples. The number of porosities in the nHAp samples lower compared to samples without the addition of nHAp, indicating the effectiveness of nHAp to enhance the remineralization of the teeth by maintain the pH and the number of bacteria.

Acknowledgment

This research was funded by a grant from Prospective Technology-Based Starter Companies 2019 (No. 001/F1/PPK.2/Kp/III/2019), Ministry of Research, Technology and Higher Education, Republic of Indonesia.

References

[1] Pitts, Nigel B., et al. "Dental caries." *Nature Reviews Disease Primers 3*: 17030. 2017, pp. 1-16

[2] Lemos, J. A., et al. "The biology of Streptococcus mutans." *Microbiology Spectrum 7.1* (2019).

[3] Kakanur, Madhu, et al. "Exploring the multitude of risk factors associated with early childhood caries." *Indian Journal of Dental Research 28.1* : 27 (2017)

[4] Rajesh, K. S., Shashikanth Hegde Zareena, and MS Arun Kumar. "Assessment of salivary calcium, phosphate, magnesium, pH, and flow rate in healthy subjects, periodontitis, and dental caries." *Contemporary Clinical Dentistry 6.4* : 461 (2015)

[5] Lakshmi Bolla, Vijaya, et al. Correlation between the PH of saliva, plaque and buffering capacity of saliva. *International Journal of Applied Dental Sciences*. 2017, pp. 48-50.
[6] Shahmoradi, M., et al. "Synthesis of stabilized hydroxyapatite nanosuspensions for enamel caries remineralization." *Australian Dental Journal 63.3*. 2018, pp. 356-364.
[7] Roveri, Norberto, et al. "Surface enamel remineralization: biomimetic apatite nanocrystals and fluoride ions different effects." *Journal of Nanomaterials*. 2009 : 8 (2009)
[8] Walsh, L.J. "Dental Plaque Fermentation and Its Role in Caries Risk Assessment". *International Dentistry SA*., 8 (5), 2006, pp. 34-40.
[9] Mescher, A L. *Junqueira's Basic Histology Text and Atlas*. New York : Mc Graw Hill Education Medical. (2013)
[10] Goodson, J. Max. "Salivary Glucose: A Metabolic Disease Marker And a Risk Factor for Oral Disease Development." *Adv Dent & Oral Health 4.3* (2017).
[11] Rosihan, Adhani, et al. "Effect pH on demineralization dental erosion." *International Journal of Chemical Engineering and Applications 6.2* (2015): 138-141.
[12] Bowen WH. "The Stephan Curve revisited". *Odontology 101.1* (2013), pp. 2-8
[13] Manarelli, Michele M., et al. "Protective effect of phosphates and fluoride on the dissolution of hydroxyapatite and their interactions with saliva." *Caries Research 51.2*. 2017, pp. 96-101.
[14] Neel, Ensanya Ali Abou, et al. "Demineralization–remineralization dynamics in teeth and bone." International journal of nanomedicine 11 (2016): 4743.
[15] Loskill, P. et al. "Reduced adhesion of oral bacteria on hydroxyapatite by fluoride treatment". *Langmuir*. 29 (18), 2016. pp. 5528-5533.
[16] Jespersen, N.D., Brady, J.E., Hyslop, A. *Chemistry: The Molecular Nature of Matter*. USA: John Wiley & Sons (2012)
[17] Amaechi, Bennett T. "Protocols to Study Dental Caries In Vitro: pH Cycling Models." *Odontogenesis*. Humana Press, New York, NY, 2019. pp. 379-392.
[18] Ballardini, A., et al. "New hydroxyapatite nanophases with enhanced osteogenic and anti-bacterial activity." *Journal of Biomedical Materials Research Part A* 106.2. 2018, pp 521-530.
[19] Kensche, A., et al. "Efficacy of a mouthrinse based on hydroxyapatite to reduce initial bacterial colonisation in situ." *Archives of Oral Biology* 80. 2017, pp 18-26.
[20] Park, Miseon, John B. Sutherland, and Fatemeh Rafii. "Effects of nano-hydroxyapatite on the formation of biofilms by Streptococcus mutans in two different media." *Archives of oral biology* 107. 2019, pp 104484.
[21] Zaharia, Agripina, et al. "Biomimetic chitosan-hydroxyapatite hybrid biocoatings for enamel remineralization." *Ceramics International* 43.14. 2017, pp 11390-11402.
[22] Pepla, E., Besharat, L.K., Palaia, G., Tenore, G., & Migliau, G. "Nano-Hydroxyapatite and its applications in preventive, restorative, and regeneratve dentistry : a review of literature." *A Journal Of Odontostomatology Sciences,* 5 (3). 2014, pp. 108-114
[23] Amaechi, Bennett T. "Remineralization therapies for initial caries lesions." *Current Oral Health Reports 2.2*. 2015, pp. 95-101.
[24] Juntavee, N., Juntavee, A., and Plongniras, P. "Remineralization potential of nano-hydroxyapatite on enamel and cementum surrounding margin of computer-aided design and computer-aided manufacturing ceramic restoration". *International Journal of Nanomedicine: 13*. 2018, pp. 2755-2765.

Policy Implementation in Water Services

Harry Nenobais*, Sarbini
University of Prof. Dr. Moestopo (Beragama), Jakarta, Jl. Hang Lekir 1/8, Jakarta Pusat 10270, Indonesia

*harryestinenobais@gmail.com

Abstract. This study analyzes the factors that influence the implementation of policies in clean water management based on participatory management of the Sukoharjo community. This research was conducted at the Tirta Makmur Regional Public Water Company in Sukoharjo Regency. The research method used is a qualitative method using the theory of implementation of the policies of Donald Van Meter and Carl Van Horn (1975). The research results conclude that the factors that influence the implementation of clean water service policy based on participatory management by the Tirta Makmur Regional Public Water Company in Sukoharjo Regency are clear policy and standard objectives, consensus on changes in policy implementation, human resource capacity of implementing organizations, financing, the quality of interorganizational relations, a healthy political, social and economic environment, and participatory management. The results of this study suggest that the Regional House of Representatives, Regency Government, and Management of Regional Public Companies have a large consensus of change, government and community participation, investment in service improvement, preparing raw water sources.

Keywords: Concensus of Changes, Finance, Capacity of Human Resources, Execellent Services, Participatory Management.

1. Introduction

The need of clean water is a vital requirement for society as water is related to mass fullfilment of people, so it should be managed by the nation. In line with the Constitution of UUD 1945 Chapter 33, verse 3 that the monopoly of earth, water as well as natural riches contained in it should be profited optimally for the well-being of the people. This study analyzes the factors that influence the implementation of policies in clean water management based on participatory management of the Sukoharjo community.

Considering the importance of the benefits and functions of clean water for human life, so the management will be in the authority of the Central Government through the Regional Government by submitting the authority of clean water management to the Regional Government in a district corporation namely Regional Public Water Company. Constitution 23 Year 2014 On Regional Government Chapter 9 verse 3 states: "the affairs concurrent with verse 1 is the affairs of the Government which are divided between the Central Government and Regional Province and District or City."

This is in line with the Indonesian Commitment in the paradigm of global development namely *Sustainable Development Goals* (SDGs) or the goal of Sustained Development (President Decree no 59 Year 2017) which is aimed at Work Unit of The Goal of Sustained Development to the General Assembly of United Nations [1].

One of the goals of SDGs is to ensure the adequacy of clean water and proper sanitation, (www.*uclg.org*). The government through Regional Public Water Company that serves the requirement of water for society is not only capable of conferring clean water service but also can do better than that to enhance services by providing standard quality drinking water [2].

Through Regional Public Water Company Drinking Water Tirta Makmur, Sukoharjo is one of the districts in Indonesia that keeps doing efforts to improve the fullfilment of clean water for society. The goal that has to be carried out by Regional Public Water Company Drinking Water Tirta Makmur in line with the Regional Regulation No 20 Year 2016 is a heavy task. *Firstly,* to give benefits for the development of regional economy namely the gaining of company profits that can contribute to Regional Income (PAD). *Secondly,* to carry out social functions namely to serve people who need items as well as services as extensively as possible.

The two goals that can be simply regarded as contradictive, in which the function of a company to achieve as wide margins as possible and to avoid factors that can effect minimal profits. On the other side, it has a social function which ignores the Principle of Profit and Loss in the calculations of accountancy. When a company has implemented the social function in line with the regulation and reduced or even experienced a loss, it will become the risk of the policy of the regional government.

The consequence of the two goals that can be simply regarded as contradictive will finally defeat the profit goal or vice versa where the social function yields to profit goals. This can lead to the profit loss of regional companies and service to society can turn unsatisfactory. Certainly, these two goals should be harmonized so they can go side by side. At this point, the managerial needs are vital so the two goals can be implemented in line with the prevailing regulations in order to gain profits to the maximum and the social function to serve the needs of the society for clean water can be carried out optimally.

By using the theory of Donald Van Meter and Carl Van Horn (1975), so the questions of the study area: (1) How is the implementation of the policy of clean water service based on the participated management at Perumda Air Minum Tirta Makmur Sukoharjo District? (2) What are the obstacles faced by Perumda Drinking Water Tirta Makmur DisSukoharjo District? (3) What are the efforts exerted by Perumda Drinking Water Tirta Makmur Sukoharjo District in the running of their tasks in servicing society? [3].

2. Methodology

This study uses a qualitative research method namely describing or explaining something as it is. The data obtained will be analyzed as well as described based on the findings from the field. The tools used in this study includes a camera, a recorder and detailed interview guidance sheets [4].

3. Result and Discussion

Based on the result of the previous study [5,6], and this study of implementation of the policy of drinking water service based on the participated management at Regional Public Water Company Drinking Water Tirta Makmur District Sukoharjo in line with the implemntation theory of Donald Van Meter and Carl Van Horn, it was found that the implementation goal of the policy of Local Regulation No 20 Tahun 2016, it was found *Firstly,* Regional Public Water Company Drinking Water Tirta Makmur has caterred good service, despite some obstacles. Regional Public Water Company Drinking Water Tirta Makmur has managed to achieve profits. *Secondly,* resources in the form of funds and human resources. Regional Public Water Company Drinking Water Tirta Makmur has yet to develop the company owing to the allocation of the profits to regional income. The capacity and profesionality of the employees need to be enhanced despite the facts that Regional Public Water Company Drinking Water has conducted a variety of steps to increase the quality of the human resources. *Third,* the quality of international relations. The relation between Regional Public Water Company Drinking Water Tirta Makmur and local government and regional parliament Sukoharjo Regency is quite harmonious, so is it with other related instances. *Fourth,* the characteristics of the operational organisation of Regional Public Water Company Drinking Water Tirta Makmur uses the participated strategies from planning, organizing, implementing and supervising. It also involves the employees in the activities of the company in giving incentives and rewards. Regional Public Water Company has also set up the standard procedures of services that can be monitored by local government as well as Regional Parliament. *Fifth,* political, social and economic environments. As the Director of Regional Public Water Company Drinking Water Tirta Makmur has a good relation with the political power that is dominant in Sukoharjo, so the political, social and economic environments are quite conducive. Regional Public Water Company Drinking Water also established a good relationship with clientele (all policies are socialized first) and also the social role. *Sixth.* The attitude of the operational staff. Participated management that is carried out by the highranking chiefs of Regional Public Water Company and Appreciation and Rewards for the employees in addition with the guidance of Standard Procedure, so the employees can run their functions well. *Seventh.* Implementation of policies requires a modification consensus of the Regional Parliament. The Local Government of Sukoharjo and Regional Public Water Company Drinking Water Tirta Makmur dan Society. *Eighth.* The obstacles faced by Regional Public Water Company Tirta Makmur are among others: the tariff of clean water that is relatively lower; the equipment and technology that have become obsolete due to aging; delayed development of the company owing to the allocation of the profits to regional income; a lack of standard water sources because of the dependence on streams and reservoirs that go dry in the dry season.

4. Conclusion

Implementation of the policy of Local Regulation No 20/2016 concludes that factors that influence the success of the implementation of the policy of clean water service based on participated management done by Regional Public Water Company Drinking Water Tirta Makmur Sukoharjo Regency encompasses (1) Goals of Policies: Regional Public Water Company Drinking Water Tirta Makmur has carried out clean water service with the quality and water standards in line with the prevailing regulations in spite of obstacles. Nevertheless, Regional Public Water Company Drinking Water has managed to generate profits. (2) Resources and Funds: Regional Public Water Company Drinking Water Tirta Makmur has yet to develop the company optimally due to the allocation of the company profits to Regional Income, capacity and professionalism of the employees need to be sustainedly enhanced even though Regional Public Water Company has conducted a variety of steps to increase the quality of the human resources [7]. (3) Quality of interorganisational relation: Relation between Regional Public Water Company and Regional Public Water Company Drinking Water Tirta Makmur with the Regional Government and Regional Parliament of Sukoharjo District is quite stable as is it with other related (4) organization of the Operators: Regional Public Water Company Drinking Water Tirta Makmur has used participated strategies from planning, organizing, implementing and supervising. The involvement of the employees in the granting of incentives and rewards is assured. Regional Public Water Company has set up the Standards of Operations of services that can be monitored by the Regional Government and the Regional Parliament [8]. (5) political, social and economy: Since the director of Regional Public Water Company Drinking Water Drinking Water Tirta Makmur has established a well woven relation with the political power that is dominant in Sukoharjo District, so the political, social and economic environments in Sukoharjo District is adequately conducive and Regional Public Water Company also has a good rapport with clientele (all policies are socialized first) and social roles [9]. (6). Attitude of Operators: Participated Management done by the high ranking chiefs of Regional Public Water Company as well as Appreciation and Rewards for the employees in addition of the Guidance of the Standard Procedures to ensure that the employees can carry out their main jobs and functions well. The need for a massive support from the Regional Government and Regional Parliament of Sukoharjo District and Regional Public Water Company Drinking Water Tirta Makmur Sukoharjo District with all the limitations has attempted to establish and run the participated management in the management of Regional Public Water Company Drinking Water Tirta Makmuar with the following strategies: *First,* management of the organisation based on participated principles [10]. *Second,* to build a clean water service with standard procedures that have been set up. *Third*, service product namely clean water which is in line with the regulations (Minister of Health Regulation No 492/2010 and Minister of general of Public Work and Public House No 27/2016), *Fourth*, to encourage the participation of employees into the policy making as well as services. *Fifth*, granting of incentives to the employees with good achievements in the form of material and nonmaterial rewards. *Sixth,* quick and online services.

References

[1] Angmor, Emmanuel Narteh, et al, "Management of Community Water Project: Assessing the Challenges of Indigenous Arrangements, " *Journal of Geography and Regional Planning,* V 9 (6) pp.122-127, June, 2016.
[2] Damayanti, Lina, "Sustainability Analysis of Community Managed Rural Water Supply System in Sindanggalih Village and Bayongbong Village in District of Garut that Received The Community Based Water Supply and Sanitation Program Pamsimas," Thesis, University of Indonesia, Jakarta, 2016.
[3] Van Meter, S. Donald & Carl E. Van Horn. "The Policy Implementation Process: A Conceptual Framework. *Sage Journals Administration & Society*, 1974.
[4] Creswell, W John & J. David Creswell, "Reseach Design Qualitative, Quantitative, and Mixed Methods Approach," 5th Edition. Thousand Oaks, United State, Sage Publications, 2018.
[5] Natalia, Benita Merry, "The Implementation Program Zona Air Minum Prima (ZAMP) to Meet The Drinking Water Society (Studi on PDAM Malang City)," *Journal Administrasi Publik*, Vol 2, No.1, 2014.
[6] Rofiana Vifin, "Policy Implementation Program Pamsimas (Supply Drinking Water and Sanitation on Base Society)," *The Indonesian Journal of Public Administration*, Volume 1, No 2, 2015.
[7] Khan, Anisur Rahman, "Policy Implementation: Some Aspects and Issues," *Journal of Community Positive Practice*, XVI (3), 3-12, 2016.
[8] Jimenez, Alejandro & Hakan Tropp, "Addresing the Policy Implementation Gaps in Water Service; The Key Role of Meso-Instituions. *Journal Water International*, Volume 43, 2018.
[9] Dye, R. Thomas, "Understanding Public Policy," Thirteenth Edition, Boston: Pearson, 2011.
[10] Olajuyigbe, Ayo Emmanuel, "Community Participation and Sustainability Issue: An Evaluation of a Donnor-Driven Water Sector in Ikaram Millennium Village Project, Nigeria, *Open Journal of Social Science*, 4, 90-103, 2016.

IOP Conf. Series: Materials Science and Engineering **924** (2020) 012012 doi:10.1088/1757-899X/924/1/012012

Design of Frangible Composite Production Process and Plant Layout with Powder Metallurgy Method and Capacity of 50.000.000 Rounds/Year

Widyastuti*, F F Sulistya, V M Pratiwi, W Jatimurti, and H Ardyananta
Department of Materials Engineering, Faculty of Industrial Technology, Sepuluh Nopember Institute of Technology
Jl. Raya ITS, Keputih, Sukolilo, Surabaya, Jawa Timur, Indonesia

*wiwid@mat-eng.its.ac.id

Abstract. Frangible composite material is one of the types of advanced material utilized by defense technology application which has been certified. In fact, this material is in development for mass production. In order to realize the mass production stage, it is necessary to have a comprehensive design on process production, specification, machine capacity and the blueprint of plant's layout. Production process planning would be conducted by the target up to 50 million rounds per year. This process uses powder metallurgy method which consists of mixing, compaction, sintering, finishing, quality control, and packaging. This research would result various variable and constants in each processes, plant design, specification, and number of machines needed by using data from literature review and previous research. Then, the number of machines is going to be calculated in order to find the certain number of machines that are needed. The calculation implies the needed of production machines which are 1 unit of mixer, 3 units of automatic mechanical compaction machine, 1 unit of continuous sintering furnace, 1 unit of finishing machine, and 3 units of quality control machines. For the production process simulation, we use Tecnomatix Plant Simulation in order to simulate the processes to get the data. The simulation process would be started by inputting the total number of machines that are used by and the variables and constants in each process which had been determined before. By calibrating the running time for 365 days 3 hours 15 minutes, the result stated that the simulated production process produced the number of outputs of 50,037,120 rounds/year and 45,696 rounds per shift.

Keywords **:** Composite, Frangible, Powder Metallurgy, Production Process, Plant Layout

1. Introduction

One the main requirement of defense equipment is ammunition. Recently, ITS developed frangible composite based ammunition which is planned for mass production, but currently has no integrated production line. The production process of frangible composite uses powder metallurgy method due to the frangibility factor which is only obtained by this method. Generally, the stage of processes consists powder mixing, powder compaction, sintering, finishing, and quality control.

The powder mixing is conducted in order to get the powder mixture by a double cone mixer. Then, the next process is powder compaction process by using automated compaction machine. The next step is sintering process which uses continuous sintering furnace. The last step is quality control process using conveyor scale and optical micrometer sensor. This research intends to determine the variables and constants of each processes and also to calculate the exact number of machine needed, all to achieve the goal of a 50 million rounds/year production capacity.

After determining the variables, constants, and number of machine needed for each processes, the plant layout design is conducted using CORELAP algorithm to reduce the material handling cost [1]. First, the production department layout is determined according to the process flowchart and it will be fixated because the flow of process cannot be rerolled. Then, CORELAP algorithm is conducted using supporting department and the fixated production department. Total Closeness Rating (TCR), which determines the degree of closeness for each department with each other, is determined and calculated. The departments are then ordered based on their TCR descending. Then, the departments are placed in the plant layout according to the order. This will result in an Area Allocation Diagram (AAD) which serves as a base for designing the final layout. The AAD is then given allowance for material handling in each processing department, which will improve production output, reduce bottleneck and delay, and reduce material handling cost [2].

For the final part, the production process is then simulated using Tecnomatix Plant Simulation, with the inputs are running time, number of machines, type of processes (batch or continuous), and processing capacity. The simulation process will give the designed production process its production capacity for its given running time, which is a year full of production. The result is then analyzed, whether it reached the production target of 50.000.000 rounds/year or not. If the result did not meet the expected target, a new design is iterated by recalculating number of machines and determining the variables and constants for each processes.

2. Experimental method

2.1. Research Location

The plant of frangible composite would be established within the building of 12 meters x 12 meters and the area of 144 m^2. This mini plant has 3 main sections of manufacturing which are production line of frangible composite material, production line of atomization to yield the powder that would be used for frangible composite, and the office section for administration and non-technical activity.

2.2. Determination of Stages within Production Process

The production process uses powder metallurgy method, which is mixing, powder compaction process, sintering, finishing, quality control and the last one, packaging [3]. This research would be carried out of fixation upon the variable and constant in each process that the production process of frangible projectile can work properly and obtain the product who met the specifications. These stages consist of but not limited to: 1. Mixing, 2. Powder compaction, 3. Sintering, 4. Finishing, 5. Quality Control, 6. Packaging.

2.3 Processing Time

In this stage, the timing process would be determined for each process to decide the specification and the total number of machines working for, in order to achieve production target. Assuming the failure rate is about 10% on compaction process, sintering and quality control, the production target in each process could be determined by:

$$\text{Initial production target} = \frac{\text{Final production target}}{(1-\mathit{failure\ rate})} \tag{1}$$

2.4 Specification and Quantity of Machine

The total number of machines from machine's specification and production target can be determined by:

$$\text{Number of machines} = (\text{machine capacity})/(\text{plant rate}) \quad (2)$$

Generally, there are six section of facility and production service within company which are Production department, Warehouse, Meeting room, Office, Prayer room, Toilet, and Parking lane.

2.5 Activity Relationship Chart (ARC)

Activity Relationship Chart (ARC) is a diagram used for obtaining relationship from specific activities in order to determine where the activity which have to be adjacent or otherwise within the layout design of a facility. ARC consists of production activity and service activity.

In explaining of the adjacency relation among departments, we use Activity Relationship Chart by inserting several symbols as follows on Table 1:

Table 1. Degree of closeness in ARC.

Degree of Closeness	Description
A	*Absolutely Necessary*
E	*Especially important*
I	*Important*
O	*Ordinary*
U	*Unimportant*
X	*Undesireable*

2.6 Total Closeness Rating (TCR)

In this step, the calculation process of TCR value would be conducted. The assessment is conducted by adding the total closeness rating of an activity or department based on criterion within Table 2 and ARC diagram that has been created [3].

Table 2. Closeness rating value for ARC diagram.

Degree of Closeness		Description	Color Code	Weight Closeness Rating	Closeness Rating
A	*Absolutely Necessary*	Absolutely necessary to be adjacent	Red	243	6
E	*Especially important*	Very Important to be adjacent	Orange	81	5
I	*Important*	Important to be adjacent	Green	27	4
O	*Ordinary*	Ordinary adjacent	Blue	9	3
U	*Unimportant*	Unecessary to be adjacent	White	1	2
X	*Undesireable*	Unexpected to be adjacent	Brown	0	1

2.7 CORELAP Analysis

After the value of TCR is obtained, each production or facility section have to be conducted CORELAP algorithm calculation. The step of CORELAP algorithm can be started by the calculation of Total Closeness Rating (TCR) for each departments. The first step is to choose the department with the highest value of TCR and putting it in the center of layout. Then, if there are equal value, we take the department with the largest area. We place the departments corelating with A and already selected, then the correlation of E, I, O, U and X. If there are same criteria, we use the department with the largest area. If the department has been chosen, the placement is decided based on Placing Rating which is the sum of weight closeness rating among departments that has been inserted and that will be inserted. If the placing rating is equal, then the border length or total units per square that are side by side will be compared.

2.8 Area Allocation Diagram
After we get the layout with the most optimum layout score, we can proceed to the Area Allocation Diagram (AAD) scheming stage. AAD is a description of layout in a global scope representing closeness relationship among sections and the area scale of exact floor.

The AAD process will be conducted twice and separately between production section and production facility. The production section will be organized first. Then. It is combined by production facility yielding the layout of a whole mini plant that want to be constructed [3].

2.9 Production Line Simulation
Production line simulation is conducted by using software Tecnomatix Plant Simulation. In simulation method, we can assume that the machine is operating, corresponding with the type of process whether it is batch or continuous. The Variable that will be inputted are type of process, production capacity in a cycle, processing time, and total of machines.

3. Results and discussion

3.1 Production Process
The method used for producing frangible composite material is powder metallurgy method. The parameter determination in each process in this method is organized by considering previous research and related literature.

The powder mixing is conducted by fixation process upon various variable to be used as a constant within production process design of frangible composite material Cu-10% Sn. We used wet mixing process with mixing media of ethanol, due to its low boiling point temperature (78.37°C) [4]. That variable produced frangible composite material with the best result therefore it is used as a constant in powder mixing process.

The type of compaction process that is used is single action because of its ability to give high strength and high density to the part which is directly contacting with moving punch [5]. The temperature while compaction processing is cold compaction, without heating process. The reason that we used cold compaction method is because tin, as the matrix of the composite, has low melting point which is 232°C. Accordingly, it does not need to combine sintering process with high pressure. The utilization of cold compaction is to reduce the working time because there is no heating and cooling process and to reduce the cost of die because hot compaction needs a die with resistance of high temperature [4]. The variable constant of compaction pressure for production can be obtained from previous research where the mechanic test implied that the best result was resulted by compaction pressure of 700 psi.

Liquid sintering process is able to create intermetallic compounds, Cu_3Sn and Cu_6Sn_5 which increase the mechanical properties of Cu-Sn [6]. Sintering process within industry, is carried out on liquid state because the solubility of solid within liquid can be filled in the porosity, so it can improve the mechanical properties [7]. Pre-sintering temperature that is used is about 100°C in 3-5 minutes [8]. Sintering temperature used is 260°C with 30 minutes of working process [9].

One of the types of finishing process which are compatible to be used for this frangible composite is vibro-abrasive. It is also known as rumbling or tumbling because it is able to process many objects instantly [10]. The finishing media used is porcelain VZ EB 0610 weighting 25 kilograms and liquid booster FEL120-B32 weighting about 50 grams [10]. The duration of the process is 180 minutes [10].

Previous research [9] stated that the mass of composite is 5.23 grams with tolerance ± 0,1 grams and the diameter is 9.02 mm with the tolerance ± 0.02 mm. Meanwhile, the simulation result shows that the optimum length of composite material is 16.0 mm with a tolerance of ± 0.8 mm.

Based on benchmarking result in various manufacturing company, shows that the aspect that is controlled by QC is weight and dimension because both of them affect the mechanical properties. In previous research also showed that the optimum frangibility factor of the frangible composite can be obtained on composite mass which are 5.23 grams with the tolerance ± 0,1 grams and diameter of 9.02 mm with the tolerance ± 0,02 grams. Meanwhile, the simulation result shows that the optimum length of composite is about 16.0 mm with the tolerance ± 0.8 mm.

The frangible composite material which do not meet specification will be rejected while QC process. Then, the rejected material will be collected for recycling. This will improve the salvage value from rejected material so it can be more efficient for production cost.

In packaging process, 100 (one hundred) composites are packaged into one box. Then, a silica gel is inserted and a plastic case is tightly wrapped to the package.

3.2 Production Target in Each Processes

Assuming the failure rate is about 10% in compaction process, sintering, and QC, initial production target can be calculated as follows in the Table 3.

Table 3. Production target in each processes.

Processes	Initial Production Target (rounds/shift)
Quality control	50778
Sintering	56420
Compaction	62688

3.3 Specification and Quantity of Machines

For the next step to establish a production line is by deciding the number of machines which are needed and use the number of production target, production capacity, and processing time. To determine the number of machines which are needed use the control variable that consist of production target/shift (8 hours), working time, and plant rate. The calculation of the production target/shift (8 hours) determined by :

$$Production\ target\ \left(\frac{rounds}{shift}\right) = \frac{Production\ target\ \left({}^{rounds}/_{year}\right)}{\left(\left({}^{total\ days}/_{year}\right)\Big/\left({}^{number\ of\ shift}/_{day}\right)\right)} \quad (3)$$

The results of the calculation of production target (rounds/shift) is 45700 grain/shift.

The calculation is conducted in order to find the effective working time in 1 shift. Expected performance can be assumed 100%. The effective working time can be calculated as follows:

$$Work\ time = 1\ shift\ (mins)\ x\ Exp.Performance\ \% \quad (4)$$

The results of the calculation of work time are 28.800 seconds.

Plant rate is the target of production capacity that has to be achieved by machine. The value can be calculated as follows:

$$Plant\ rate\ \left({}^{second}/_{pcs}\right) = \frac{working\ time\ (second)}{Production\ target\ per\ shift\ (pcs)} \quad (5)$$

The results of the calculation of plant rate (second/pcs) is (28800 seconds)/(45700 pcs). The spesification and quantity of machines which needed in this line production as shown in Table 4.

Table 4. Spesification and quantity of machines.

Processes	Machine Capacity (seconds/rounds)	Total of Machines (units)
Continous sintering furnace	0,466	1
Compaction	1,333	3
Mixer	48.000	1
Finishing	0,276	1
Quality control	1,7	3

3.4 Departments of the Plant

Deriving from the number of machine needed, we can calculate the needed area of each department. Therefore, we can calculate the production section area and supporting production area. The data has been attached as shown in Table 5 for Production Section and Table 6 for Supporting Section.

Table 5. Production Section Data.

No.	Section	Dimension (meter)		
		Length	Width	Area
1.	Loading Scrub Atomization	1,5	1,5	2,25
2.	Scrub Preparation Atomization	1,5	3	4,5
3.	Scrub Feeding Atomization	3	3	9
4	Atomization	3	3	9
5	Mixing	2	5	10
6	Compaction	5	3	15
7	Sintering	5	3	15
8	Finishing	2	3	6
9	QC	3	3	9
11	Packaging	2	3	6
Total Area				61

Table 6. Supporting Production Section Data.

No.	Section	Dimension (meter)		
		Length	Width	Area
1.	Office	6	3	18
2.	Manager Room	3	3	9
3.	Warehouse	1	3	3
4	Toilet	3	3	9
Total Area				39

3.5 ARC Diagram of the Plant

Each sections of production is combined into one production section because the placement is based on the sequence of process. Then, the ARC diagram is made based on all supporting department and the production department, as shown in Figure 1.

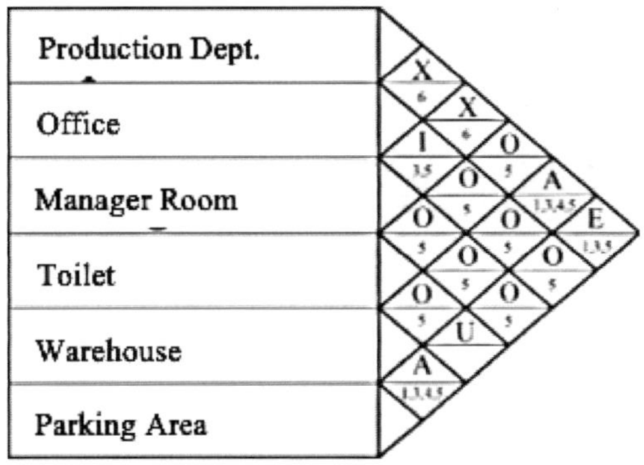

Figure 1. ARC Diagram of frangible composite Plant.

Then, we can calculate TCR value in each section. Therefore, we get the sequence of placements which is shown in Table 7.

Table 7. TCR Value and Allocation Order

No	Section Name	Total TCR	Area Section (m^2)
1	Warehouse	76	3
2	Parking Area	58	24
3	Toilet	18	9
4	Production Dept.	-12	61
5	Office	-12	18
6	Manager Room	-12	9

3.6 Layout Design of Plant

Based on the sequence of placement in Table 5, the departments can then be placed according to the CORELAP algorithm rules. The results are as shown in Figure 2 and Figure 3 below.

3.7 Line Production Simulation

To simulate the production line, the variable and constant of each processes and total machines is inputted into the software Tecnomatix Plant Simulation 14. By this method, every machines can operate either continuously or batchly. The constant inputted is machine production process (continuous or batch), production capacity (pcs/cycle) and process time. Then, the model is simulated for 1 year 3 hours 15 minutes (365 days 3 hours 15 minutes). The Production output in 1 year is 50,037,120 rounds and .production output in 1 hour is 5712 rounds.

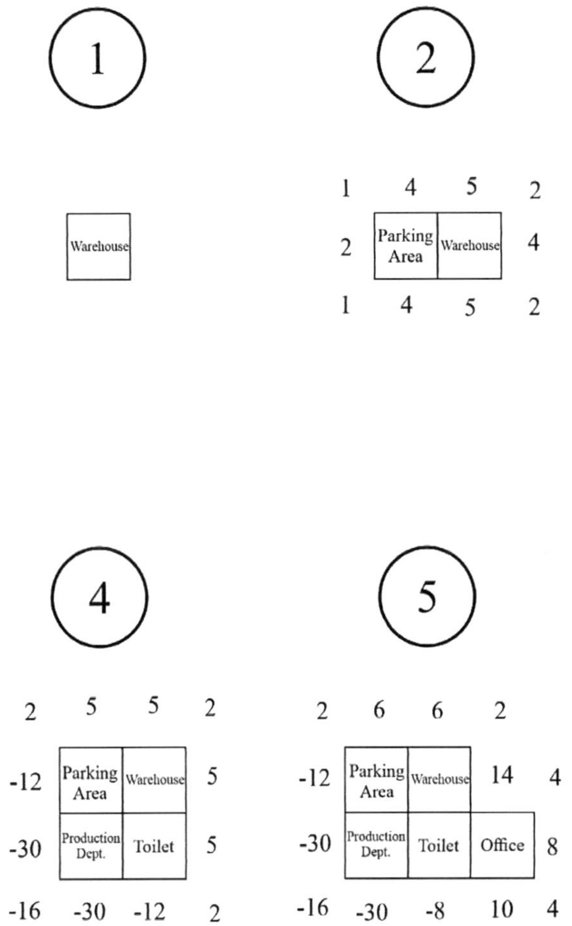

Figure 2. Sequence of department placing by CORELAP Algorithm.

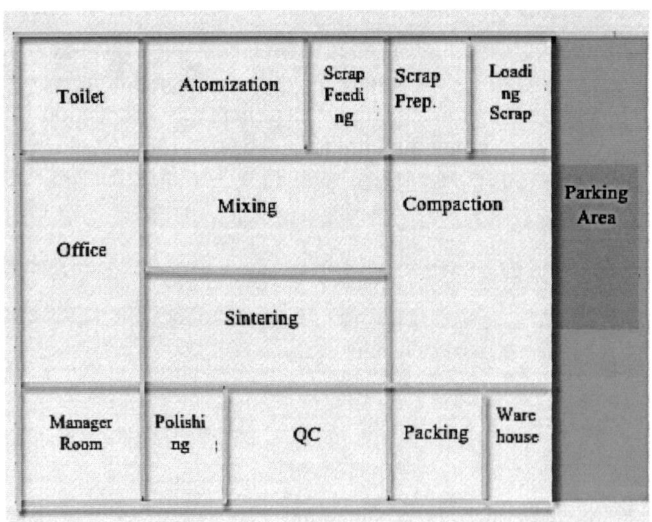

Figure 3. Final layout design.

IOP Conf. Series: Materials Science and Engineering **924** (2020) 012012 doi:10.1088/1757-899X/924/1/012012

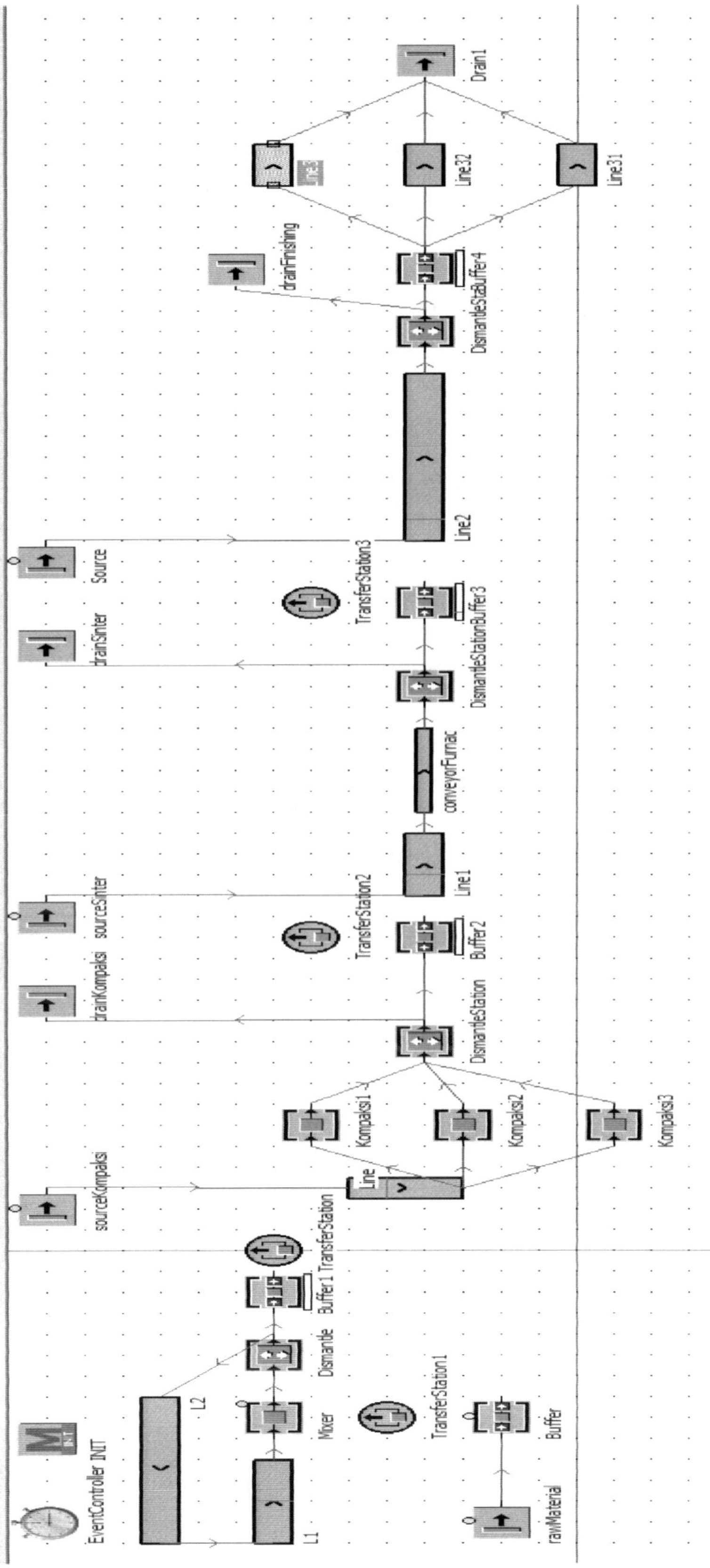

Figure 4. Frangible composite production scheme as simulated in Tecnomatix Plant Simulation.

4. Conclusion

The result of this research implied that the plant layout design of *frangible* composite plant has been arranged and proven to produce 50 million rounds/year with powder metallurgy method. The total machine that are needed for this mini plant is as follows, 1 unit of mixer, 3 units of compaction machine, 1 unit of sintering furnace, 1 unit of finishing machine, and 3 units of QC conveyor. From the simulation of production line using Tecnomatix Plant Simulation, we get that the production output of the designed plant layout is 50,037,120 rounds per year with a running time of 365 days 3 hours 15 minutes.

Acknowledgment

We would like to thank ITS for their contribution in funding this research through the PENPRO (Penelitian Prototype).

References

[1] N. Aribowo, S. Wignjosoebroto, and S. Sudiarno. *Redesign Facility Production Layout With Corelap Algorithm 2-Opt Algorithm And Planar Graph Algorithm To Minimize Material Handling Cost Study Case At Pt. Swadaya Graha-Gresik*. Surabaya:Institut Teknologi Sepuluh Nopember, 2009. (Indonesian)

[2] D. Christyono, and M.L. Singgih. *Plant Layout Improvement with CORELAP, MST, CRAFT Algorithm and Material Handling Consideration*. Surabaya:Institut Teknologi Sepuluh Nopember, 2015. (Indonesian)

[3] S.S. Heragu. *Facilities Design 4th Edition*. CRC Press, 2016.

[4] J.A. Tompkins. *Facilities Planning*. Michigan: John Wiley Inc., 1996.

[5] S.A. Tsukerman. *Powder Metallurgy 1st Edition*. Elsevier, 1965.

[6] ASM International. *ASM Handbook Volume 7 Powder Metal Technologies and Applications*. ASM International, 1998.

[7] P. Jhony and Widyastuti. *Analysis of the Effect of Sintering Temperature and Holding Time on Interface Bonding for Metal Matrix Composites (MMCs) Cu-10%wtSn Made by Powder Metallurgy*. Surabaya: Institut Teknologi Sepuluh Nopember, 2014. (Indonesian)

[8] R. M. German. *Sintering: from Empirical Observations to Scientific Principles*. Oxford: Elsevier Inc, 2014.

[9] J.M. Tambunan, Widyastuti, and H. Ardyananta. *Study Design of Body and Temperature Continuous Furnace Sintering Material Frangible Cu-5wt%Sn*. Surabaya: Institut Teknologi Sepuluh Nopember, 2017. (Indonesian)

[10] Firmansyah and Widyastuti, "Effect of Sintering Temperature in Powder Metallurgy Method to Frangibility and Ballistic Performance of Frangible Composite Cu-10%wtSn", *Jurnal Teknik ITS*. 1-6. 2015. (Indonesian)

IOP Conf. Series: Materials Science and Engineering **924** (2020) 012013 doi:10.1088/1757-899X/924/1/012013

Shape Modification of Manganese Oxide Prepared by Solvothermal: Effect of Precipitation Agent

Retno Kusumaningrum[1], Faustina De Yesu Prisila Abi[2], Galuh Sukmarani[1], Alvin Muhammad Habieb[1], Fiqhi Fauzi[1], Wahyu Bambang Widayatno[3], Agus Sukarto Wismogroho[3], Muhamad Ikhlasul Amal[4], Nurul Taufiqu Rochman[4], and Alfian Noviyanto[1,5,*]

[1] Nano Center Indonesia, Jl. PUSPIPTEK, South Tangerang, Banten 15314 Indonesia
[2] Department of Chemistry, Faculty of Mathematics and Sciences, Universitas Katolik Widya Mandira, Kupang, East of Nusa Tenggara 71152 Indonesia
[3] Research Center for Physics, Indonesian Institute of Sciences, PUSPIPTEK, South Tangerang, Banten 15314, Indonesia
[4] Research Center for Metallurgy and Materials, Indonesian Institute of Sciences, PUSPIPTEK, South Tangerang, Banten 15314, Indonesia
[5] Department of Mechanical Engineering, Mercu Buana University, Jl. Meruya Selatan, Kebun Jeruk, Jakarta 11650, Indonesia

*a.noviyanto@nano.or.id

Abstract. Nano-sized manganese oxide was synthesized by the solvothermal method. Manganese sulfate was used as a precursor, which is obtained from the Sumbawa manganese ore. Two kinds of the precipitating agent, ammonium persulfate and sodium hydroxide, were used in this study and the effect on the solvothermal product were examined thoroughly. The solvothermal temperature and time were 120°C for 18 hours, respectively, for both precipitating agents. The results showed that the phase of manganese oxide was influenced by the precipitating agent, as indicated by X-ray diffraction analysis. Phase α MnO_2 was obtained from the reaction between $MnSO_4$ and ammonium persulfate, while Mn_3O_4 for the sodium hydroxide. The formation of α-MnO_2 was influenced by the excess of NH_4^+ which came from the manganese ore leaching into $MnSO_4$. One notable finding in this study is that the morphology of manganese oxide was also affected by the precipitating agent. For instance, the shape of MnO_2 and Mn_3O_4 was needle and sphere, respectively. Moreover, the strict long thinner needle of α-MnO_2 was formed and the aspect ratio increased at a higher temperature. While the particle size of Mn_3O_4 decreased with the increased temperature. These results imply that variation of a precipitating agent is imperative to obtain the specific manganese oxide product, including the shape.

***Keywords*:** α-MnO_2, Mn_3O_4, solvothermal, needle, sphere

1. Introduction

Manganese has wide application in human life and industry, indicated by the extensive study to the synthesis of manganese oxide and its application. Moreover, manganese is one of element which interesting and unique properties because of its various oxidation state, which influenced the different structure. The most common oxidation numbers of manganese are +2, +3, +4, +6 and +7.

Nanomaterial of manganese oxide is one of important functional metal oxide as a catalyst [1,2], supercapacitor [3], waste treatment [4,5] and pharmaceutical/ medical [6] owing to excellent physical and chemical properties. Two of manganese oxides are extensive use, i.e., manganese dioxide (MnO_2) and tri manganese tetroxide (Mn_3O_4) [7]. MnO_2 is a polymorph, which has α, β, γ, δ and ε structure [8]. In order to control the structure of MnO_2, stabilize the tunnel structure is important. Several factors that influence the tunnel structure are pH, counteraction, temperature and concentration and the existing of the inorganic cation. Moreover, the size of MnO_2 is the results of a different tunnel [9]. The tunnel structure of α-MnO_2, β- MnO_2, γ- MnO_2 are 2×2, 1×1 and 1×2 Mn octahedral tunnels (1D), respectively, while δ-MnO_2 is a 2D layered structure [10,11]. Tri manganese tetroxide (Mn_3O_4) that found in stable temperature form is tetragonal hausmannite and a metal oxide which has two cations, Mn^{2+} and Mn^{3+} that occupied the tetrahedral and octahedral position, respectively. This counterpart is important to encourage its pseudo-capacitance supercapacitors [12,13].

From the above perspective, in order to obtain a desirable manganese oxide, attention needs to pay in the controlling reaction system. The soft chemical route is a solution process with easy controllable reaction condition and provides many alternative sources of reactant, also more suitable for large-scale production of nanomaterial with specific morphology [14]. Solvothermal is the simple method which most favorable to synthesis manganese oxide. This method also provides the nanostructure and phase convenient to control.

The objectives of this study are synthesis manganese oxide with different precipitating agents using a solvothermal method. Moreover, we use manganese ore instead of commercial one to synthesis manganese sulfate as a precursor to synthesis manganese oxide. The precipitating agents used in this study are sodium hydroxide and ammonium persulfate. The effect of solvothermal temperature on the phase and shape of manganese oxide was examined thoroughly.

2. Experimental method

This research used manganese ore from Sumbawa, West of Nusa Tenggara, Indonesia as starting material. All chemical reagents which used in this research are analytically pure grade. The sulfuric acid 95% (H_2SO_4), hydrogen peroxide 30% (H_2O_2) and ammonium sulfate 25 % ($(NH_4)_2SO_4$) from Merck were used in synthesis manganese hydroxide from its ore. The 510 mmol of H_2SO_4 solution was heated at 70 °C using a hot plate that equipped with a magnetic stirrer. Ten grams of manganese ore was added slowly to the solution. Subsequently, 147 mmol of H_2O_2 was added carefully drop by drop to ignore the bumping solution. The leaching process was continued for 90 minutes. After the leaching solution cooling down, the slurry and filtrate were separated used Buchner funnel that connected with a vacuum pump. The red wine solution was added by 25% of NH_4OH till brown precipitate formed and the pH of the solution was 8. The clear manganese sulfate ($MnSO_4$) solution was ready to be used as a precursor.

The precipitating agents used in this study were sodium hydroxide (NaOH, Merck) and ammonium persulfate NH_4OH, XiLong Scientific). Manganese oxide was synthesis by solvothermal method. The process was started by pouring 45 mL of manganese sulfate into beaker glass and added the 180 mmol of NaOH or four mmol of ammonium persulfate. The total volume was adapted with the volume Teflon jar in the solvothermal reactor. The solution then was mixed for 15 minutes at room temperature. Then, the solution was poured into Teflon jar-lined stainless-steel autoclave. The solvothermal reaction was done for 18 hours in a variety of temperature; 90, 120 and 150 °C. The final product from NaOH and ammonium persulfate precipitating agent was dark brown and black, respectively. Then, the powder was washed till pH 7 using demineralization water and ethyl alcohol.

The phase of the final product was characterized by room temperature X-ray diffraction (XRD: Shimadzu 7000 Maxima-X tube, Japan) using CuKα line. The morphology of the manganese oxide powder was observed by field emission scanning electron microscope (Fe-SEM: JIB-4610F *Multi Beam System*, JEOL, Japan). The average particle size and aspect ratio were determined using SEM image by calculate 100 particles and analyzed statistically.

3. Results and discussion

Figure 1 displays the XRD patterns of manganese oxide from the solvothermal using different precipitating agents. Although our previous study shows that many types of manganese oxide products might be formed due to the reaction of $MnSO_4$ and NaOH [15], the spontaneous reaction only occurred for the reaction that the product was Mn_3O_4 based on the Gibbs free energy calculation. This calculation was verified by the experimental. Fortunately, our theoretical calculation matches the experimental results, as shown in Figure 1 (a). However, α-MnO_2 was obtained after a solvothermal reaction between $MnSO_4$ with $(NH_4)_2S_2O_8$. The difference of manganese oxide product phase that obtained after the solvothermal reaction is interesting, hence, it seems the formation of manganese cation plays an important role to determine the manganese product.

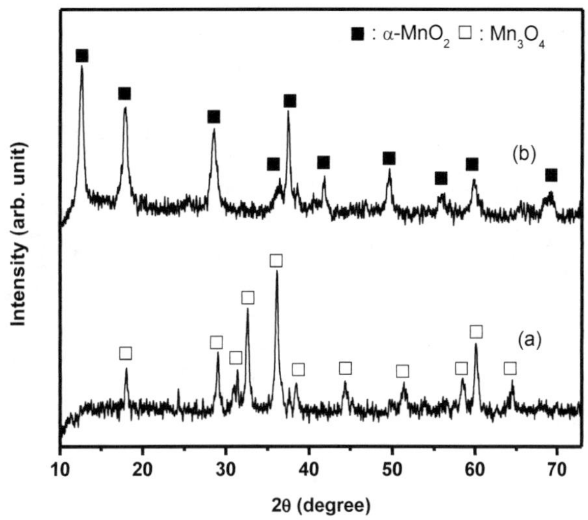

Figure 1. X-ray Diffraction of manganese oxide in different precipitating agent.

The formation of Mn_3O_4 from the reaction between $MnSO_4$ and NaOH can be seen in Equation (1). However, several steps occurred in the formation of Mn_3O_4. By nature, manganese sulfate dissolves in the water to form Mn^{2+}, while NaOH imparts OH^- ion in the system. The OH^- oxidized the Mn^{2+} to form MnO, as shown in Equation (2). Simultaneously, the formation of Mn_2O_3 is also appeared, as shown in Equation (3). Hence, the presence of MnO and Mn_2O_3 is essential for the formation of Mn_3O_4 (Equation (4)), which is having a role to oxidizing Mn^{2+} from manganese sulfate. Hydroxide ion oxidized Mn^{2+} into Mn^{3+} and together with Mn^{2+} formed Mn_3O_4. However, there is a possible direct reaction to form Mn_3O_4, as shown in Equation (5). Regardless of the direct or indirect reaction, Mn_3O_4 is the product of the reaction between $MnSO_4$ and NaOH.

$$MnSO_4 + NaOH \rightarrow Mn_3O_4 + Na_2SO_4 + H_2O \quad (1)$$

$$Mn^{2+} + OH^- \rightarrow MnO + H^+ \quad (2)$$

$$2Mn^{2+} + 3OH^- \rightarrow Mn_2O_3 + H^+ \quad (3)$$

$$MnO + Mn_2O_3 \rightarrow Mn_3O_4 \quad (4)$$

$$3Mn^{2+} + 4OH^- \rightarrow Mn_3O_4 + 2H^+ \quad (5)$$

IOP Conf. Series: Materials Science and Engineering **924** (2020) 012013 doi:10.1088/1757-899X/924/1/012013

Figure 2. SEM images of the α-MnO_2 after solvothermal at (a) 90°C, (c) 120°C and (e) 150°C, Mn_3O_4 after solvothermal at (b) 90°C, (d) 120°C and (f) 150°C.

IOP Conf. Series: Materials Science and Engineering **924** (2020) 012013 doi:10.1088/1757-899X/924/1/012013

On the other hand, the oxidation of Mn^{2+} into higher oxidation number, i.e., Mn^{4+}, happens when $(NH_4)_2S_2O_8$ uses as an oxidizing agent, resulting in the formation of α-MnO_2 (Equation (6) and (7)). Other study showed the formation of α-MnO_2 along with β-MnO_2 using analytical-high purity grade $MnSO_4$ and $(NH_4)_2S_2O_8$ [16]. Moreover, the excess of $(NH_4)_2SO_4$ in our system occurs due to reaction of NH_4OH with excess SO_4^{2-} from leaching agent during iron removal procedure, it seems responsible for the formation of α-MnO_2 without β-MnO_2. The existing of NH_4^+ ion gives effect to stabilize 2 × 2 tunnels forming significantly, thus enhance the formation of SO_4^{2-}, which decreases the formation of Mn^{4+} [16].

$$MnSO_4 + (NH_4)_2S_2O_8 \rightarrow MnO_2 + (NH_4)_2SO_4 \quad (6)$$

$$Mn^{2+} + S_2O_8^{2-} \rightarrow Mn^{4+} + 2SO_4^{2-} \quad (7)$$

Figure 2 presents the SEM images of the manganese oxide product after the solvothermal reaction at various temperature. It should be noted that the shape of manganese oxide depends on the phase of manganese oxide. For instance, the needle shape was obtained when the phase was α-MnO_2 (Figure 2 (a)), while spherical shape for the Mn_3O_4 (Figure 2 (b)). The needle shape of α-MnO_2 is more likely due to an excess of SO_4^{2-} in the system, which suppresses the formation of β-MnO_2 [16]. α-MnO_2 has a double chain of MnO_6 octahedral structure and lead to the needle shape formation. In contrast, as a transition metal oxide with the normal spinel structure, Mn_3O_4 was found in tetragonal, which Mn^{3+} and Mn^{2+} ions fill in the octahedral and tetrahedral position of the spinel structure [17].

Interestingly, in our study, we found that increase in the temperature lead to the decrease of particle size, in particular for the Mn_3O_4, as shown in Figures 2 (b), (d) and (f). The average particle size of Mn_3O_4 at 90, 120 and 150°C was 63.14, 66.74 and 42.04 nm, respectively. The shape of Mn_3O_4 improves with the increasing temperature, while the perfect sphere was observed in the solvothermal temperature of 150°C (Figure 2 (f)). Generally, the particle size increase with the increasing temperature, however, in our study, we found opposing results. This is more likely due to the nucleation rate is the function of the temperature. In our system, the nucleation rate is faster than crystallization growth, therefore, the crystallite size decrease with increasing temperature. This phenomenon can be understood due to the increasing temperature, i.e., 150 °C, increase the thermal energy, which is related to the faster Brownian motion [18]. Moreover, the rate of oxidation of Mn^{2+} appeared sooner in this system. By considering that the cation formation initiated the nucleation, it can be confirmed that at a higher temperature the systems prefer to form the nuclei rather than the growth of particle. The crystallite size of Mn_3O_4 at 90 and 150°C was 39.3 and 27.2 nm, respectively, showing that the crystallinity of Mn_3O_4 decrease with increasing temperature. In order to have a better understanding, we performed an investigation to elucidate the effect of crystallization. Solvothermal at 150°C for 24 h were done and the results showed that the average particle size of Mn_3O_4 at 150°C was 59.3 nm, indicating that the particle size increase with prolongs the time of solvothermal. This can be explained that after the nucleation phase finished, the Ostwald ripening phenomenon occurs and result the large particle.

In the case of the α-MnO_2, longer and thinner needle shape enhance with the increasing temperature. This result similar to the other study, which defined that aspect ratio of α-MnO_2 and β-MnO_2 is influenced the product morphology. From their study, high temperature affected the growth anisotropic of crystal and gave a higher aspect ratio product [10]. The aspect ratio of α-MnO_2 at 90, 120 and 150°C was 12.41, 18.33 and 32.88, respectively, indicating that at 150°C the diameter of the needle thinner compared to 90°C.

4. Conclusion

Manganese oxide with different phase and shape were successfully synthesized using different precipitating agents and solvothermal method. In the first step, the precursor, i.e., manganese sulfate, for the synthesis of the manganese oxide was prepared from the local manganese ore. Then, α-MnO_2 and Mn_3O_4 was obtained from the reaction of manganese sulfate with sodium hydroxide or ammonium persulfate, respectively, for 18 h of solvothermal. Indeed, the formation of Mn_3O_4 by the reaction of manganese sulfate with sodium hydroxide can be understood due to its has the lowest Gibbs free energy, therefore the reaction is spontaneous. Moreover, the presence of OH^- is imperative to oxidize the Mn^{2+} into Mn^{3+} while there is still Mn^{2+} and led to the formation of Mn_3O_4. Meanwhile, the formation of α-MnO_2 due to the existence of ammonium persulfate as a precipitating agent that forms Mn^{4+}. The phase also affects the morphology of the particle. For instance, two types of shape were obtained after the solvothermal method for 18 h; needle and sphere which belonged to α-MnO_2 and Mn_3O_4, respectively. One notable finding in this study is the crystallinity of the Mn_3O_4 decreased with increasing the solvothermal temperature, as indicated by decreasing the crystallite size from 39.3 nm at 90°C of solvothermal to 27.2 nm at 150°C of solvothermal. Moreover, the average particle size also decreased with increasing temperature, which is 63.14, 66.74 and 42.04 nm were obtained after solvothermal reaction at 90, 120 and 150°C, respectively. This phenomenon is more likely due to the competition of the nucleation rate and crystallization rate, where the nucleation rate is dominant in our case. The needle shape of α-MnO_2 was thinner and longer at 150°C of solvothermal reaction compared to 90°C. The aspect ratio of α-MnO_2 was 12.41, 18.33 and 32.88 after solvothermal reaction at 90, 120 and 150°C, respectively.

Acknowledgment
The funding for this research was provided by the National Intensive Research System Innovation Program 2019 (INSINAS), Ministry of Research and Technology of Indonesia. This research also supported and facilitated by Research Centre of Physics, Indonesian of Science Institute.

References

[1] C. Liu, J. W. Shi, C. Gao, and C. Niu, "Manganese oxide-based catalysts for low-temperature selective catalytic reduction of NOx with NH3: A review," *Appl. Catal. A Gen.*, vol. 522, pp. 54–69, 2016.
[2] Z. B. Jildeh, J. Oberländer, P. Kirchner, P. H. Wagner, and M. J. Schöning, "Thermocatalytic behavior of manganese (IV) oxide as nanoporous material on the dissociation of a gas mixture containing hydrogen peroxide," *Nanomaterials*, vol. 8, no. 4, 2018.
[3] P. A. Shinde and S. Korea, "Hydrothermal synthesis of manganese oxide thin films using different oxidizing agents for supercapacitor application," *Int. J. Eng. Res. Technol.*, vol. 10, no. 1, pp. 532–537, 2017.
[4] H. Chen and J. He, "Facile synthesis of monodisperse manganese oxide nanostructures and their application in water treatment," *J. Phys. Chem. C*, vol. 112, no. 45, pp. 17540–17545, 2008.
[5] S. Chakrabarti, B. K. Dutta, and R. Apak, "Active manganese oxide: A novel adsorbent for treatment of wastewater containing azo dye," *Water Sci. Technol.*, vol. 60, no. 12, pp. 3017–3024, 2009.
[6] J. Shin, R. M. Anisur, M. K. Ko, G. H. Im, J. H. Lee, and I. S. Lee, "Hollow manganese oxide nanoparticles as multifunctional agents for magnetic resonance imaging and drug delivery," *Angew. Chemie - Int. Ed.*, vol. 48, no. 2, pp. 321–324, 2009.
[7] N. Wang *et al.*, "Facile synthesis of monodisperse Mn_3O_4 tetragonal nanoparticles and their large-scale assembly into highly regular walls by a simple solution route," *Small*, vol. 3, no. 4, pp. 606–610, 2007.
[8] G. Pistoia, A. Antonini, D. Zane, and M. Pasquali, "Synthesis of Mn spinels from different polymorphs of MnO2," *J. Power Sources*, vol. 56, no. 1, pp. 37–43, 1995.
[9] Y. F. Shen, S. L. Suib, and C. L. O'Young, "Effects of Inorganic Cation Templates on Octahedral Molecular Sieves of Manganese Oxide," *J. Am. Chem. Soc.*, vol. 116, no. 24, pp. 11020–11029, 1994.
[10] X. Wang and Y. Li, "Synthesis and formation mechanism of manganese dioxide nanowires/nanorods," *Chem. - A Eur. J.*, vol. 9, no. 1, pp. 300–306, 2003.
[11] S. Devaraj and N. Munichandraiah, "Effect of crystallographic structure of MnO_2 on its

IOP Conf. Series: Materials Science and Engineering **924** (2020) 012013 doi:10.1088/1757-899X/924/1/012013

electrochemical capacitance properties," *J. Phys. Chem. C*, vol. 112, no. 11, pp. 4406–4417, 2008.

[12] O. Y. Gorbenko *et al.*, "The structure and properties of Mn_3O_4 thin films grown by MOCVD," *Solid State Commun.*, vol. 124, no. 1–2, pp. 15–20, 2002.

[13] V. C. Bose and V. Biju, "Mixed valence nanostructured Mn_3O_4 for supercapacitor applications," *Bull. Mater. Sci.*, vol. 38, no. 4, pp. 865–873, 2015.

[14] F. Cheng *et al.*, "Facile controlled synthesis of MnO_2 nanostructures of novel shapes and their application in batteries," *Inorg. Chem.*, vol. 45, no. 5, pp. 2038–2044, 2006.

[15] R. Kusumaningrum *et al.*, "Reactivity of Manganese Sulphate From Sumbawa Manganese Ore With Precipitating Agent: Theoretical and Experimental Evaluation," *J. Phys. Conf. Ser.*, vol. 1191, p. 012052, 2019.

[16] X. Wang and Y. Li, "Selected-control hydrothermal synthesis of α- and β-MnO_2 single crystal nanowires," *J. Am. Chem. Soc.*, vol. 124, no. 12, pp. 2880–2881, 2002.

[17] H. Jiang, T. Zhao, C. Yan, J. Ma, and C. Li, "Hydrothermal synthesis of novel Mn_3O_4 nano-octahedrons with enhanced supercapacitors performances," *Nanoscale*, vol. 2, no. 10, pp. 2195–2198, 2010.

[18] N. T. K. Thanh, N. Maclean, and S. Mahiddine, "Mechanisms of Nucleation and Growth of Nanoparticles in Solution, " *Chem. Rev.*, vol, 114, pp. 7610-7630, 2014

Reactivating Adsorption Capacities of Spent Bleaching Earth for Using in Crude Palm Oil Industry

Kurniawan Eko Saputro[1,2], Puji Siswanti[1], Dwi Wahyu Nugroho[1,2], Radyum Ikono[1], Alfian Noviyanto[1,4*] and Nurul Taufiqu Rochman[4]

[1] Nano Center Indonesia, Jl. PUSPIPTEK, South Tangerang, Banten 15314, Indonesia
[2] Department of Industrial Engineering, Faculty of Engineering, University of Nahdlatul Ulama Indonesia, Jl. Taman Amir Hamzah No. 5, Jakarta 10430, Indonesia
[3] Department of Mechanical Engineering, Faculty of Engineering, Mercu Buana University, Jl. Meruya Selatan, Kebon Jeruk, Jakarta 11650, Indonesia
[4] Research Center for Metallurgy and Materials, Indonesia Institute of Science, Jl. PUSPIPTEK, South Tangerang, Banten 15314, Indonesia

*a.noviyanto@nano.or.id

Abstract. The spent bleaching earth from the palm oil refining industries contains various contaminants, including free fatty acids and hazardous material. The conventional procedure is harmful to the environment. In this study, spent bleaching earth (SBE) was successfully regenerated by calcination and acid-activation process. The calcination process was performed at 500, 600 and 700°C for 2, 3 and 4 h. All samples were then mixed with $CaCO_3$ and lastly activated with aqueous acid. The results showed the SBE color restored at calcination temperature higher than 500°C, even though the decomposition of the volatile compound was removed at 680°C according to thermal gravimetric analysis. Treatment affected the clay structure and mostly removed the impregnated organic matter. The primary phase of regenerated SBE is similar to virgin bleaching earth (VBE), implying the calcination and acid-activation is not alter the phase. Moreover, the adsorbent capacity of regenerated SBE was comparable with VBE, showing the regenerated SBE has promising potential for many applications, in particular for the palm oil refining industry. Regenerated SBE gave significantly larger specific surface area, microporous surface area and total pore volume with smaller the average pore diameter.

Keywords **:** spent bleaching earth, regenerated SBE, calcination, acid activation

1. Introduction

Palm oil is a highly efficient and yielding source of food and fuel. In 2017, the crude palm oil (CPO) production in a whole nation reached 38.5 million tons with 18 million tons was exported and the remains being consumed domestically [1]. In the long term, the global demand for palm oil shows a growing trend as a ballooning global population, which increases the consumption of food and cosmetic product that contain palm-derived raw material. Moreover, CPO is also a promising source for biodiesel due to abundant resources, especially in Indonesia [2].

The crude palm oil (CPO) through the refining process before it is derived from other products. Traditional palm oil refining processes generally involve four major steps which are degumming, bleaching, neutralization and deodorization [3]. Bleaching is an important process in palm oil refining. The purpose of bleaching processes is to adsorb impurities in crude palm oil mostly both saturated and unsaturated fats [4]. The bleaching step is realized by using acid-activation clay which known as bleaching clay (usually used is bentonite) because of its high specific surface [5]. The bleaching process generates massive waste of spent bleaching earth. Globally, it is estimated that about 1 million tons or more of SBE are disposed of in the refining process and it expected to grow even further with an increasing global production [6]. The spent bleaching earth contains 20-40 wt. % of the adsorbed oil with a high content of free fatty acids, which is not removed completely during the final filter pressing process. It is typically dumped in the landfill as waste and disposed of. Spontaneous auto-ignition is the main concern with the handling of SBE from the bleaching of unsaturated vegetable oils which can trigger fire hazard [7,8]. If the abundance residues can be converted into a more useful and valuable product, it will be essential for more effective production [9]. Therefore, it is important to develop new economical technologies for SBE disposal and to reduce potential contamination.

Many studies are dedicated to the conversion of these wastes and their reuse in different applications to minimize the contamination risk. The regenerated material (RSBE) has particularly been used as an adsorbent to separate pigments from crude palm oil [3,8,9], hexane micella and from aqueous solution [10], heavy metal ions [11], and to capture the carbon dioxide [5]. Currently, several methods to RSBE are: removed the oil content by the solvent extraction [12–15], acid solutions [16,17], hot alkali solutions [18] and inorganic salt solutions [17,19], thermal activation [16,20] and acid activation [17]. Due to increasingly high the cost of dispose waste to landfills, pretreatment SBE using n-hexane was the best alternative to remove the oil content. It was economically feasible for expensive and relatively stable oils [6]. To our knowledge, however, no study on the reuse of SBE regenerated by using a thermal treatment method in a furnace without chemical additives have been presented.

Hence, the main purposes of this research are as follows: regeneration and reuse spent bleaching earth in the palm oil bleaching, to reduce the production cost of oil refinery and to minimize environmental pollution from this waste. The best regenerated material (RSBE) was characterized by using FTIR, SEM, and XRD methods. On the other hand, measuring some value, such as heavy metals, pore size, and color of the bleached oil, allow us to validate our proposed regeneration method and to conclude on adsorbent capacity.

2. Experimental method

SBE was supplied by PT. SMART Tbk (Palm oil refinery industry in Indonesia). The oil was extracted from SBE by maceration method whereby SBE was mixed with an n-hexane for 4 h at room temperature with 1:5 w/v ratio. This SBE was named deoiled SBE. Then, to activate the deoiled SBE by thermal heating, calcination was performed at 500, 600, and 700°C for 2, 3, and 4 h in the furnace. 0.1 M $CaCO_3$ solution was added to calcinated SBE and then mixed for 1 h at 105°C. Then the solution was filtered, dried and leached by the acid solution. Furthermore, the SBE powder was chemically activated by mixing with 10% of sulfuric acid (H_2SO_4) for 3 h on a magnetic stirrer. Next, the samples were rinsed with distilled water until the pH 3 – 4 and dried at room temperature. This sample was named regenerated spent bleaching earth (RSBE). Commercial VBE was obtained from PT Clariant AG, Indonesia was also used for comparison with RSBE from this study.

Thermal decomposition of SBE was analyzed by thermogravimetry and differential thermal analysis (TG-DTA, STA PT1600, LINSEIS, Germany) in an air atmosphere with heating rate 5°C/min from room temperature to 1000°C. Molecular bonding of the SBE, deoiled SBE, RSBE and VBE was characterized by Fourier-transform infrared spectroscopy (FTIR, IRPrestige-21, Shimadzu, Japan). X-ray diffraction (XRD, Rigaku, Japan using CuKa as a source of radiation) was used to observe the phase in VBE, deoiled SBE and RSBE. Microstructure and elemental composition were evaluated by scanning electron microscope (SEM, SU-3500, Hitachi, Japan) equipped with energy dispersive spectroscopy (EDS, Horiba, Japan). The pores of VBE and RSBE, including the surface area, were measured by Brunauer–Emmett–Teller (BET) method using Quantachrome instrument (BET, Nova 4200e, USA).

3. Results and discussion

The decomposition temperature of the oil adsorbed on all samples was determined using TG-DTA before proceeding with calcination (Figure 1). The percentage of calcined products from SBE determined using the muffle furnace and TGA-DTG at 680°C showed about 24.8%. This result is in agreement with previous reports [6,21], that oil retention in SBE of 20–40%. Due to the removal of the hydroxyl group in silica bonding at higher temperatures, the percentage of weight loss for SBE increased by 20–40% [22,23]. The thermogram of SBE (Figure 1) showed three weight loss steps. The first weight between 40 - 200°C loss is caused by the removal of adsorbed water molecules and chemically bonded OH groups [22,23]. The second step showing a substantial endothermal weight loss is primarily associated with the decomposition and burning of the sample organic content and also suggests the carbonization of the residual oils [20]. The third weight loss is observed which corresponds to a subtle endothermic peak due to the loss of structural hydroxyl groups in silica (Si-OH) bonding [22,23].

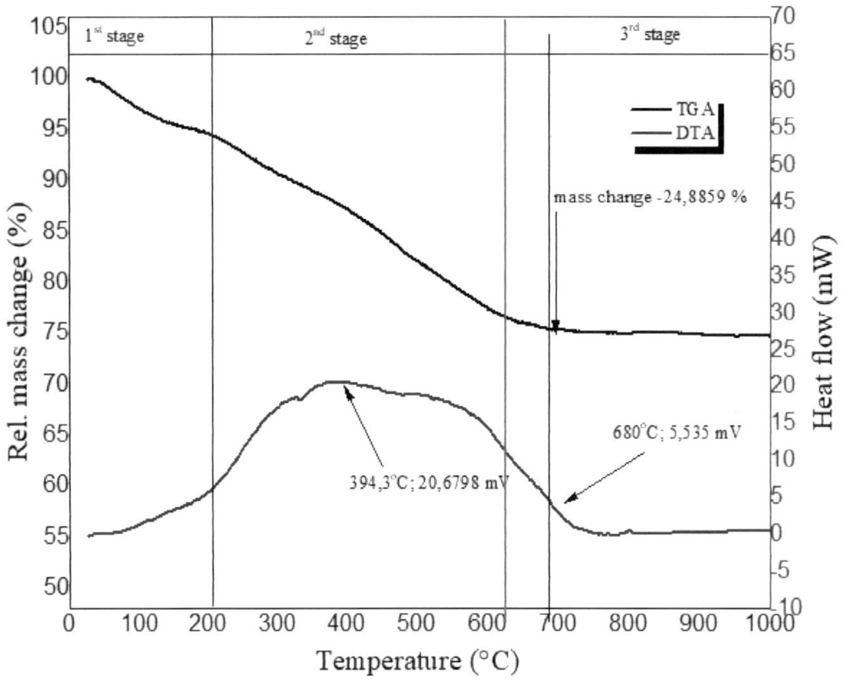

Figure 1. Thermogravimetric and DTA plots of SBE.

From our previous study, the amount of oil in SBE is 25%, which is similar to other reports [4,7,15]. Thus, the extraction of oil from SBE by the maceration method was incomplete. To confirm these results, SBE was calcinated at three different temperatures and times, as shown in Figure 2. The color of SBE is a function of temperature and time, which is the darkest color observed at 500°C for 2 h. This is in agreement with TG analysis (Figure 1), whereby the decomposition of SBE is completely at 680°C. The color of SBE powder at 600°C showed lighter color compared to 500°C due to the holding time during calcination. Furthermore, the color of SBE at 600 and 700°C did not change significantly indicating the absence of volatile compounds.

Table 1. The chemical element of SBE, deoiled SBE, and RSBE (wt. %).

Sample	Si	Al	Fe	Ca	P	Mg	S	O	C
SBE	5.98	1.44	0.84	0.33	0.3	0.21	0.19	39.46	51.25
SBE-deoiled	6.7	1.64	0.97	0.35	0.45	0.26	0.17	41.86	47.59
RSBE	28.71	3.09	1.59	0.46	0.14	0.34	0.26	65.4	-

Table 1 shows the elemental composition of SBE, deoiled SBE, and RSBE. The bleaching earth samples represent the expected typical elements such as silicon, aluminum, iron and a small amount of calcium, phosphorus, and magnesium. SBE and deoiled SBE had a significant amount of carbon, while the amount of carbon in deoiled SBE inconsiderably compared to SBE. This result is corresponding with TG analysis of deoiled SBE, whereby decomposing of deoiled SBE observed. Contrary, carbon did not observe in RSBE, which is similar to typical RSBE, as shown in Table 1.

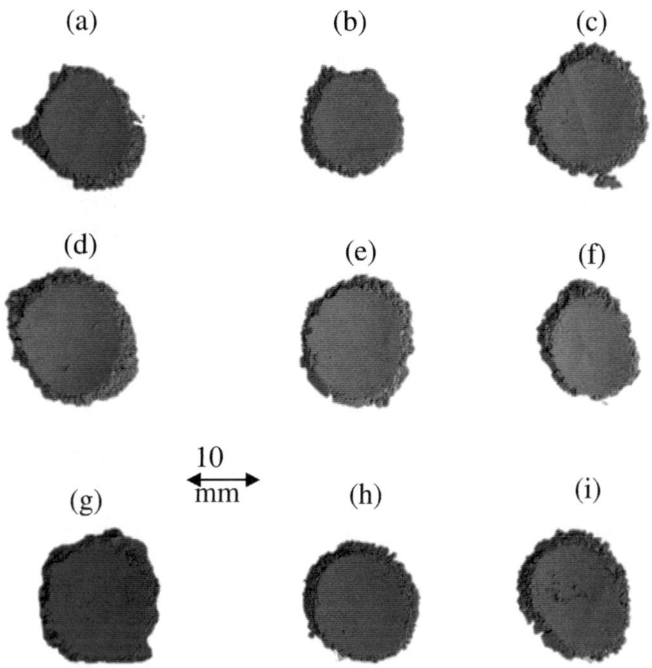

Figure 2. Images of SBE after calcination at (a) 700°C for 2 h, (b) 700°C for 3 h, (c) 700°C for 4 h, (d) 600°C for 2 h, (e) 600°C for 3 h, (f) 600°C for 4 h, (g) 500°C for 2 h, (h) 500°C for 3 h and (i) 500°C for 4 h.

Figure 3 presents the FTIR spectra of SBE, deoiled SBE, RSBE and VBE. As shown in Figure 3, a quite strong of stretching vibration of the absorption band CH_2-CH_3 was observed at 2924 cm^{-1} and 2852 cm^{-1} in SBE and deoiled SBE. Moreover, a stretch vibration of ester carbonyl (C=O) also appeared at 1747 cm^{-1}. These absorption bands which appeared only in the FTIR spectra of SBE and deoiled SBE is characteristic of edible oil [20]. These absorption bands disappeared in RSBE due to calcination process and acid activation, which removes the volatile compounds. RSBE obtain in this study was comparable to commercial VBE, as shown in Figure 3. The structure of montmorillonite indicated by the presence of absorption bands Si-O and Si-O-Al at 1047 cm^{-1} and 694 cm^{-1}, respectively. Moreover, stretching bands at 528 cm^{-1} and 468 cm^{-1} also suggested the presence of Si-O-Al. The existence of quartz was described by absorption bands ν (Si-O-Si) and δ (Si-O) [24]. A strong band at 3630 cm^{-1} and 3427 cm^{-1} showed the possibility of the hydroxyl linkage. However, a broad band at 3427 cm^{-1} and a band at 1637 cm^{-1} in the spectrum of SBE suggest the possibility of water of hydration in the adsorbent. A characteristics band reveals the interlayer hydrogen bonding in clay at 3630 cm^{-1}.

The XRD patterns of deoiled SBE, RSBE, and VBE were shown in Figure 4. The presence of montmorillonites observed in XRD patterns for all samples, which confirms the results of FTIR. Although all samples showed partly amorphous phase, the peaks of montmorillonite, quartz, and cristobalite were spotted very clear. These results informed that calcination and acid activation did not change the phases of RSBE, it demonstrates the effectiveness of this method.

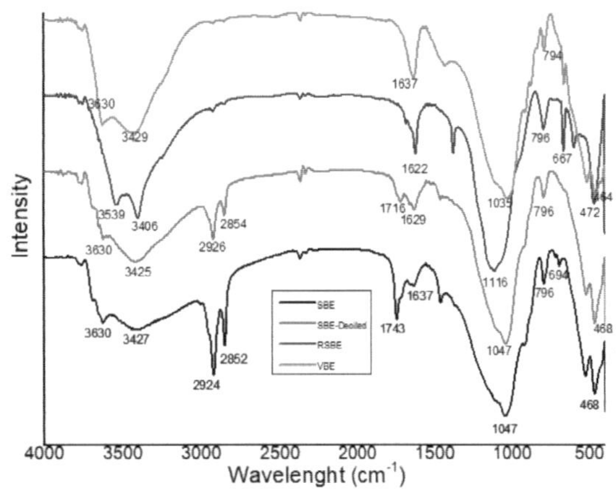

Figure 3. FTIR spectra of VBE, SBE, deoiled SBE and RSBE.

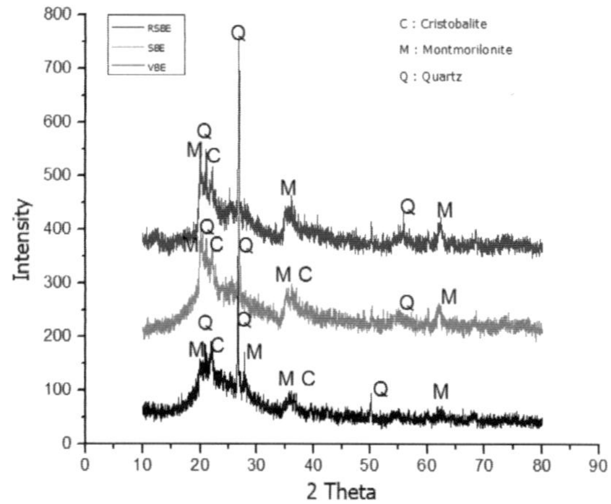

Figure 4. XRD patterns of VBE, deoiled SBE and RSBE.

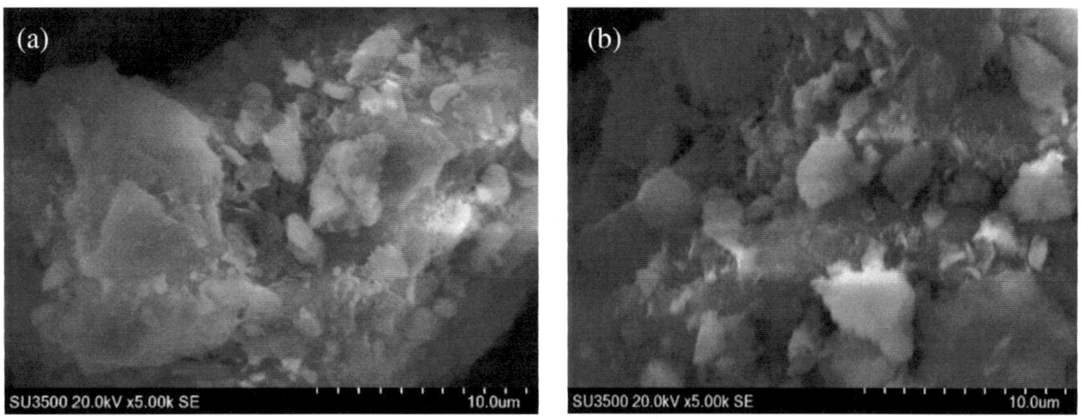

Figure 5. SEM photographs of (a) deoiled SBE and (b) RSBE.

Figure 5 shows the SEM images of the deoiled SBE and RSBE. The microstructure of deoiled SBE and RSBE showed similarity in the shapes and sizes. The calcination and acid activation did not alter the microstructure of RSBE, since it is not different compared to deoiled SBE. Although the particle size of RSBE was very coarse, the surface area of RSBE was very large, as shown in Table 2. The surface area of RSBE was 151.780 m^2/g, which is almost twice of VBE's surface area. This indicates that the particle of RSBE contains a lot of pores. Table 2 shows the average pore size of RSBE and VBE was 29.357 and 56.508 nm, respectively. The pore size of RSBE falls into a mesoporous category according to the International Union of Pure and Applied Chemistry (IUPAC) definition, while VBE had macropores. Moreover, the microporous volume of RSBE and VBE was 0.049 and 0.026 cm^3/g, respectively, indicating the VBE only contain a small amount of micropores. These results are evident in the distribution of pores, as shown in Figure 6, which shows very small of micropores in RSBE and

VBE. Figure 6 also shows that the dominant pores of VBE and RSBE were at 60.20 and 36.84 nm, respectively.

Table 2. Textural characteristics of VBE and RSBE.

Sample ID	Surface area (m^2/g)	Microporous volume (cm^3/g) *SF Method	Total pore volume (cm^3/g)	Average pore diameter (nm)
VBE	78.157	0.026	0.2208	56.508
RSBE	151.780	0.049	0.2228	29.357

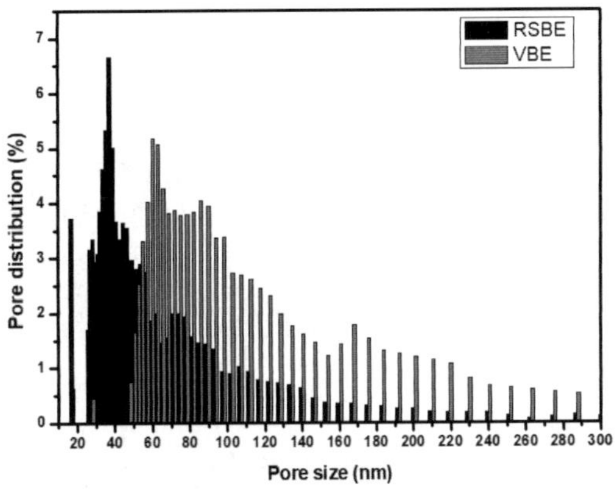

Figure 6. Pore size distribution of RSBE and VBE.

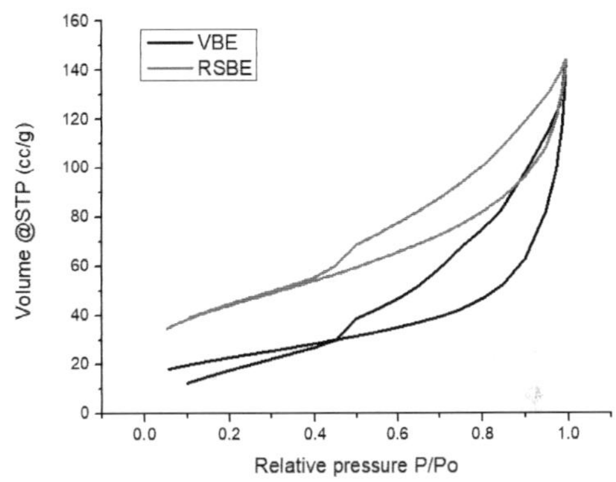

Figure 7. Nitrogen adsorption-desorption isotherm of RSBE and VBE.

Figure 7 displays the isotherms of adsorption-desorption VBE and RSBE using N_2. These materials (VBE and RSBE) gave the adsorption-desorption isotherm type V. These results revealed the broad distribution of pores, which is in agreement with Figure 6. The isotherm graphics show the reversible isotherm under a lower relative equilibrium pressure, but under higher relative pressure they exhibit a hysteresis loop. RSBE adsorbed more nitrogen than VBE about 144.03 cm^3/g, 142.7 cm^3/g, respectively, however, it is not notably different. The desorption curves indicated that hysteresis phenomena were rather small with the VBE but were greater in the RSBE.

Table 3 shows the physical characteristics of this studied material. In this test, RSBE were tested with a specific standard of PT. SMART Tbk. The highest bleaching ability for RSBE was 25% belongs to A2. Bulk density, acidity and moisture content have met industry standards. Comparing the results to the VBE, the bleaching performances were 0.2 times higher than the VBE in actual QC procedure. These results suggest that applying RSBE could reduce the application of VBE and the volume of spent bleaching clay treatment, and therefore also reduce the cleanup costs.

The study on adsorption capacities of the RSBE for CPO in commercial refinery site presents interesting reuse for a circular economy. Its efficiency in adsorption-desorption of impurities (including colored pigments, gum and trace contaminants, etc.) during CPO degumming and bleaching cannot be discussed here due to limitations of the study. Further study will be focusing on the targeted impurities removal efficacy, reuse cycle vis-à-vis commercial standards (PT SMART Tbk vs. VBE). Thus, based on the results of textural and physical characteristics, the RSBE provides better performance. This study provides a simple process and procedure to develop sustainable action for spent bleaching clay.

Table 3. Physical characteristics of regenerated RSBE.

Properties	A1	A2	B1	B2	PT. SMART Tbk
Moisture content	12,53	3.7	6,16	6,95	Max. 15
pH	2,62	2.56	2,53	2,55	2-6
Acidity	3,69	0.31	3,64	3,44	Max. 5
Bleach Ability	NA	25%	14%	NA	>20%

Note:
A1. Temperature 500 °C; 2 h, A2. Temperature 500 °C; 3 h, B1. Temperature 650 °C; 2 h, B2. Temperature 650 °C; 3 h.

4. Conclusion

The SBE was successfully regenerated by calcination at 500°C in air atmosphere for 3 hours and followed the acid activation. Although the volatile compounds were completely removed at 680°C, the SBE showed light color after calcination at 600°C, indicating the removal of volatile compounds. The primary phase of RSBE is similar to VBE, implying the calcination and acid-activation is not alter the phase. The characteristic of RSBE was a high surface area (151.780 m^2/g) with a mesoporous structure. Moreover, the adsorbent capacity of RSBE was comparable with VBE, in fact, slightly higher, showing the regenerated material (RSBE) have promising potential for many applications, such as palm oil refinery, construction and other fields in large scale for their accessible, cost-effective and high yield.

Acknowledgment

Financial support from the Ministry of Research, Technology and Higher Education through Grant No. 24/PPK/E/E4/2017 and support from the Center for Innovation, Indonesia Institute for Sciences are gratefully acknowledged. The authors are grateful for a fruitful discussion on the interpretation of XRD and SEM analysis with Mr. Nanda Hendra Pratama.

References

[1] G. McDonald, A. Rahmanulloh, Indonesia Oilseeds and Product Annual 2018, 2018.
[2] U.K. Nizar, J. Hidayatul, R. Sundari, B. Bahrizal, A. Amran, A. Putra, L. Latisma Dj, I. Dewata, The Effect of Titanium Tetrahedral Coordination of Silica-Titania Catalyst on the Physical Properties of Biodiesel, in: IOP Conf. Ser. Mater. Sci. Eng., 2018: p. 012036. doi:10.1088/1757-899X/335/1/012036.
[3] A. Boukerroui, M.S. Ouali, Regeneration of a spent bleaching earth and its reuse in the refining of an edible oil, J. Chem. Technol. Biotechnol. 75 (2000) 773–776.
[4] A.W. Nursulihatimarsyila, K.Y. Cheah, T.G. Chuah, W.L. Siew, T.S.Y. Choong, Deoiling and regeneration efficiencies of spent bleaching clay, Am. J. Appl. Sci. 7 (2010) 434–437.
[5] C. Su, L. Duan, F. Donat, E.J. Anthony, From waste to high value utilization of spent bleaching clay in synthesizing high-performance calcium-based sorbent for CO_2 capture, Appl. Energy. 210 (2018) 117–126. doi:10.1016/j.apenergy.2017.10.104.
[6] T.N.B. Kaimal, P. Vijayalaksmi, A.A. Laxmi, B. Ramalinga, Process for simultaneous conversion of adsorbed oil to alkylesters and regeneration of commercial spent bleaching earth for reuse, US 2002/0115875 A1, 2002.
[7] R.A. Fattah, N.A. Mostafa, M.S. Mahmoud, W. Abdelmoez, Recovery of oil and free fatty acids from spent bleaching earth using sub-critical water technology supported with kinetic and thermodynamic study, Adv. Biosci. Biotechnol. 05 (2014) 261–272. doi:10.4236/abb.2014.53033.
[8] S.J.T. Pollard, C.J. Sollars, R. Perry, The reuse of spent bleaching earth: A feasibility study in waste minimisation for the edible oil industry, Bioresour. Technol. 45 (1993) 53–58. doi:10.1016/0960-8524(93)90143-Y.
[9] K.Y. Foo, B.H. Hameed, Microwave-assisted preparation and adsorption performance of activated carbon from biodiesel industry solid reside: Influence of operational parameters,

Bioresour. Technol. 103 (2012) 398–404. doi:10.1016/j.biortech.2011.09.116.

[10] J.F. Toro-Vazquez, A. Rocha-Uribe, Adsorption Isotherms of Sesame Oil in a Concentrated Misceila System, J. Am. Oil Chem. Soc. 70 (1993) 589–594.

[11] J. Tang, B. Mu, M. Zheng, A. Wang, One-step calcination of the spent bleaching earth for the efficient removal of heavy metal ions, ACS Sustain. Chem. Eng. 3 (2015) 1125–1135. doi:10.1021/acssuschemeng.5b00040.

[12] E.L. Foletto, C.C.A. Alves, L.R. Sganzerla, L.M. Porto, Regeneration and utilization of spent bleaching clay, Lat. Am. Appl. Res. 32 (2002) 205–208.

[13] R.S. Nebergall, D.R. Taylor, C.J. Kucharz, Process for regenerating spent acid-activated bentonite clays and smectite catalysts, 5,468,701, 1995.

[14] Y.Z. Wang, H.L. Xu, L. Gao, M.M. Yan, H.L. Duan, C.M. Song, Regeneration of spent lubricant refining clays by solvent extraction, Int. J. Chem. Eng. 2015 (2015) 1–7. doi:10.1155/2015/207095.

[15] A.A. Al-Zahrani, M.A. Daous, Recycling of spent bleaching clay and oil recovery, Process Saf. Environ. Prot. Trans. Inst. Chem. Eng. Part B. 78 (2000) 224–228. doi:10.1205/095758200530682.

[16] M.R. Sabour, M. Shahi, G.A. Dezvareh, Reactive dye extraction utilizing regenerated bleaching earth, Glob. J. Environ. Sci. Manag. 3 (2017) 299–310. doi:10.22034/gjesm.2017.03.03.00.

[17] I. Polowczyk, P. Cyganowski, E. Lorenc-Grabowska, R. Sawicki, A. Bastrzyk, A method of thermal and chemical activation of waste bleaching earth for preparation of a low-cost carbon-mineral adsorbent for chromium(VI) removal, Sep. Sci. Technol. 53 (2018) 1142–1155. doi:10.1080/01496395.2017.1310236.

[18] J.I. Chang, H.S. Tai, T.H. Huang, Regeneration of spent bleaching earth by lye-extraction, Environ. Prog. 25 (2006) 373–378. doi:10.1002/ep.10170.

[19] C. Meziti, A. Boukerroui, Regeneration of a solid waste from an edible oil refinery, Ceram. Int. 37 (2011) 1953–1957. doi:10.1016/j.ceramint.2011.02.016.

[20] W.T. Tsai, H.P. Chen, M.F. Hsieh, H.F. Sun, S.F. Chien, Regeneration of spent bleaching earth by pyrolysis in a rotary furnace, J. Anal. Appl. Pyrolysis. 63 (2002) 157–170. doi:10.1016/S0165-2370(01)00150-4.

[21] S.K. Loh, S. James, M. Ngatiman, K.Y. Cheong, Y.M. Choo, W.S. Lim, Enhancement of palm oil refinery waste - Spent bleaching earth (SBE) into bio organic fertilizer and their effects on crop biomass growth, Ind. Crops Prod. 49 (2013) 775–781. doi:10.1016/j.indcrop.2013.06.016.

[22] M. Mana, M.S. Ouali, L.C. de Menorval, Removal of basic dyes from aqueous solutions with a treated spent bleaching earth, J. Colloid Interface Sci. 307 (2007) 9–16. doi:10.1016/j.jcis.2006.11.019.

[23] Z. Wu, C. Li, X. Sun, X. Xu, B. Dai, J. Li, H. Zhao, Characterization, acid activation and bleaching performance of bentonite from Xinjiang, Chinese J. Chem. Eng. 14 (2006) 253–258. doi:10.1016/S1004-9541(06)60067-0.

[24] P.S. Nayak, B.K. Singh, Instrumental characterization of clay by XRF, XRD and FTIR, Bull. Mater. Sci. 30 (2007) 235–238. doi:10.1007/s12034-007-0042-5.

Clean Water Programs In Oefafi Village, Kupang, Nusa Tenggara Barat

Prasetya Yoga Santoso*, Radja Erland Hamzah, and Erma Pratiwi
Department of Communication Science, Faculty of Communication Science, University of Prof. Dr. Moestopo (Beragama), St. Hang Lekir I/8, Jakarta Pusat 10270, Indonesia

*prasetyayogasantoso@gmail.com

Abstract. Oefafi is one of the driest places in Indonesia and community experiences water shortages. They need to walk up to 1-2 km to get a water source due to the unavailability of water. There are many aid programs in the Oefafi but it is unsuccessful. They were not used maximally by the community and the level of community involvement was low. Jaka Perkasa Citra Cemerlang Foundation (JPCC) is a non-profit organization. This foundation empowers the community through the clean water programs in Oefafi. The research objective is to find out how JPCC foundation's communication strategy in empowering the local community through the clean water program. It also focuses on the supporting factors as well as the inhibiting factors in implementing community empowerment communication strategies. This research uses a qualitative approach with a case study method. The results of the study showed the success case of community empowerment in Oefafi. Empowerment by JPCC Foundation already showed an active participation. The communication strategy carried out successfully to trust in the community so that the community has the awareness and willingness to change. Although there are some communities who refuse to be empowered due to the character of people.

***Keywords*:** Community Empowerment, Clean Water Program, Oefafi

1. Introduction

Strategy is a comprehensive planning formulation of how the company will achieve its mission and objectives. Strategy will maximize competitive advantage and minimize competitive limitations [1]. Humans as social beings are inseparable from communication. The communication is not only as a medium to interaction, but communication is also a tool for humans to survive. Communication has the power to make a selection of various stimuli. It will be selected and these stimuli can give society more powerful motivations [2].

Communication strategies are important role in community empowerment communication. Communication strategy consists of plan, way, pattern, position and perspective [3]. Community empowerment in the Oefafi Kupang is one of case study of this issue. Though water and sanitation are fundamental for promoting good hygiene behavior, health, and well-being, many individuals and communities remain without safe water and sanitation facilities [4]. Oefafi is unavailability of water, and it affects people's lifestyle. Some villages in Kupang find it difficult to get access to water, so people have to walk 1.5 km to get water. People spend their time just walking to get water without upgrade the way more effective and useful. Local community tends to take water to the rivers closest to the place of residence which is not suitable for consumption. The quality of water is unhealthy and unclean. In addition, certain rivers will be drought in the dry season. People have to look for other places that are farther to get water. This is frightening condition because the community's basic needs are not being fulfilled. Community empowerment is a power to improve the performance of products supply and public services by spreading the decision-making authority and responsibility to the people. It would encourage people's involvement in decision-making and responsibility.

Community empowerment is a deliberate effort to facilitate local communities in planning, deciding and managing the local resources through collective action and networking. Thus, they have the ability and independence economically, ecologically, and socially [5,6]. There are eight elements in community empowerment. First, it is the creation of atmosphere to open up opportunities to develop the potential. Second, it is supportive action by giving encouragement. Third, it is motivation. Fourth, it is raising awareness. Fifth, it is empowering by strengthening the potential. Sixth, it is the development of opportunities. Seventh, this is devolution by transferring the authority. The last, it is protection of the weak [7].

Empowerment generates community resources, opportunities, knowledge and skills to increase capacity in determining their future [8,9]. Empowerment is successful if it can provide opportunities, knowledge and skills to the community. It must change the community to live independent. There are aid institutions such as NGOs (Non-Governmental Organizations) both from private and government institutions to help local community as a project. Thus, community empowerment is not happened and there is no change in lifestyle for long term that impacts to the community. The aid is not effective.

Jaka Perkasa Citra Cemerlang Foundation or JPCC Foundation organized a clean water programs n order to empower the community. There was a change in lifestyle such as the establishment in the daily needs. Community was accustomed to a healthy lifestyle. JPCC Foundation, the actor of community empowerment in Oefafi, thinks that it is good if Oefafi community has a decent life and has a better standard of living. Local community is expected to have more improvements such as having a mindset to change, improving clean lifestyles, maintaining health, and changing in the economic sector. Through the clean water program, the villagers no longer waste their time to get water, because access to water is easy to reach. To replace the time available, JPCC Foundation provides education to the community to do suitable agricultural such as growing vegetables (spinach, kale, corn). Thus, it can be enjoyed alone and sold it to get additional income [10].

In order to empower the community, it cannot be separated from the communication strategy. Each planning and implementation of activities will be anticipated. Strategy in communication is a way to manage the implementation of communication operations. It needs to be successful. Communication strategies essentially consist of planning and management to achieve a goal. In order to achieve these goals, the strategy is not as a road map to shows direction but it must also show operational tactics [11]. Communication is important because accomplishment of community empowerment depends on the management of communication methods and techniques to give information and knowledge to the community [12]. Community empowerment as noble personality requires certain learning processes and this process will not run well without communication between policy makers and clients [13]. From this theoretical explanation, the communicator must be able to make a communication strategy. Thus, the message can influence the communication target. The aim of this research is to find out how JPCC foundation's communication strategy in empowering the local community through the clean water program. It also focuses on the supporting factors as well as the inhibiting factors in implementing community empowerment communication strategies.

2. Methodology

This research uses the constructivism paradigm. There are various definitions of constructivism by experts. According to Guba and Lincoln in Denzin [14], constructivism adopts the relativism ontology, transactional epistemology, and hermeneutic or dialectical methodology. The purpose of this research is to find out the implement of JPCC Foundation's communication strategy in community empowerment through the clean water program in Oefafi, Kupang, Nusa Tenggara Timur.

IOP Conf. Series: Materials Science and Engineering **924** (2020) 012015 doi:10.1088/1757-899X/924/1/012015

Constructivism is a paradigm or antithesis about the importance of observation and objectivity in the reality of science. Ontologically, reality exists in various forms of mental construction based on social, local and specific experiences, and agents. Therefore, reality cannot be generalized as a positivist paradigm. On this philosophical basis, this paradigm states that the epistemological relationship between the observer and the object is an entity, subjective and a combination of their interactions [15].

This research is a qualitative research in a case study approach. This research is designed as a dynamic form to obtain an overview of various interesting problems in social life. Case studies are comprehensive descriptions and explanations of various aspects of individuals, groups, communities, programs, or social situations [16].

In communication strategy, it focuses on the formulation of clear objectives. It also focuses on the conditions and situations of the audience. Thus, the main step is to identify the audience and target [17]. It can be based on the introduction between the NGO as anactor and community as an agent. It also needs to consider the selected communicator due to the conditions and situations. There are steps in formulating a communication strategy such as get to know the audiences, prepare messages, set methods, and use a suitable media [13].

3. Results and Discussion

Based on observations, problem solving, is needed to design relevant development communication strategy designs [18]. The situation and conditions of local community shows low level of community empowerment to encourage equitable development. It is reflected on the mindset of people who prefer to take water from rivers. Although the rivers are far from their house and the quality of the river is not good. This empowerment has not yet produced equitable change among all Oefafi society.

There are some issues about the successfulness of empowerment. Realize that poverty is a structural problem and multi-dimensional. The complexity of the various dimensions of poverty related social, political, economic, and asset [19]. Based on the observation of the important role of JPCC foundation in the empowerment of rural communities, it would certainly not be separated from the ability of the government due to the community empowerment communication strategies.

It is important to consider the understanding and implementing of communication strategy in empowerment. It will be an important ability especially if the empowerment is in a different cultural context such as JPCC foundation clean water program in Oefafi, Kupang. Community involvement is one of the main keys in clean water program. Community engagement creates a sense of ownership of the facilities. Thus, cooperation between JPCC foundation and the local community is important. Oefafi community is not easy to accept the changes and the presence of foreigners, their culture is also different from the JPCC foundation internal workers. There may be differences in interpreting messages. Another problem in the society was the lack of public trust. Local community cannot trust the message by people outside the area of Oefafi. The following are the steps in formulating a communication strategy based on the results of the study:

The communication strategy based on four point, first is Audience, JPCC foundation need to know audiences thoroughly and intimately. JPCC foundation does not only know the names of local community but also have merged with the local community in Oefafi. The second is construct the message, the message is delivered to the local community of Oefafi. It consists of 3 forms of messages, such as: Informative, persuasive, and coercive, JPCC foundation conveys messages in good information about the objectives of the clean water program, the benefits of healthy living, and lessons of hygiene. JPCC foundation also provides information on life values. Thus, local community can organize good and right lifestyle. The third is Method, JPCC foundation uses a repetition method in empowering the community to instill a new mindset. The messages of the clean water program in Oefafi were delivered repeatedly and continuously. And the fourth is media selection, JPCC FOUNDATION uses the local resident's kitchen as place to gather. They cook together while they tell stories to increasing closeness and togetherness. JPCC FOUNDATION can identify the conflicts in the process of empowerment through clean water works. Closeness is considered appropriate because it influences behavior and touches emotions between individuals, Because the JPCC FOUNDATION is able to provide a role outside the use of resources and to engage in social activities designed to increase community profits based on existing regulations [20].

4. Conclusion

This research shows the changes in the lifestyle of Oefafi community especially in public health and sanitation. Local community shows an active participation. It reflects the successful of the empowerment by JPCC foundation. JPCC foundation has succeeded to get public trust. Local community has awareness and willingness to change although there are some people who refuse to be empowered due to the nature of the local community. They do not like to live in groups. To explain JPCC foundation s communication activities, researcher uses the communication strategy by Harold Lasswell communication model. It also focuses on the importance of the goal factors by JPCC foundation in communication activities. Thus, JPCC foundation's communication strategy of community empowerment to approach and build relationships focus on the audience, composing messages, establishing methods and selection in using media.

Reference

[1] J. D. Hunger and T. L. Wheelen, *Strategic Management. Manajemen Strategis*. Yogyakarta: Andi, 2003.
[2] D. Priatama, "Communication Strategy Of Community Empowerment Institutions In The Socialization Of Community Development Programs In Loa Bakung, Samarinda City" *eJournal Ilmu Komnunikasi*, vol. 1, no. 2, pp. 70–84, 2013. *(Indonesian)*
[3] R. D. Smith, *Strategic Planning for Public Relations*, Second Edi. London: Lawrence Erlbaum Associates, 2005.
[4] F. Dery, E. Bisung, S. Dickin, and M. Dyer, "Understanding empowerment in water, sanitation, and hygiene (WASH): a scoping review," *J. Water, Sanit. Hyg. Dev.*, pp. 1–11, 2019.
[5] A. N. Masar, "Communication Strategy of Community Empowerment Institutions (LPM) in Disseminating the City Park Program in Sanga-Sanga, Kutai Kartanegara." *J. Ilmu Komun. Univ. Mulawarman*, vol. 2, no. 4, pp. 88–101, 2014. *(Indonesian)*
[6] G. Dalziel, "Community Empowerment Through Development of Locally Sustainable Drinking Water Sources," in *Pipelines Specialty Conference 2009*, 2009, pp. 1289–1298.
[7] A. I. Indrawijaya, *Organizational Behavior*. Bandung: Sinar Baru Algensindo, 2002. *(Indonesian)*
[8] Suparjan and H. Suyatno, *Community Development: from Development to Empowerment*. Yogyakarta: Aditya Media, 2003. *(Indonesian)*
[9] G. Vitri and H. Herman, "Infrastructure maintenance system for community development projects to improve the quality of infrastructure services in West Sumatra Province," in *IOP Conference Series: Materials Science and Engineering PAPER*, 2019, pp. 1–11.
[10] "JPCC Foundation." [Online]. Available: https://www.jpccfoundation.org/.
[11] Y. Z. Abidin, *Communication Management: Philosophy, Concepts and Applications*. Bandung: Pustaka Setia, 2005. *(Indonesian)*
[12] Jumrana and M. A. Tawulo, "Facilitator in Community Empowerment Communication" *J. Komun. Profetik*, vol. 8, no. 1, pp. 19–30, 2015. *(Indonesian)*
[13] M. Nasor, "Community Empowerment Through Interpersonal Communication" *Al-Adyan J. Stud. Lintas Agama*, vol. 11, no. 1, pp. 17–35, 2016. *(Indonesian)*
[14] N. K. Denzin and Y. S. Lincoln, Eds., *The SAGE Handbook of Qualitative Research*, Fifth Edit. Los Angeles: Sage Publications, 2018.
[15] A. Salim, *Theory and Paradigm: Social Research*. Yogyakarta: Tiara Wacana, 2006. *(Indonesian)*
[16] D. Mulyana, *Qualitative Research Methodology: New Paradigms of Communication and Other Social Sciences*. Bandung: Remaja Rosdakarya, 2008. *(Indonesian)*

[17] A. Arifin, *Communication Strategy: A Brief Introduction*. Bandung: Armico, 1984. *(Indonesian)*
[18] S. Amanah, "The Role of Development Communication in Empowering Coastal Communities." *J. Komun. Pembang.*, vol. 8, no. 1, pp. 1–19, 2010. *(Indonesian)*
[19] R. Fatriadi, F. Asteriani, and C. Cahyaningsih, "Effectiveness of the National Program for Community Empowerment (PNPM) for Infrastructure Development Accelerated and Geoplanology in District of Marpoyan Damai, Pekanbaru," *JGEET (Journal Geosci. Eng. Environ. Technol. Eff.*, vol. 2, no. 1, pp. 53–63, 2017.
[20] Y. Risa, "Community Empowerment as The Corporate ' s Social Responsibility : Case Study of PT . Tirta Investama Plant Solok , Indonesia," *Varia Justicia*, vol. 15, no. 2, pp. 43–50, 2019.

Bond Strength, Hardness, and Microstructure Analysis of Stellite Coating Applied on 410 Steel Surface Using Flame Spray, Plasma Spray, and High-Velocity Oxyfuel Spray Process

Nail Widya Satya, Sunoto Mudiantoro, and Winarto Winarto*
Metallurgy and Materials Engineering, Faculty of Engineering, Universitas Indonesia, Depok 16424, Indonesia

*winarto@metal.ui.ac.id

Abstract. Three thermal spray methods often used in industry are flame spray, plasma spray, and high-velocity oxyfuel (HVOF) spray. This research is intended to compare the properties of those methods in depositing Stellite coating on 410 martensitic stainless steel. The results showed that both plasma spray and HVOF coating show an even deposition, which could not be achieved by a flame spray method. Those three coatings meet manufacture minimum bond strength requirements. The highest bond strength and hardness were provided by the HVOF process with a value of 33.1 MPa and 719 HV, respectively. According to bond strength and standard hardness deviation, the HVOF process gives the most homogeneous coating. Substrate hardness just below the coating interface after flame spray, plasma spray, and HVOF process are raised by 236%, 56%, and 65% each from the specification. HVOF coating has the best cross section compared to others. Smallest porosity percentage, porosity size, and average interface unbonding is got by the HVOF process, with a value of 0.2%, 7.2 μm, and 31%, respectively. Coating microstructure after etching shows phases related to heat input during application. The dendritic structure is observed on flame spray and plasma spray coating after etching but not on HVOF coating. Oxides and carbides of both cobalt and chrome are formed in the coating.

Keywords: Stellite; 410 Steel; Flame Spray; Plasma Spray; HVOF

1. Introduction

Thermal spray is often applied to increase abrasion resistance. Stellite is a commonly used material in the thermal spray process as it has excellent resistance against abrasion, oxidation, and corrosion at elevated temperatures. This resistance is gained from cobalt content as its primary alloy. The service temperature of a Stellite coating is around 538°C to 843°C. Three thermal spray methods often used in industry are flame spray, plasma spray, and HVOF. Flame spray generally used in a small industry because of its simple system and relatively cheap equipment. Plasma spray and HVOF equipment usually owed by a bigger shop or industry as the price is quite high. In this research, the property of these three methods in depositing Stellite coating on 410 martensitic stainless steel will be compared. It can give some idea to engineer to choose the suitable thermal spray methods based on the coating produced.

Recent researches in thermal spray focus on advance technology involving self-made equipment, whereas basic knowledge is still needed by the beginners new to thermal spray. This research is performed using industrial equipment and powder available in the market. Coating application and characterization methods used can be easily emulated by practitioners or researchers for a different type of base material, powder, method, or parameter. Coating color gradation after etching rarely observed in other studies even though it can provide valuable information about heat input of the spraying process.

2. Experimental method

The substrate used is 410 stainless steel classified as martensitic. This steel has 142 HV hardness. The specimen is cut to a dimension of 20x20x10 mm. Two types of Stellite, i.e., Stellite 1 and Stellite 6, are used as a coating. Substrate and powder chemical composition can be seen in Table 1.

Table 1. Substrate and powder chemical composition (in % weight).

Element	Co	Fe	Cr	Mn	Si	Ni	C	S	Mo	W	Nb
S S 410	0.018	85.2	13.3	0.487	0.451	0.199	0.105	<0.005	0.069	<0.02	0.014
Stellite 6	63	1.5	28	-	1	1.5	1	-	-	4	-
Stellite 1	50.5	2	30	-	1.1	1.5	2.4	-	-	12.5	-

Before the coating application, the substrate surface is blasted with 96 psi pressure to an average roughness of 9.2 μm using Oerlikon Metco Metcolite C grit. This grit is composed of 94 wt% Al_2O_3, 3.5 wt% TiO_2, and 2.5 wt% other compounds. The blasting particle has a size of –2 +0.6 mm with 1.92 g/cm^3 density. Stellite 6 powder used is a CoCrW alloy Praxair CO-106-1. This powder has a size of -45 +15 μm with 4.41 g/cm^3 density. The powder has 23.4 MPa minimum bond strength and 285-361 HV average hardness. Besides Stellite 6, Stellite 1 with brand Dura-Metal DM H ST 1 is used for the HVOF process. This powder has a size of -53 +20 μm, 4.7 g/cm^3 density, and 694 HV hardness.

Flame spray process is performed using SUPERJET-S Eutalloy equipment. Oxygen and acetylene gas sets are 60 psi and 15 psi, respectively. Both gas valves are regulated manually until the oxidation flame, as well as the setting of the powder deposition rate. Besides, the coating is applied from a 25-30 mm distance. The gas used in the system is argon, hydrogen, and air. The coating is applied from a 100–125 mm distance with a 3.5 kg/hour rate. The plasma spray parameter can be seen in Table 2.

Table 2. Plasma spray parameter.

Argon Pressure (Psig)	Argon Flow (FMR)	Hydrogen Pressure (Psig)	Hydrogen Flow (FMR)	Ampere	Voltage (V)	Carrier Gas (SCFH)	Vibrator (psi)
70	150	70	20	500	72	37	25

HVOF process is performed using Diamond Jet DJC Control Unit, 9MP-DJ Closed Loop Powder Feed Unit, water-cooled DJ Spray Gun, and ABB IRB 2600 robot. A robot is used since the manual application is impossible due to the high gas pressure of the system. The gas used in the system is oxygen, hydrogen, and air. The coating is applied from a 250 mm distance with a 2.5 kg/hour rate and gas rate of 3,000 m/s. The spray parameter can be seen in Table 3.

Table 3. HVOF spray parameter.

Oxygen Pressure (Psig)	Hydrogen Pressure (Psig)	Air Pressure (Psig)	Oxygen Flowmeter (FMR)	Hydrogen Flowmeter (FMR)	Air Flowmeter (FMR)	Oxygen Flow (SCFH)	Hydrogen Flow (SCFH)	Air Flow (SCFH)
170	140	100	26	57.5	37	398.35	1345.49	663

All coating process is applied from 90° angle or perpendicular to the substrate surface. The preheat of 100°C is applied to the substrate surface before the coating application. Specifically, for flame spray process, preheat variation of 100°C, 200°C, 300°C, and without preheat is used. Figure 1 shows the thermal spray application for each method.

Characterization is conducted to know and compare properties for each process variation. Coating properties to be observed are bond strength, hardness, microstructure, morphology, and chemical composition. In figures that follow, some abbreviations are used, i.e., FS for flame spray, PS for plasma spray, NP for without preheat, P for preheating, S-1 for Stellite 1 coating, S-6 for Stellite 6 coating, and S for 410 stainless steel substrate.

The bond test is performed using a universal testing machine to know the bond strength of each specimen. Microstructure examination before and after etching is carried out using an optical microscope to analyze the coating structure and make measurements such as porosity, unbonded interface, and coating thickness. Vickers hardness test was performed from the specimen cross section to measure the hardness of the coating and substrate. SEM and EDS test was used to check morphology and identify elements present, respectively.

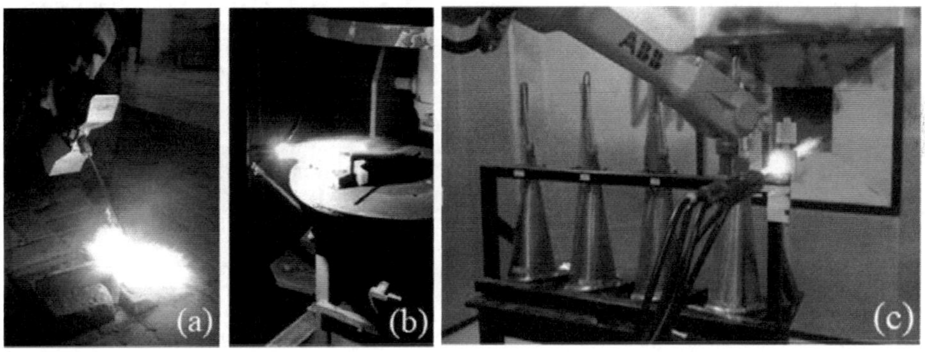

Figure 1. Thermal spray application (a) flame spray (b) plasma spray (c) HVOF spray.

3. Results and discussion

3.1. Coating

The result for some coating applied can be seen in Figure 2. All specimens are in as-sprayed condition.

Figure 2. Coating result (a) flame spray without preheat (b) flame spray 100°C preheat (c) flame spray 200°C preheat (d) flame spray 300°C preheat (e) plasma spray (f) HVOF spray.

In Figure 2 (a), it can be seen that adhesion between coating and substrate is not excellent. As preheat temperature is increased, as shown in Figure 2(b), (c), and (d), even coating started to be formed. In Figure 2(d), some vesicles are formed on the surface. In general, flame spray coating in Figure 2(a)-(d) shows greenish color. Figure 2(e) and (f) from plasma spray and HVOF show even coating deposition. HVOF coating has brighter color than plasma spray that probably caused by lower temperature in the HVOF process and higher tungsten content in Stellite 1, which in this case plasma spray is applied using Stellite 6 while HVOF is applied using Stellite 1.

3.2. Bond test

The bond test is performed on 19 specimens consisted of 4 flame spray specimens, three plasma spray specimens, 4 HVOF specimens with Stellite 6 coating, and 8 HVOF specimens with Stellite 1 coating. In the flame spray process, higher preheat temperature leads to higher bond strength with the highest value of 36.5 MPa. The average bond strength for the plasma spray process is 27.4 MPa, with a 59% standard deviation. HVOF spray process has an average bond strength of 35.9 MPa and 33.1 MPa for Stellite 1 and Stellite 6, respectively. The result can be caused by higher tungsten and less cobalt in Stellite 1, where tungsten has more significant Young modulus than cobalt resulted in higher strength. The standard deviation for the HVOF process is around 33%. Based on this value, HVOF produces the most homogeneous coating. The three methods satisfy manufacture minimum bond strength specification of 23.4 MPa.

3.3. Hardness test

The indenter is positioned in the middle of the coating thickness. The highest average hardness is obtained by HVOF spray Stellite 1 with a value of 719 HV with an 8% standard deviation. Apart from the process itself, the high hardness may be caused by the addition of 8.5% tungsten compared to Stellite 6, where tungsten is harder than cobalt. The second hardest coating is got from the flame spray process of Stellite 6 with a value of 459 HV with an 11% standard deviation. Different from the bond test, preheat variation relatively not influence coating hardness. Plasma spray of Stellite 6 produces 356 HV hardness with a 12% standard deviation. Same as the bond test result, based on standard deviation, HVOF has the most homogeneous coating.

In addition to coating hardness, substrate hardness is also examined with an indentation point located about 50 μm below coating. Average substrate hardness of 477 HV, 222 HV, and 234 HV are obtained from the flame spray, plasma spray, and HVOF, respectively. Compared with 410 steel, substrate hardness after application of the flame spray, plasma spray, and HVOF each are raised by 236%, 56%, and 65%. Increasement for plasma spray and HVOF is considered normal due to the coating or cutting process. High substrate hardness for the flame spray process could be the result of contact with the oxidation flame during coating application. The hardness of 477 HV is similar to 410 steel after normalizing, which is about 522 HV [1]. Normalized 410 steel has a martensitic lath structure that resulted in a hard, low toughness material in nature [1]. Tempering is required to decrease hardness and increase toughness [1].

3.4. Metallographic examination

The microstructure of a thermal spray coating generally consisted of structure built up of splats, oxide, inter-lamellar pores, and interface voids [2]. To investigate the porosity, unmelted particle, contaminant, and coating interface, the coating cross-section is first observed without etching. The microstructure without etching of the flame spray process is shown in Figure 3. It can be seen that the majority of the area is filled by porosity, especially for the specimen without preheating. The flame spray process produces the most significant amount of porosity, both in terms of quantity and size. Porosity could be caused by a lack of powder impact speed, unmelted particle side effect, wrong application angle, or shrinkage [3,4]. In the flame spray process, higher preheat temperature resulted in a less porosity generated. The existence of an unmelted particle can be caused by a lack of heat to melt the particle [5]. Splats are not visible to be formed, and part of the white structure produced is as dense as weld structure.

It can be due to direct heat input to the coating during the process so that powder in some area are entirely fused. This condition is different from plasma spray as well as HVOF, where heat is concentrated inside the gun, and in-flight period reduces the particle temperature, which eventually triggers splats formation along with the thickness of the coating.

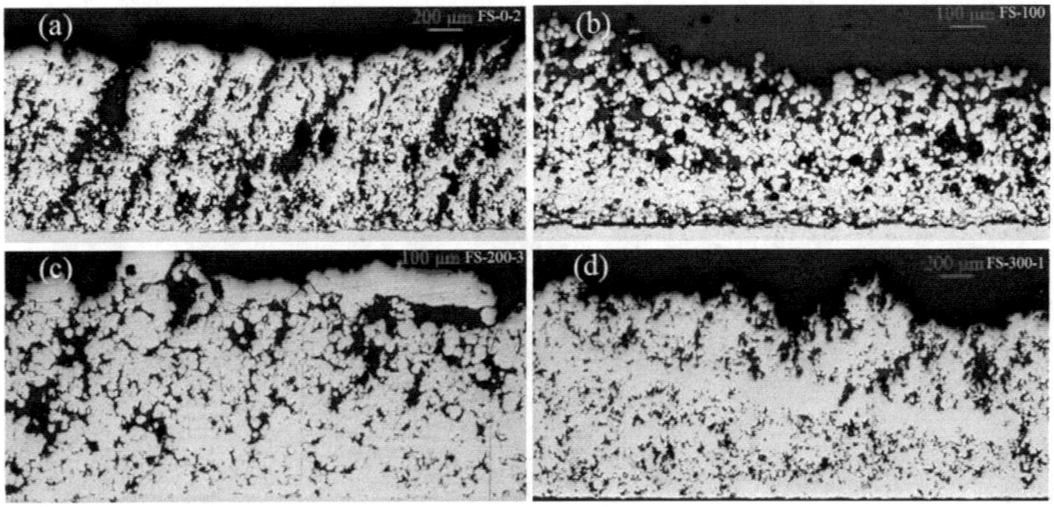

Figure 3. Flame spray microstructure without etching (a) without preheat, 50x magnification (b) 100°C preheat, 100x magnification (c) 200°C preheat, 100x magnification (d) 300°C preheat, 50x magnification.

The microstructure without etching of the plasma spray process is shown in Figure 4. Coating cross-section is dominated by grey splats that evenly spread. Splats formation indicates a melted coating powder [3]. These flat structure also denotes that the powder particle has perfectly melted before deposition [6]. Porosity is much less compared to the flame spray coating with a more uniform thickness.

Figure 4. Plasma spray microstructure without etching 200x magnification.

IOP Conf. Series: Materials Science and Engineering **924** (2020) 012016 doi:10.1088/1757-899X/924/1/012016

Microstructure without etching of HVOF spray is shown in Figure 5. The cross-section of the HVOF spray microstructure showed a dense coating with little splats and porosity. A higher density is followed by higher hardness. The HVOF process produced the lowest porosity percentage and size with a value of 0.2% and 7.2 μm each. Besides porosity, unbonding length at the coating interface is an important criterion to be evaluated. Unbonding has a significant effect on bond strength results. The shortest unbonding is provided by the HVOF process with a value of 31%.

Figure 5. HVOF spray microstructure without etching 200x magnification.

Microstructure after etching for every process is shown in Figure 6–8. The etching is performed by dipping the specimen for 5 minutes into the solution made of 30 ml HCl 36% and 10 ml HNO_3 65%.

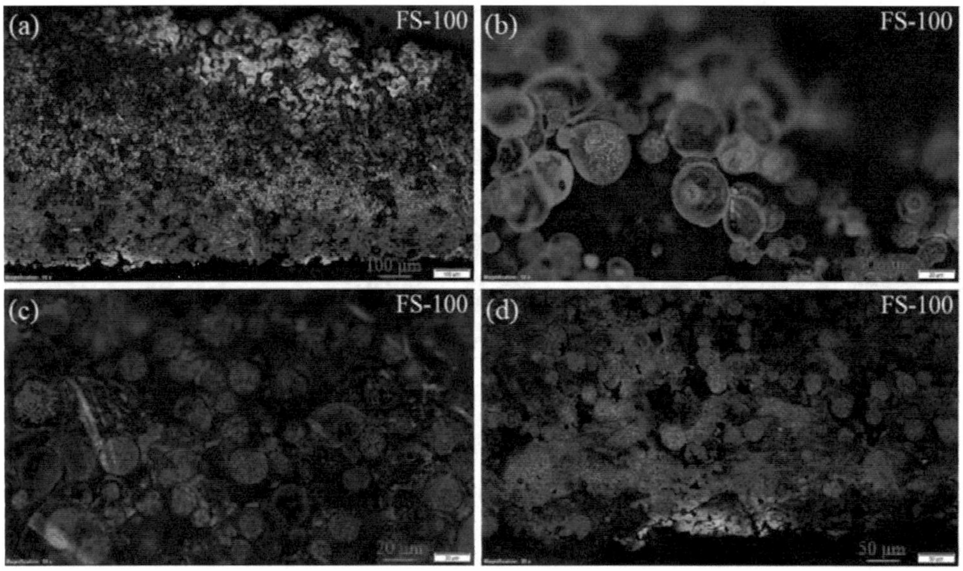

Figure 6. Flame spray microstructure after etching 100°C preheat (a) 100x magnification (b) 500x magnification (c) 500x magnification (d) 200x magnification.

In Figure 6 (a), color gradations appear with brown color at coating interface, blue in the middle, and white at the top of the coating. As magnification is increased to 500x, as seen in Figure 6 (b) and 6 (c), a dendritic structure are visible inside the unmelted powder particle. The existence of this structure signifies that the etching process was done correctly. Figure 6 (d) shows an area with heat concentration after etching. By looking at this color gradation, it is understood that the areas with the highest heat will turn yellow, followed by orange, purple, blue, green, and brown as the heat gradually reduced.

Figure 7 shows the plasma spray microstructure after etching. In Figure 7 (a), it can be seen that color gradation is more evenly spread compared to flame spray result. This result indicates a more homogeneous heat deployment. Similar to flame spray, the area near the interface has a brown color, followed by purple in the middle, and white at the top. In Figure 7 (b-d), a dendritic structure is seen to be formed. The dendritic structure formed after etching in flame spray and plasma spray are the same as found from the deposition of Stellite 6 using the cold spray in another research [7,8]. These structures are also identical to the structure produced from Stellite 6 welding [9].

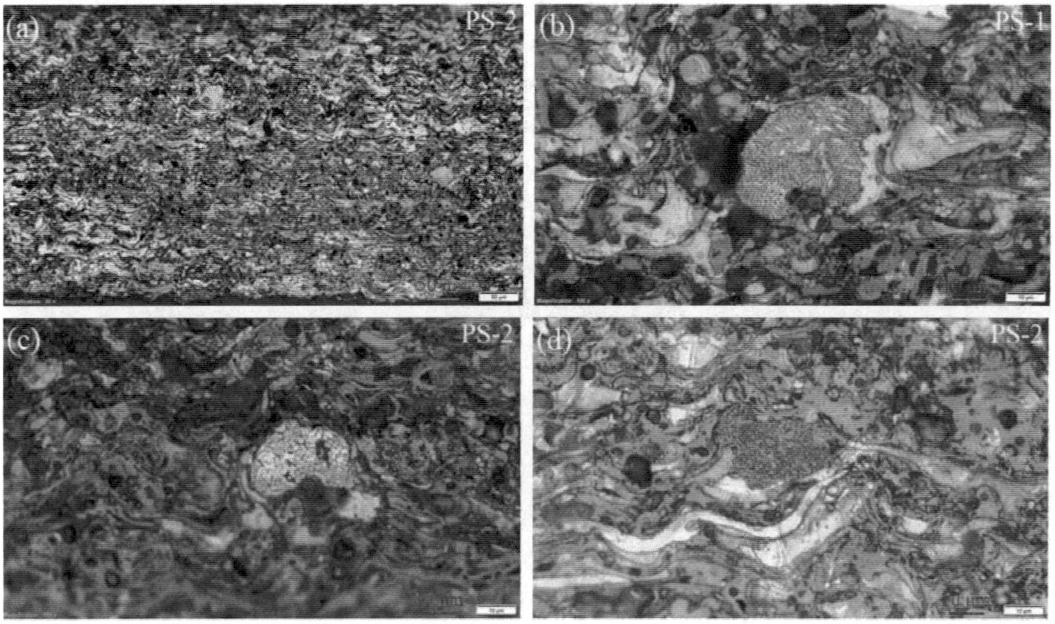

Figure 7. Plasma spray microstructure after etching (a) 200x magnification (b) 1,000x magnification (c) 1,000x magnification (d) 1,000x magnification.

HVOF spray microstructure after etching is shown in Figure 8. A white and brown area seen signifies a low heat input during coating application. This low heat input, together with the highest system pressure from the HVOF process resulting in a dense, hard, high bond strength and low porosity coating. The dendritic structure is not found even with high magnification. It can be due to the density of the coating so that the acid solution cannot penetrate through the coating pores.

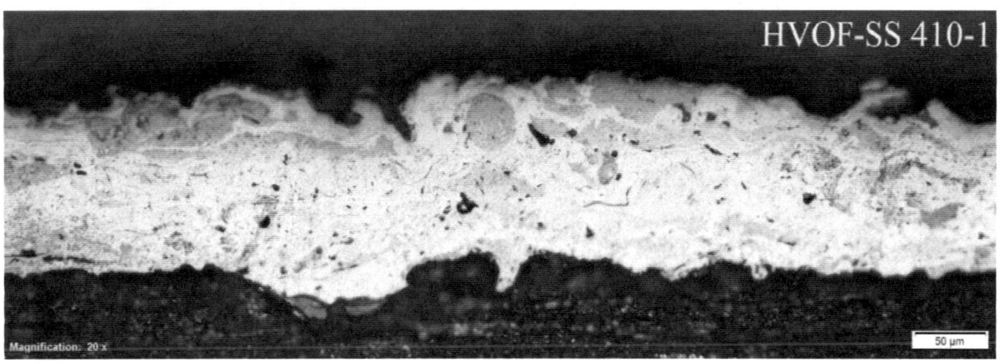

Figure 8. HVOF microstructure after etching 200x magnification.

3.5. SEM & EDS

SEM and EDS tests are performed to observe morphology, change in chemical composition, and phase of the coating. An example of the coating analyzed is shown in Figure 9. SEM result confirms the common black area at the interface to be a cavity. Based on increasing and decreasing of elements found by EDS result for three thermal spray process, oxides like CoO, Cr_2O_3, and $CoCr_2O_4$ may be formed within the coating [10-12]. The amount of these oxides will be greater for longer oxidation time [10]. Besides, carbides like CoC, Cr_7C_3, Co_6W_6C, or $Cr_{23}C_6$ also might formed [7,10,11]. In some areas, silicon oxide and the iron oxide may also be formed at the coating interface. Porosity and cavity existence at the interface could be marked by increasing carbon and oxygen content.

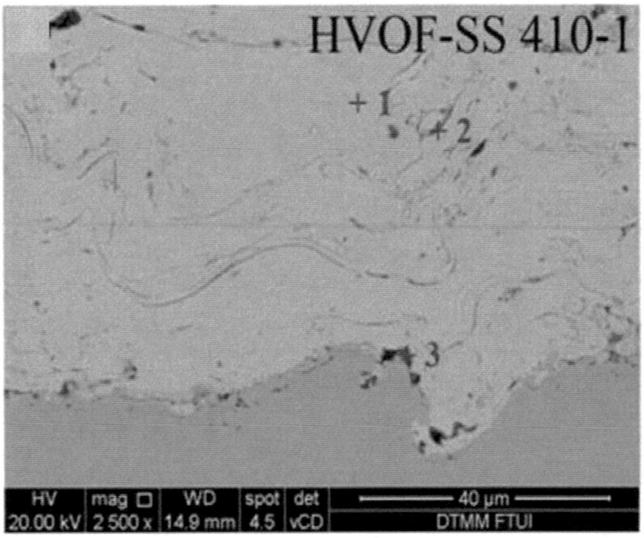

Element	1	2	3
	Wt%	Wt%	Wt%
CK	2.8	3.04	3.31
OK	1.2	2.52	1.57
AlK	0	0.19	0.51
SiK	5.72	6.15	1.83
CrK	31.82	34.44	18.24
FeK	1	1.32	55.15
CoK	57.47	52.33	19.39

Figure 9. SEM 2,500x magnification and EDS of HVOF spray coating.

4. Conclusion

All three thermal spray processes satisfy manufacture minimum bond strength specifications. HVOF spray produces the highest bond strength and hardness with most homogeneous coating. Flame spray process could reduce substrate toughness while plasma spray and HVOF spray not. The HVOF process produced the lowest porosity percentage and size. The HVOF process also produced the shortest unbonding length. Color gradation at plasma spray structure after etching indicates a more homogeneous heat deployment than flame spray. HVOF spray microstructure after etching signifies a low heat input during coating application. Oxides and carbides are formed within coating for all three thermal spray processes.

Acknowledgment

The authors would like to acknowledge the DRPM, Universitas Indonesia, for the financial support through the PIT9 Research Grant 2019 with the contract No: NKB-0082/UN2.R3.1/HKP.05.00/2019.

References

[1] G. Chakraborty, C. Das, S. Albert, A. Bhaduri, V. T. Paul, G. Panneerselvam, and A. Dasgupta, "Study on tempering behavior of AISI 410 stainless steel," *Materials Characterization*, vol. 100, pp. 81–87, 2015.

[2] W. Winarto, D. Priadi, N. Sofyan, and A. Wicaksono, "Wear Resistance and Surface Hardness of Carbon Nanotube Reinforced Alumina Matrix Nanocomposite by Cold Sprayed Process," *Procedia Engineering*, vol. 170, pp. 108 – 112, 2017.

[3] S. Vignesh, K. Shanmugam, V. Balasubramanian, and K. Sridhar, "Identifying the optimal HVOF spray parameters to attain minimum porosity and maximum hardness in iron-based amorphous metallic coatings," *Defence Technology*, vol. 13, pp. 101-110, 2017.

[4] M. Axente and V. Geaman, "Characterization of Ni-5Al and Sprasteel Coatings Deposited by Electric Arc Wire Thermal Spraying," *Metalurgia International*, vol. XVI no. 5, pp. 89, 2011.

[5] R. Paredes, S. Amico, and A. d'Oliveira, "The effect of roughness and pre-heating of the substrate on the morphology of aluminum coatings deposited by thermal spraying," *Surface & Coatings Technology*, vol. 200, pp. 3049 – 3055, 2006.

[6] S. Saeidi, K. Voisey, and D. McCartney, "Mechanical Properties and Microstructure of VPS and HVOF CoNiCrAlY Coatings," *Journal of Thermal Spray Tech.*, vol. 20, pp. 1231–1243, 2011.

[7] B. Li, Y. Jin, J. Yao, Z. Li, Q. Zhang and X. Zhang, "Influence of laser irradiation on deposition characteristics of cold sprayed Stellite-6 coatings," *Optics and Laser Technology*, vol. 100, pp. 27–39, 2018.

[8] N. Cinca, E. López, S. Dosta and J. Guilemany, "Study of stellite-6 deposition by cold gas spraying," *Surface & Coatings Technology*, vol. 232, pp. 891–898, 2013.

[9] L. Baiamonte, M. Tului, C. Bartuli, D. Marini, A. Marino, F. Menchetti, R. Pileggi, G. Pulci and F. Marra, "Tribological and high-temperature mechanical characterization of cold sprayed and PTA-deposited Stellite coatings," *Surface & Coatings Technology*, vol. 371, pp. 322–332, 2019.

[10] R. Shoja and Razavi, "Laser Surface Treatment of Stellite 6 Coating Deposited by HVOF on 316L Alloy," *Journal of Materials Engineering and Performance*, vol. 25, pp. 2583–2595, 2016.

[11] Z. Pala, M. Bai, F. Lukac and T. Hussain, "Laser Clad and HVOF-Sprayed Stellite 6 Coating in Chlorine-Rich Environment with KCl at 700 C," *Oxidation of Metals*, vol. 88, pp. 749–771, 2017.

[12] S. Houdkova´, E. Smazalova´ and Z. Pala, "Effect of Heat Treatment on the Microstructure and Properties of HVOF-Sprayed Co-Cr-W Coating," *Journal of Thermal Spray Technology*, vol. 25, pp. 546–557, 2015.

IOP Conf. Series: Materials Science and Engineering **924** (2020) 012017 doi:10.1088/1757-899X/924/1/012017

Fracture Toughness and Microstructure of Gas Metal Arc Welded Plates of Aluminium Alloy 5083 Using Different Filler Wires and Heat Inputs

Arif Hidayat, Arif Junaidi, Sunoto Mudiantoro, and Winarto Winarto*
Department of Metallurgical and Materials Engineering, Faculty of Engineering, Universitas Indonesia,
Kampus UI Depok 16424 Jawa Barat – Indonesia

*winarto@metal.ui.ac.id

Abstract. Aluminum alloys AA5083 is alloys which mostly used in the hull components of the ship and high-speed craft. Commonly gas metal arc welding (GMAW) is used for joining the construction. Welding strength especially cracks toughness, is very important on this welding due to the regular operation of the ship or high-speed craft. In this research, welding of aluminum alloys AA5083 with dimension 150 mm x 350 mm and thickness 15 mm and 20 mm, were carried out by gas metal arc welding using two difference filler metal ER5183 and ER5356, and 99% argon is used as shielding gas. The result of the experiment that the maximum value of excellent toughness was shown on welding with thickness plate 15 mm using filler metal ER5356 and heat input 7.8 kJ/cm.

***Keywords*:** Aluminium Alloy AA5083 Welding, Filler Metal, Mechanical Properties, Microstructure, Crack tip Opening Displacement

1. Introduction

The Republic of Indonesia is the largest archipelagic country in the world, consisting of 17,508 islands, including 9,638 unnamed islands and 6,000 uninhabited islands. As an archipelagic country often referred to as the emerald of the equator, Indonesia has a unique and challenging configuration; besides, it has a strategic archipelago position between two continents and two oceans. This position has a high level of vulnerability to the influences of neighbour countries due to the opening and spread of Indonesian territories, which have caused disruption to the security stability and can destabilize the Indonesian integrity as NKRI. Therefore, sea transportation (ships & speed boats) is needed to support all trades and security activities [1]. Aluminum is the right choice of engineering material to be used for the hull of sea transportation because Aluminium has a lightweight and a good strength that is almost the same as the strength of low carbon steel [2]. One type of aluminum that is widely used in shipbuilding fabrication is Aluminium Alloy 5083, an aluminum that can be annealed with an alloy composition of 4.5% Mg, has good strength, and is easily welded, and also has high corrosion resistance [3]. In addition to the necessary materials used must also take into account the development process so that the quality of the product can meet the design plan.

Fabrication is one of the processes or stages that are important in shipbuilding, where the connection process between parts or components is carried out using welding at this stage [4]. The AA5083 aluminum welding process using the GTAW welding method is more reliable than the GMAW method, from the results of research conducted by Liu, et al. Found that the mechanical properties of the welding results using the GTAW method are better than using the GMAW method. Besides, no welding defects were found in the microstructure of welds using the GTAW method [5]. However, welding of aluminum alloy AA5083 especially in shipyards uses the GMAW method and generally uses multi-pass, the results of research that have been done show that after a long time and after experiencing improvements due to the age of the weld, the mechanical properties of the weld may have decreased and the microstructure has changed [6, 7]. Welding with GTAW is a welding method that uses protective gas. Commonly used protective gases are argon and helium with varying protective gas flows. Based on the results of the investigation showed that welding GTAW for aluminum AA5083 using argon shielding gas would produce a smaller grain size compared to using helium gas. Small grain size may produce better mechanical properties, so the use of argon gas as a protective gas is better than using helium [8, 9].

The welding procedure is essential and has a significant influence on the quality of the welds, mechanical properties, and microstructure of the welding results. For aluminum alloy material A5083, which is generally used for the maritime industry, testing and analysis have been carried out using aluminum alloy A5083-O and A5083-H321 with numerical simulations. From the results of the investigation found that there are differences in resistance behaviour to crack growth and the validity of simulation figures from fatigue crack growth based on the RPG stress criteria [10, 11].

Another thing that influences welding is the heat treatment process before welding. The pre-heat treatment influences the mechanical properties of aluminum welding. The Aluminium alloys 6061-T6 and 7075-T651 were welded using the MIG method before welding samples were preheated (preheating aging heat treatment), the results of the investigation showed that the welds that were subjected to heat treatment before welded had better mechanical properties compared to samples without heat treatment [12]. The effect of Butt-joint connection design, with variations in the distance of the magnesium alloy welding joint using the GTAW, has been carried out to determine the mechanical properties and microstructure. From the results of the investigation found no porosity in the microstructure and other types of weld joint found porosity [13].

The purpose of this research was to determine the toughness of aluminum material in its application as a body-hull of fast patrol boats. Right now, the toughness measurement of welding results using CTOD for Aluminum material is uncommonly done. So it is necessary to know the mechanical properties, microstructure, and CTOD value generated from the welding process using existing procedures in order to obtain an analysis of whether the welding procedures used in the construction of fast patrol boats using aluminum currently meet the technical requirements or standards.

2. Experimental method

2.1. Materials

The material used for base metal is Aluminium AA5083, while the filler metal used is based on the AWS A5.10 classifications, Blue Mig 5183 and Blue Mig 5356 with diameter size 1.20 mm and using argon 99% shielding gas. The material composition can be seen in Table 1 and Table 2. The thickness of the base metal used in this research is 15 mm and 20 mm. The material width is 300 mm, and the length is 350 mm.

Table 1. Material Composition of Base Metal (%).

Material	Si	Fe	Cu	Mn	Mg	Cr	Zn	Ti
Base Metal Al AA5083	0.4	0.4	0.1	0.4~1	4.0~4.9	0.05~0.25	0.25	0.15

Table 2. Material Composition of Electrode Wire (%).

Materials	Si	Cu	Fe	Mn	Mg	Cr	Ti	Al
Electrode 5183	0.40	0.1	0.4	0.75	4.75	0.15	0.15	Rem
Electrode 5356	0.25	0.1	0.4	0.12	5.0	0.12	0.13	Rem

2.2. Sample preparation

The type of connection used in welding is a butt joint welding, with a single V-groove. The distance from the connection is 2 ~ 3 mm, with a bevel angle of 60°, the surface of the joint in fine grinding, and cleaned before the welding process, the connection details such as Figure 1.

The welding process is using the gas metal arc welding and carried out by certified welders in the shipyard using two kinds of welding machines. The MIG Welding brand welding machine for material thickness of 15 mm and OTC type CPVE 400 for materials with a thickness of 20 mm. The voltage is set at 20-26 V and using the welding current between 180-200 A on the first pass and 200-270 A for the next pass.

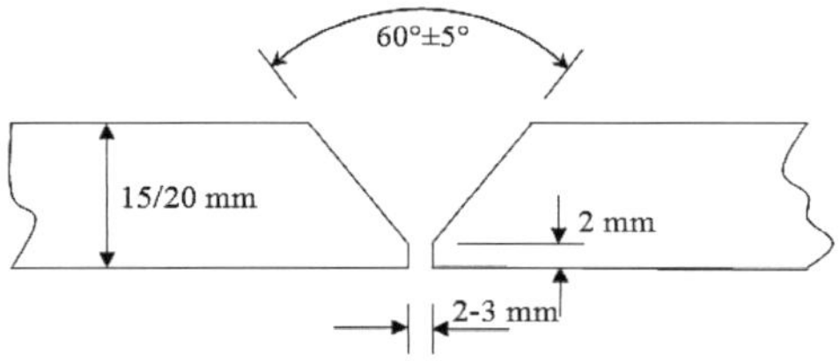

Figure 1. Detail of Weld Join.

Visual testing is done to determine the results of welding in general, especially in the surface area; the other test is a radiography examination to determine the results of the welding process in the weld metal. The observation of the microstructure was taken using the Nikon Model Epiphot 200 optical microscope, with a magnification of 500 times on weld metal and HAZ. For mechanical testing, there is three kinds of testing: Tensile Testing, Impact Testing, and Hardness Testing. Tensile testing is carried out with specimen preparation standards and testing using AWS D1.2 or ISO 4136 standards. While impact testing is carried out with sample preparation and testing using the standard of AWS D1.1 or ISO 9016 standards, each welding sample is taken three impact test specimens on weld metal, and the test is carried out at room temperature and the Hardness testing using the Vickers. Each welding sample is tested hardness on the base metal, weld metal, and HAZ (heat affected zone) area. Crack Tip Opening Displacement (CTOD) testing is one part of mechanical fracture testing conducted to determine and measure the resistance of a material to crack growth [14]. Sample preparation and testing is carried out using the BS 7448 standard or ASTM E1290, and each welding sample is taken 1 CTOD test specimen on a weld metal, and the test is carried out at room temperature. The fractographic observation was carried out with SEM on the weld test specimen with magnification up to 50 times. SEM-EDS testing aims to analyze the chemical composition by scanning methods.

3. Results and discussion

The welding parameters used in the welding process in this study are the heat input and filler metals or electrode of the welding process. The heat input required in the welding process can be seen in Figure 2, in general, it can be seen that the heat input required for welding with the metal arc welding (GMAW) method using ER5183 and ER5356 fillers for materials with a thickness of 20 mm require more significant heat input than materials with a thickness of 15 mm.

Welding results from the entire sample can be seen in Figure 2, both 15 mm and 20 mm thickness, and based on the electrodes used, showed no defects on the surface either surface porosity or undercut from the results of radiographic testing conducted by PT. Gong Samudra International in Batam, shows that all welding results are quite good and meet the requirements of existing standards. However, welding defects obtained in the form of inclusions in the sample 20-5356, where the maximum inclusion length of 2.0 mm that still meets the requirements of existing standards.

IOP Conf. Series: Materials Science and Engineering **924** (2020) 012017 doi:10.1088/1757-899X/924/1/012017

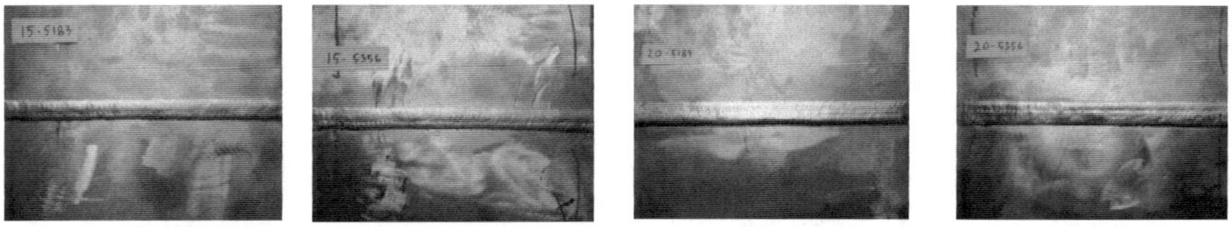

(a) 15-5183 (b) 15-5356 (c) 20-5183 (d) 20-5356

Figure 2. Welding Result of Al5083 with each parameter (Thickness – Electrode).

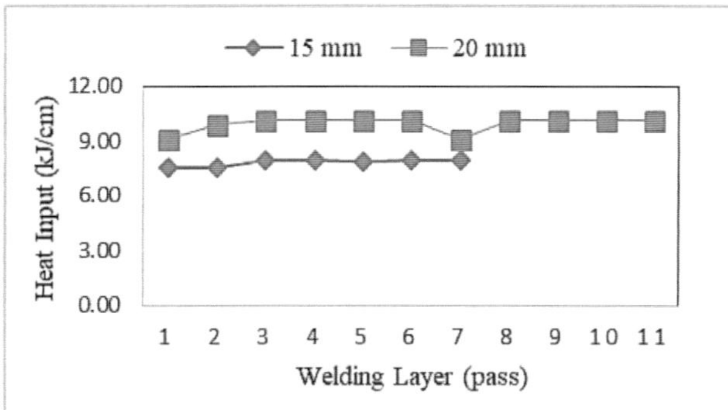

Figure 3. Heat Input on Welding Process.

From the results of macro and micro metallographic observations in Figure 3, it can be seen that the weld has been wholly integrated (homogeneous) between the base metal and the filler metal. Moreover, the results of metallographic microstructure observations made using an optical microscopy outline can be seen in the grain structure formed in the HAZ region and weld metal produced in the welding process of AA5083 aluminum alloy material using the gas metal arc welding and two metals different fillers.

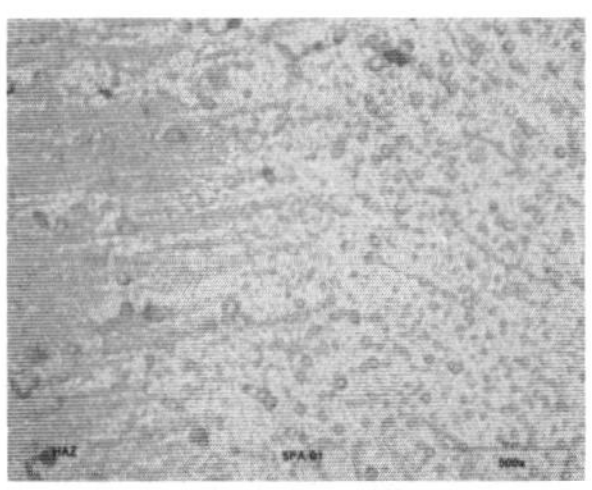

(a) 15-5183

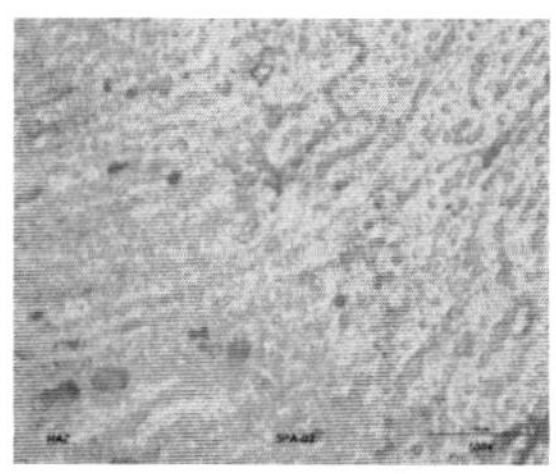

(b) 15-5356

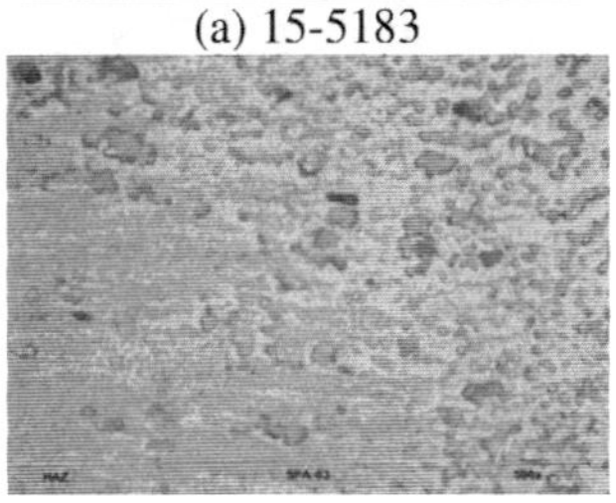

(c) 20-5183

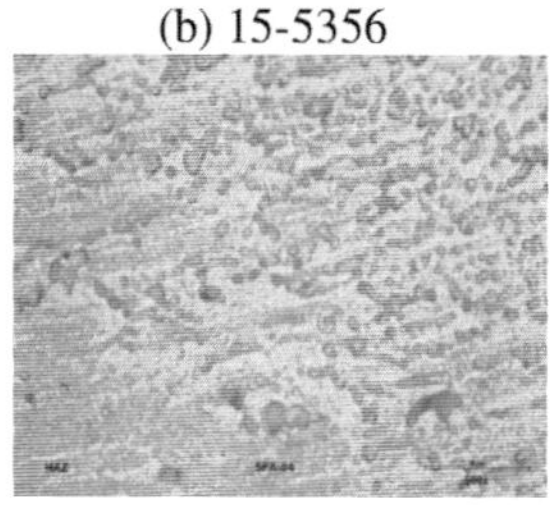

(d) 20-5356

Figure 4. Microstructures of HAZ and Weld Metal of every specimen (Material thickness – Electrode).

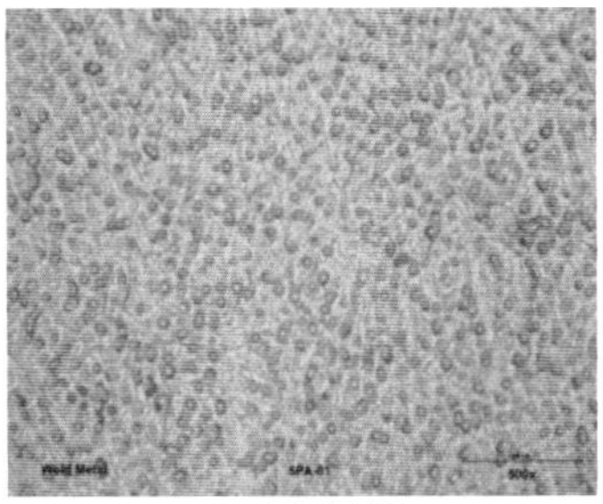

(a) 15-5183

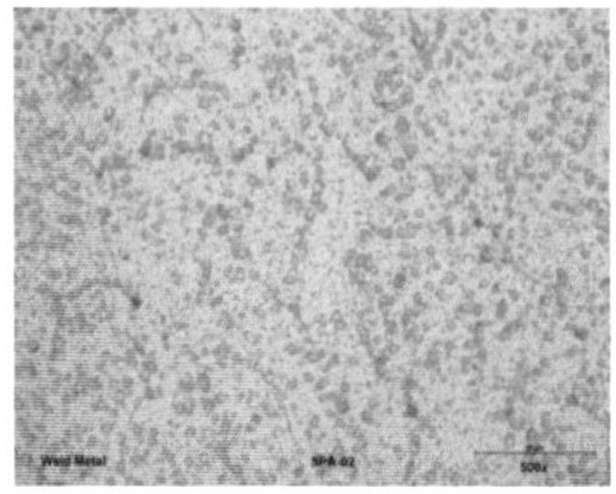

(b) 15-5356

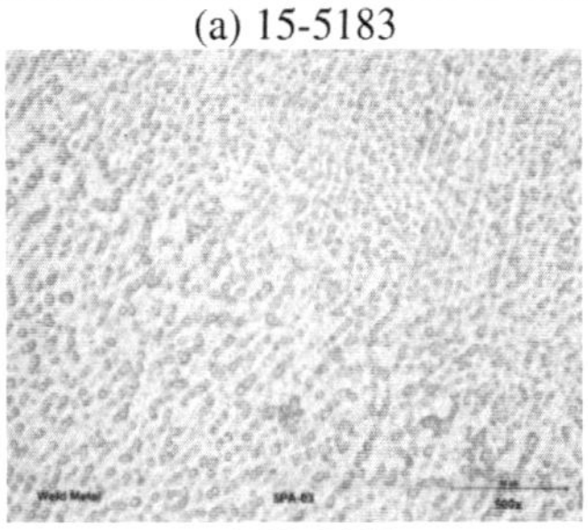

(c) 20-5183

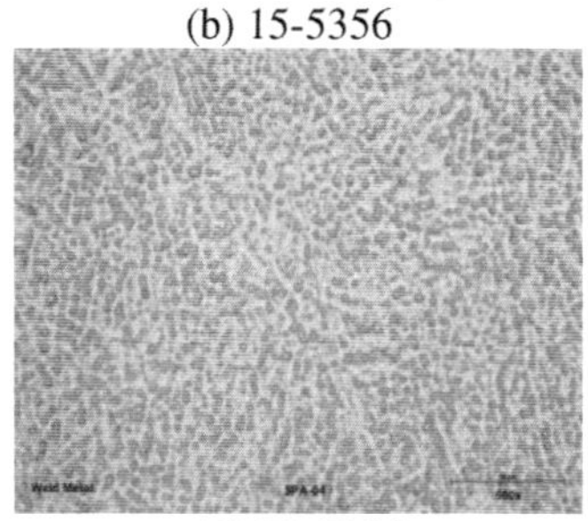

(d) 20-5356

Figure 5. Microstructures of Base Metal of every specimen (Material thickness – Electrode).

From Figure 4 which is a photograph of observations of microstructure analysis, it can be seen that the grain structure in the HAZ region varies in the form of lanes and reasonable grain structure with sizes varying from 40 μm to 60 μm, this occurs due to changes in heat cycles that occur in the area HAZ during the welding process. While in Figure 5, the microstructure of the weld metal area shows some reasonable round granules with grain sizes varying from 20 μm to 40 μm.

From the tensile test result in Figure 6 results, showed that the most significant yield strength is 172.82 N / mm^2, which is the result of welding plates with a thickness of 20 mm using ER5183 filler metal and heat input of 10 kJ/cm while the welding results that produce the highest tensile strength is welding on 15 mm material using ER5356 filler metal and heat input 7-8 kJ/cm that is 272.59 N/mm^2. It can happen because this process creates the finest/smallest grains compare to the other after the welding (Figure 5).

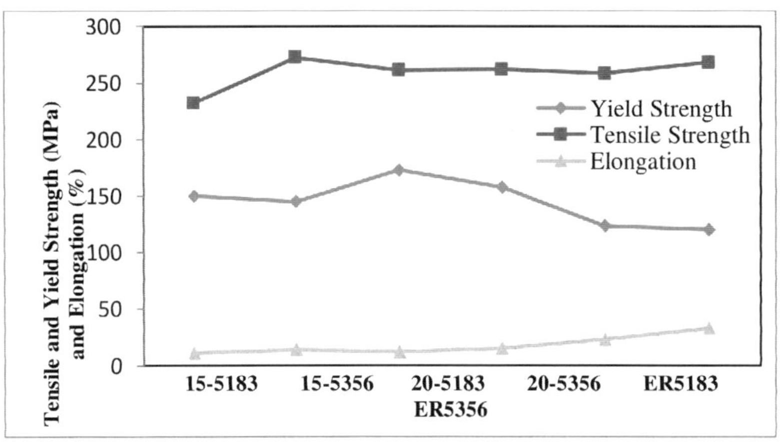

Figure 6. Comparison of the yield strength, tensile strength, and elongation of the tensile test results.

The result of the Impact test can be seen in Table 3. From this Table can be seen that welding of materials with a thickness of 15 mm produces higher impact energy compared to materials with a thickness of 20 mm. The highest average impact energy is obtained at 15 mm welding using ER5356 filler metal, which is 31.67 J, while the lowest is produced by welding 20 mm material using ER5356 filler metal that is 21.67 J.

Table 3. Result of Impact Test in every sample.

No.	Sample (Thickness – Electrode)	Impact Energy (j)
1	15-5183	30.00
2	15-5356	31.67
3	20-5183	30.33
4	20-5356	21.67

From the results of hardness testing conducted both in the base metal area, HAZ and weld metal. Based on the hardness result of all welding samples showed that the highest hardness is in the base metal, and the lowest hardness value is in the weld metal. The hardness value is higher for the welds with high heat input and ER-5183 filler wire and with a low heat input using ER-5356 filler wire. In general, the results of a hardness test can be seen in Figure 7.

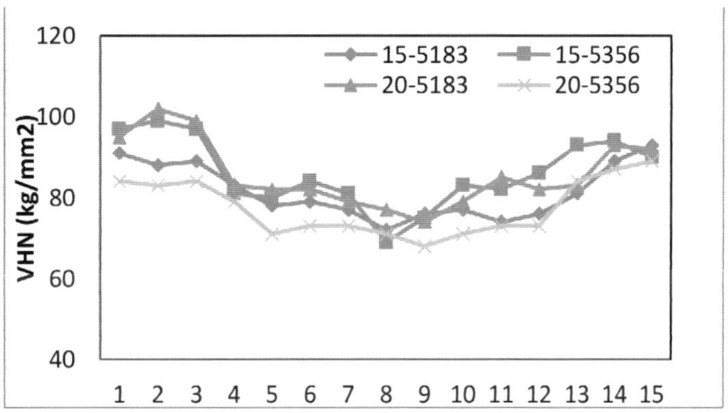

Figure 7. Hardness Distribution of AA5083 Welding.

Table 4. CTOD Test Result.

No.	Sample	K_f (Pre cracking) ($N/mm^{3/2}$)	K_f (Fracture) ($N/mm^{3/2}$)	CTOD (mm)
1	15-5183	288	1333	0.26
2	15-5356	288	1439	0.31
3	20-5183	394	1037	0.23
4	20-5356	394	1052	0.28

CTOD test results on aluminum welding AA5083 by using plates with a thickness of 15 mm and 20 mm, and by using filler metal ER5183 and ER5356 are shown in Table 4. The Table shows that the CTOD values produced by welding using ER5356 filler metal are higher compared to welding using ER5183 filler metal both for welding materials with a thickness of 15 mm or 20 mm. In welding materials with a thickness of 15 mm, a CTOD value of 0.31 was obtained when using filler metal ER5356 and a CTOD value of 0.26 when welded using filler metal ER5183. As for materials with a thickness of 20 mm, the resulting CTOD value is 0.28 with the fill metal ER5356 and only 0.23 if the fill metal used is ER5183. For welding carried out using the same filler metal and different plate, thicknesses show that the CTOD value on the 15 mm material welding results in a higher CTOD value when compared to welding on a 20 mm thick material.

Figure 8 shows a graph of the relationship between CTOD values and the heat input generated on welding using two different filler metals and different plate thicknesses. Welding results using ER5183 filler metal may result in higher CTOD values and lower tensile strength, but CTOD values may be lower if the tensile strength produced is higher. In contrast to welding with ER5356 filler metals where the CTOD value may be higher if the tensile strength is also high.

Whereas welding with 15 mm thick material may produce a low CTOD value if the tensile strength produced is also low and the CTOD value may be high if the tensile strength produced is also high, and the CTOD value may be very high when welding using a material with a thickness of 20 mm as shown in Figure 8, which shows the relationship between the CTOD value and the tensile strength of the AA5083 aluminum welding using two different filler metals and plate thicknesses [15].

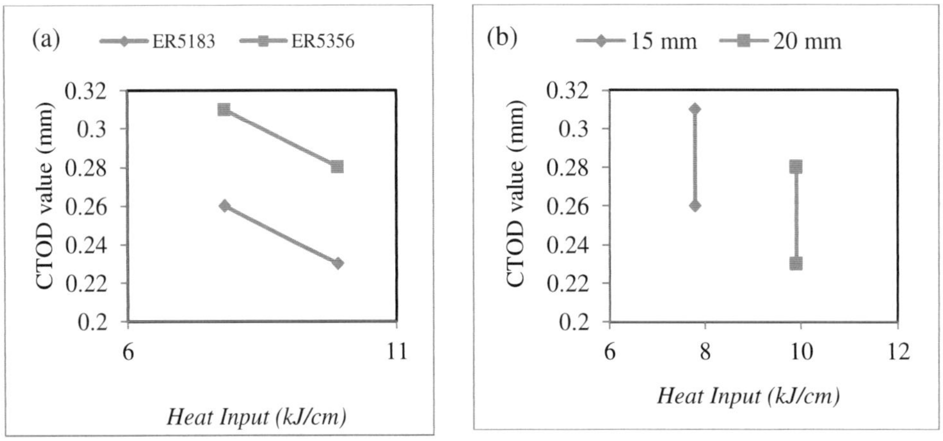

b

Figure 8. Comparison between CTOD value and heat input based on filler metal (a) and plate thickness (b).

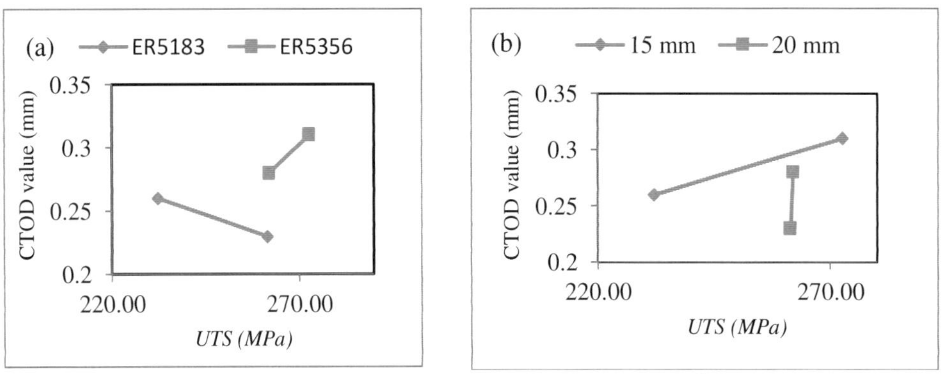

b

Figure 9. Comparison between CTOD value and Tensile Strength on filler metal (a) and plate thickness (b).

Furthermore, the relationship between CTOD value and the impact energy absorbed in aluminum AA5083 welding using two filler metals and different plate thickness can be seen in Figure 10, for welding with ER5183 filler metal, the CTOD value may decrease significantly if the impact energy produced increases, whereas for ER5356 filler metal CTOD values may increase according to the increase in impact energy produced, whereas for welding with 15 mm thick material CTOD values may increase if the impact energy also produced increases, but decreases when welding with 20 mm thick material.

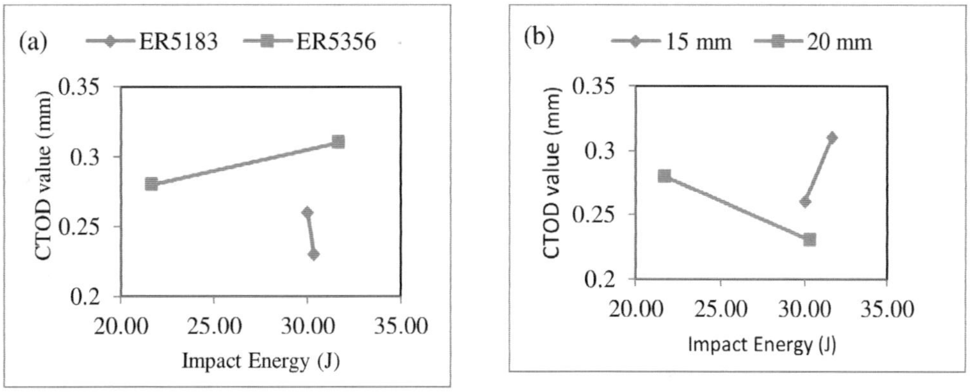

Figure 10. Comparison between CTOD value and Impact Energy on filler metal (a) and plate thickness (b).

Figure 11 is the microstructure of the SEM test results in the HAZ and weld metal regions. Based on photographs with a magnification of 2000 times, it can be seen that the grains formed in the HAZ region are more significant than the grains in the weld area. From these pictures, it can also be concluded that the 5183 electrode has a smaller size of grains compare to 5356 electrodes. SEM photos also show that solid magnesium deposits formed in the HAZ region with a larger size compared to deposits in the weld metal area. This conclusion from the research done by Jakobsen that the increasing of the Silicon may reduce the grain size [16].

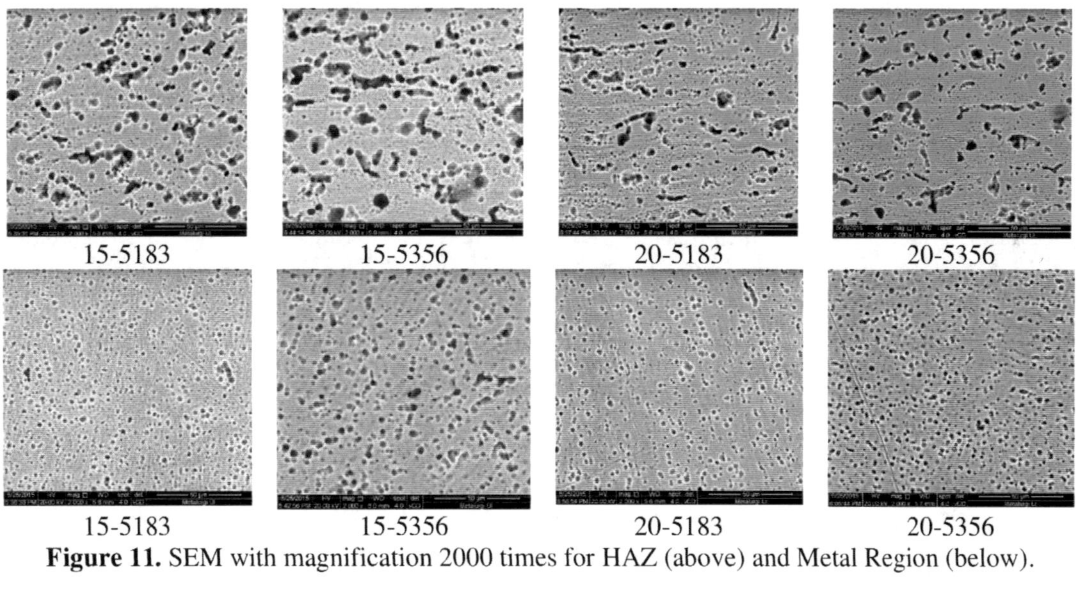

Figure 11. SEM with magnification 2000 times for HAZ (above) and Metal Region (below).

The results of the EDS examination conducted in the weld area showed the chemical composition formed from the filler metal after smelting due to the welding process. Welding with filler metal ER5183 on a 15 mm plate resulted in a 4.92% Mg element and 95.08% Mg element, while welding with ER5356 filler metal showed maximum Mg elements 5.37% and Al 94.63%. Welding of materials with a thickness of 20 mm with ER5183 shows a maximum of 4.82% Mg element and 95.18% Mg element, while welding with ER5356 filler metal shows a maximum of 5.19% Mg and 94.81% Al. This EDS result is related to the effect of heat number and the material properties. From the results of tests conducted by Vargas et al., one of the factors that influence the welding results of aluminum with the GMAW method is heat input. The heat input may affect the mechanical properties of 6061-T6 aluminum welding. It was concluded that heat input would affect the temperature of the welding process, but the amount of energy absorbed by the material depends on the alloy contained in the aluminum. So the maximum temperature of this process may affect the cooling process, which may affect the resulting microstructure changes [17].

SEM testing is also carried out on the surface of the fracture results from the CTOD test to determine the type of fault that occurs, the results of SEM testing on the surface of the fracture can be seen in Figure 12. That figure shows a dimple type surface fracture. Thus it can be concluded that the type of fracture that occurs for the welding results of AA5083 aluminum alloy material with two different filler metals is elastic (ductile). Moreover, from two types of filler metals used welding with ER5183 shows dimple surface fractures, compared to welding with filler metals ER5356.

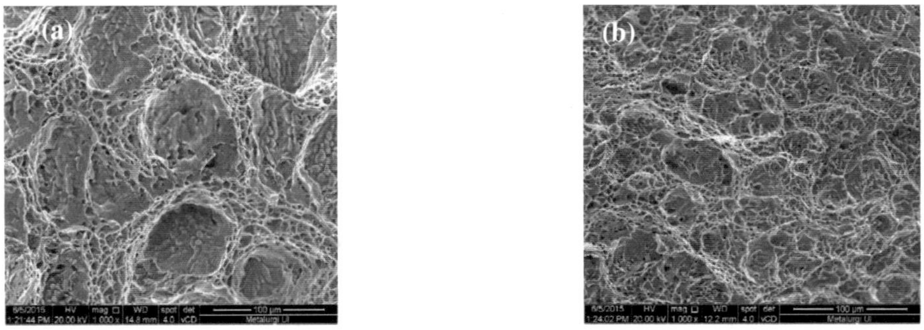

Figure 12. SEM photos of CTOD fracture surfaces of welding area for different filler metal (a) ER5183 and (b) ER5356.

4. Conclusion

From the results of the research and discussion that has been carried out, it can be concluded that the results of welding in general, both visual, NDT, and other tests can be accepted according to the classification rule standards (classification rule and regulation). It also found that the influence of plate thickness on the heat input indicates that the thicker plate used may require a higher heat input. Based on the results of the tensile test, the higher the heat input may produce a high tensile strength for welding with ER5183 filler metal, but the high heat input for welding with ER5356 produced a smaller tensile strength. While the higher hardness value happens on the welding with high heat input and ER5183 filler metals and with low heat input when using ER5356. Otherwise, impact energy is higher if the heat input on welding is lower for welding with ER5356 filler metal, but not much different when welding with ER5183 filler metal. The higher heat input on welding may produce a lower CTOD value, and with the same heat input, the CTOD value may be higher on welding with ER5356 filler metal compared to ER5183. The shape of the surface fracture in the CTOD test that occurs from the SEM results shows the simplified form of welding with ER5183 and ER5356 filler metals with a larger surface fracture diameter for ER5183. Optimization of research results obtained by welding results with a 15 mm thick plate with ER5356 and 7.8 kJ/ cm heat input obtained the highest value of fracture toughness.

Acknowledgment

The authors would like to acknowledge the DRPM, Universitas Indonesia, for the financial support through the PIT9 Research Grant 2019 with the contract No: NKB-0082/UN2.R3.1/HKP.05.00/2019.

References

[1] Lemhanas, "Structuring the Security of the Maritime Region To Maintain Security Stability in the Context of Safeguarding the Sovereignty of the Republic of Indonesia," *Jurnal Kajian Lemhanas RI,* pp. 74-75., 2012 (in Indonesia).

[2] J. Dwight, Aluminium Design and Construction, London: Taylor & Francis Group, 1999.

[3] W. D. C. Jr and D. G. Rethwisch, Fundamental of Material Science and Engineering, New York: John Wiley & Sons, 2013.

[4] H. Wiryosumarto and T. Okumura, Metal Welding Technology, Jakarta: PT Pradnya Paramita, 1996 (in Indonesia).

[5] Y. Liu, W. Wang, J. Xiea, S. Sunb, L. Wang, Y. Qian, Y. Meng and Y. Wei, "Microstructure and mechanical properties of aluminum 5083 weldments by gas tungsten arc and gas metal arc welding," *Materials Science and Engineering,* vol. A 549, pp. 7-13, 2012.

[6] M. Hakem, L. S., M. J., B. A., and T. S., "Welding and Characterization Of 5083 Alumunium alloy," in *Metal 2012*, Brno, Czech Republic, May 2, 2012.

[7] S. Katsas, J. Nikolaou, and G. Papadimitriou, "Microstructural changes accompanying repair welding in 5xxx aluminum alloys and their effect on the mechanical properties," *Materials and Design,* pp. 968-975, 2006.

[8] P. Peasura and A. Watanapa, "Influence of Shielding Gas on Aluminum Alloy 5083 in Gas Tungsten Arc Welding," *Procedia Engineering,* vol. 29, pp. 2465-2469, 2012.

[9] W.-b. Ding, W. Liu, and S.-y. Zhang, "The Properties of Pulsed Gas Metal Arc Welding for A5083 in LNG Cryogenic Storage Vessel," *The International Society of Offshore and Polar Engineers (ISOPE),* Vols. ISBN 978-1-880653-77-7 (Set), ISSN 1098-6189 (Set), pp. 242-245, 2010.

[10] J. J. Coronado and C. Cerón, "Fracture mechanisms of CTOD samples of submerged and flux-cored arc welding," *Theoretical and Applied Fracture Mechanics,* vol. 53, pp. 145-151, 2010.

[11] C. RC, W. MA, V. DM, S. GO, P. RH, and A. NU, "The development of the welding procedures and fatigue of butt-welded structures of aluminium-AA5383," *Journal Material Process Technology,* Vols. 60-5, p. 92– 93, 1999.

[12] I. Sevim, F. Hayat, Y. Kaya, N. Kahraman, and S. Şahin, "The study of MIG weldability of heat-treated aluminum alloys," *International Journal of Advanced Manufacturing Technology,* vol. 66, pp. 1825-1834, 2013.

[13] H. Dong, C. Liao, and L. Yang, "Effects of Butt Design and Gap Width on the Microstructure and Mechanical Properties of Magnesium Alloy Welds," *Journal of Material Science and Technology,* vol. 28(6), pp. 537-542, 2012.

[14] Jarvie, Tom; Stork Southwestern Laboratories, "Crack Tip Opening Displacement: Using Materials Testing to Control Cracks," *Stork Materials Technology,* pp. 1-8, 2004.

[15] J. A., J. Xian-Kui Zhu, "Review of fracture toughness (G, K, J, CTOD, CTOA) testing, and standardization," *Engineering Fracture Mechanics,* vol. 85, pp. 1-46, 2012.

[16] J. V. Jakobsen, "Microstructure and Mechanical Properties of Welded AA6082 Aluminium Alloys," Department of Materials Science and Engineering, Norwegian University of Science and Technology, Trondheim, 2016.

[17] J. A. Vargas, J. E. Torres, J. A. Pacheco, and R. J. Hernandez, "Analysis of heat input effect on the mechanical properties of Al-6061-T6 alloy weld joints", *Materials and Design,* vol. 52, pp. 556-564, 2013.

Influence the Filler Metal Containing Zirconium on the Weld Metal Porosity of The MIG Welded Aluminum Alloy 5083

Wahidun Adam, Winarto Winarto*, and Sunoto Mudiantoro
Metallurgy & Materials Engineering Dept., Universitas Indonesia, Kampus Baru UI Depok, Indonesia

*winarto@metal.ui.ac.id

Abstract. The aluminum alloy series 5083 has high corrosion resistance and high specific strength properties; hence, it is widely used for ship structures, marine components, cryogenics of LNG tanks, and unfired pressure vessels. This study was undertaken to investigate the effect of zirconium on weld metal porosity and hardness of MIG welded aluminum Alloy 5083. Zirconium is one of the alloying elements that are mostly used for grain refining in the aluminum casting process. Zirconium element was previously added in the electrode wire as a filler metal. The experiments were carried out by MIG welding for several aluminum alloy plates to produce the full penetration weldments. Two different filler wires used in the welding: one is containing zirconium, and another is no containing zirconium. The grain size of weld microstructures was measured using a Linear Intercept Method (ASTM E 112), and the hardness measurement is conducted using micro-Vickers along the cross section of AA-5083 weldment. Results showed that the grain size of AA-5083 weldment using the filler metal containing zirconium was considerably finer size than that of using the filler metal without zirconium. Moreover, the hardness of AA-5083 weldment using the filler metal containing zirconium is generally higher than that of using the filler metal without zirconium.

Keywords: Aluminium Alloys 5083; Zirconium, Weld Metal Porosity, Grain Size, and Hardness.

1. Introduction

Aluminum alloy 5083 is one of the groups of aluminum-magnesium (5XXX series) that are strengthened by alloying elements in solid solution, a fine dispersion as well as by strain hardening and grain size control [1]. They are work hardenable, have quite high strength, excellent corrosion resistance even in seawater, and very high cryogenic toughness. This alloy contains magnesium as a major alloying element with 4.5 wt. % magnesium, and also often includes small addition of transition elements such as chromium and manganese to control the grain and subgrain structure and the small impurities such as iron and silicon that are present in the form of intermetallic particles [2,3].

Aluminum Alloy - 5083 plates is generally welded by Metal Inert Gas (MIG) welding process, due to this process is more efficient and suitable for welding in industries. However, a significant hindrance in the weakness of the heat-affected zone (HAZ) and a presence of porosity in the weld fusion zone have been potential for degradation of mechanical properties as a result of welding. Peasura and Watanapa carried out MIG welding shielded by argon and helium gas on aluminum alloy AA 5083. They found that shielding gas argon provided smaller grain size, resulting in higher hardness both in the weld metal and HAZ, while helium, which has high thermal conductivity, provided larger grain size due to a large amount of heat [4]. One of the solutions for avoiding the welding degradation in the aluminum alloy is the use of micro alloying which has proven to be one of the effective mechanisms in enhancing mechanical properties of aluminum alloys through its microstructure improvement [5-7].

Seshagiri et al. [8] conducted intensive research on the effect of Zircon (Zr) addition to the mechanical properties of Al-Cu alloys welded with Gas Tungsten Arc Welding (GTAW). They found that better mechanical properties were achieved by adding zirconium to the molten zone during welding due to the finer grain size. Zhao et al. [9] examined characteristics of the Al-Mg-Sc alloy resulted from Tungsten Inert Gas (TIG) and Friction Stir Welding (FSW). They found that the A13(Sc, Zr) particles refine the weld metals of both joints effectively. Yang Dongxia et al. [10] showed that the small alloying of zirconium give the small grain size in the weld metal structure, due to the existence of Al_3Zr. Micro-hardness of the center of the fusion zone goes up from 74 $HV_{0.1}$ to 84 $HV_{0.1}$ since the reduction of grain size by small alloying of Zr.

Moreover, the small alloying of Zr gave improvement on the tensile strength. Yin Deng et al. [11] also found that the microalloying of Zr and longer the aging period at 120 °C, both can reduce the corrosion of Al-Zn-Mg alloys. This effect is due to the grain size refiner of Zr and limiting the formation of precipitates.

This work aimed to examine the grain size, weld porosity, and mechanical properties in two types of welded plates after applying different filler wires (ER5556A and ER5087) on the mechanized MIG weldments of aluminum alloy 5083.

2. Experimental method

2.1. Materials

2.1.1. Base Material and Specification. AA 5083 was used as the base material in the present investigation. This alloy is in the annealed condition or can be written as AA5083-0 and was supplied in plate form with a 5mm thickness, 300mm long, and 100 mm wide. The chemical composition of this alloy is shown in Table 1.

Table 1. Chemical composition of base metal AA-5083.

AA-5083	%Si	%Fe	%Cu	% Mn	% Mg	% Cr	%Zn	%Ti	Other	%Al
Standard	0.40	0.40	0.10	0.40-1.00	4.00-4.90	0.05-0.25	0.25	0.15	0.15	Rem
Tested	0.90	0.24	0.01	0.60	4.70	0.10	0.02	0.01	0.15	Rem

Table 2. Chemical composition of filler wire.

ER-5556A	%Si	%Fe	%Cu	% Mn	% Mg	% Cr	%Zn	%Ti	%Be	Other	%Al
Standard	0.25	0.40	0.100	0.6-1.0	5.0-5.5	0.05-0.20	0.20	0.05-0.20	0.0008	0.15	Rem
Tested	0.07	0.19	0.002	0.72	5.40	0.093	0.001	0.118	0.0007	0.0042	Rem
ER-5087	**%Si**	**%Fe**	**%Cu**	**% Mn**	**% Mg**	**% Cr**	**%Zn**	**%Ti**	**%Be**	**%Zr**	**%Al**
Standard	0.25	0.40	0.05	0.7-1.1	4.5-5.2	0.05-0.25	0.25	0.15	0.0008	0.1-0.2	Rem
Tested	0.08	0.20	0.015	0.68	4.61	0.096	0.011	0.128	0.0004	0.12	Rem

2.1.2. Filler Wire. The recommended filler alloys for welding AA 5083 are ER5356, ER5183, and ER5556 (ASM, 1993). Two filler alloys used in this investigation were ER5556A and ER5087. The last one, ER5087, is a new development of filler alloy that contains zirconium as a grain refiner. The certificate analysis for the chemical composition of filler wires can be seen in Table 2.

2.2. Weldment Preparation.

There were 12 pieces of full-penetration aluminum alloy 5083 weldments produced using a mechanized MIG welding with the following welding parameters, as presented in Table 3.

Table 3. Welding parameter of AA-5083 plate using two types of filler wires.

CODE	Filler Wires	Voltage (U) (Volt)	Current (I) (Ampere)	W.Speed (S) (mm/s)	Power (P) (kW)	Linear Energy (Q) (J/mm)
SU-01	ER 5556A	22.3	250	8.33	5.6	673
SU-02		22.5	256	9.46	5.8	610
SU-03		23.7	260	9.32	6.2	653
SU-04		22.8	266	9.43	6.1	640
SU-05		23.4	270	10.49	6.3	600
SU-06		24.0	281	11.24	6.7	596
SU-07	ER 5087	23.1	252	8.22	5.8	703
SU-08		23.5	257	9.49	6.0	633
SU-09		24.2	261	9.32	6.3	673
SU-10		24.5	267	9.55	6.5	683
SU-11		24.0	271	10.68	6.5	603
SU-12		25.8	281	10.60	7.2	686

2.3. Material Characterization.

The microstructure of weldments was investigated under an optical microscope to reveal their microstructure of grains, phases, and precipitates. The measurement of weldment grain size was carried out using the optical microscope. The grain size was determined by the Linear Intercept Method (ASTM E 112) [11]. Micro-hardness tests were conducted on a cross-section of weldment specimens with a Vickers micro-hardness tester of the 100-gr load. Hardness measurements were made across the weld interface parallel and 1.5 mm below the face surface of the sample. The indentations were made with increments of 1 mm from the middle of weld metal. The process of calculating the percentage of porosity of welds performed on the cross-sectional area of the weld of AA 5083 by using Image Analysis.

3. Results and discussion

3.1. Microstructure of Weldments

Photographs of the HAZ structure of AA-5083 weldment, as seen in Figure 1, show that the grain morphology of weldment is irregularly shaped grains and the HAZ grain size of weldment in SU-06 is coarser than the HAZ grain size of weldments in SU-09. Also, some of the precipitates are distributed in the matrix with the black cluster.

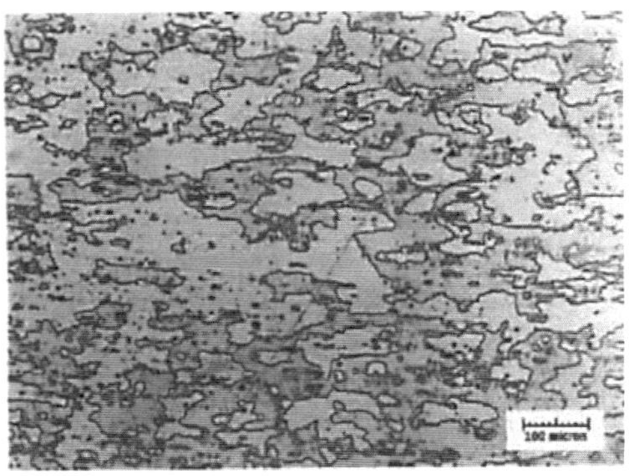

(a)

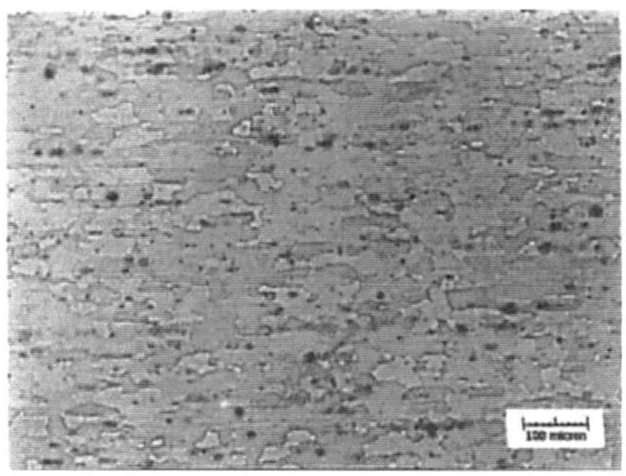

(b)

Figure 1. The microstructure of HAZ in weldments of (a) SU-06 using the filler wire of ER-5556A, (b) SU-09 using the filler wire of ER-5087, etched with Keller's Reagent.

3.2. Grain Size Measurement

The result of grain size measurement can be shown in Figure 1 (a) and (b). The grain size of both weldments, as in Figure 2, shows that the grain size for weldments using ER-5556A (Figure 2-a) is different value compared to the weldments using ER-5087 (Figure 2-b), especially at HAZ and at the interface (fusion line) area. The HAZ grain size of weldments using ER-5556A is in the range of 6.0 to 6.3 (ASTM Grain Size). This value is coarser than the grain size of weldments using ER-5087 which is the value between 7.1 and 7.4 (ASTM Grain Size). Besides, the interface grain size of weldments using ER-5556A seems to be the same trend-line with the HAZ grain size. The interface grain size in the weldments using ER 5556A is quite coarser than the interface grain size of weldments using ER-5087. Those grain refinement would affect not only the mechanical properties improvement but also hot crack susceptibility reduction of the weldments. Grain refiner element able to provide many efficient heterogeneous nucleation sites in the weldment [12].

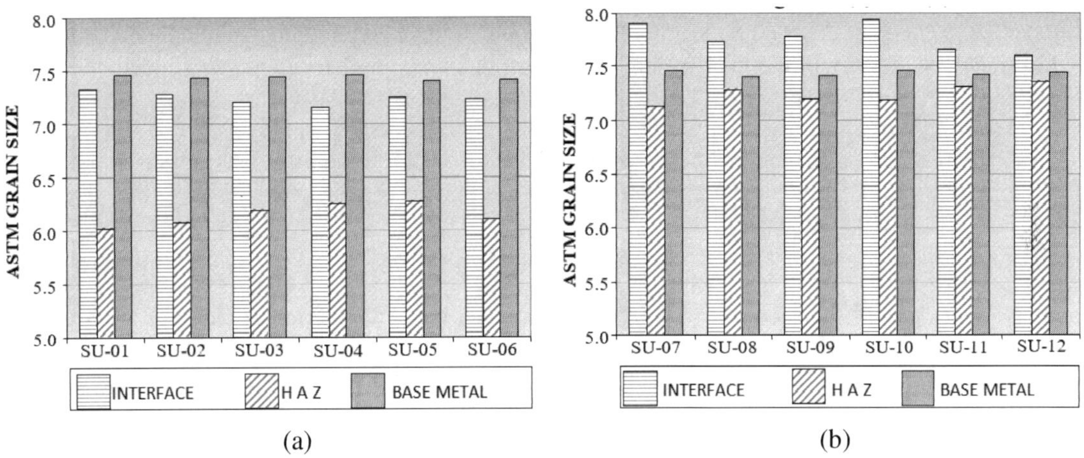

Figure 2. The result of grain size measurement for weldments using filler wire (a) ER-5556A and (b) ER-5087.

3.3. Microhardness Distribution of Weldments

Analysis of the hardness profiles was performed for both weldments. Hardness values in the weld metal, HAZ, and the base metal are plotted as a function of the distance (D) from the middle of weld metal, as shown in Figures 3 and 4. An average hardness value in the base metal (H_{BM} *ave*) equals 76 HV, with maximum fluctuations from average being approximately 3%. These fluctuations may be a result of the uneven distribution of the second phase particles in the base metal microstructure.

Figure 3(a) shows a hardness distribution of a weld produced using filler wire ER-5556A. There is a flat trend line in weld metal hardness from about 66 to 74 HV in the location between 0 mm to 4 mm from the middle of weld metal. Following that between 4 to 5 mm weld interfaces, the hardness increases slightly from 74 HV to 76 HV. After that decrease in HAZ hardness occurs in the distance between 5 mm to 9 mm from 76 HV to between 40 HV and 60 HV. This hardness is significantly the lowest value across the weldments. Subsequently, from 9 mm to 15 mm, the HAZ hardness increases sharply to approximately 76 HV, and this value remains constant at the base metal of 76 HV.

Figure 3(b) shows a hardness distribution of a weld produced using filler wire ER-5087. There is a flat trend line in weld metal hardness from about 66 to 75 HV in the location between 0 mm to 4 mm from the middle of weld metal. Then between 4 to 5 mm weld interfaces, the hardness decreases slightly from 75 HV to 70 HV. After that, the decreasing in HAZ hardness occurs in the distance between 5 mm to 9 mm from 59 HV to between 66 HV and 60 HV. This value is slightly decreasing in HAZ. Subsequently, from 9 mm to 15 mm, the base metal hardness is approximately constant at 76 HV on average.

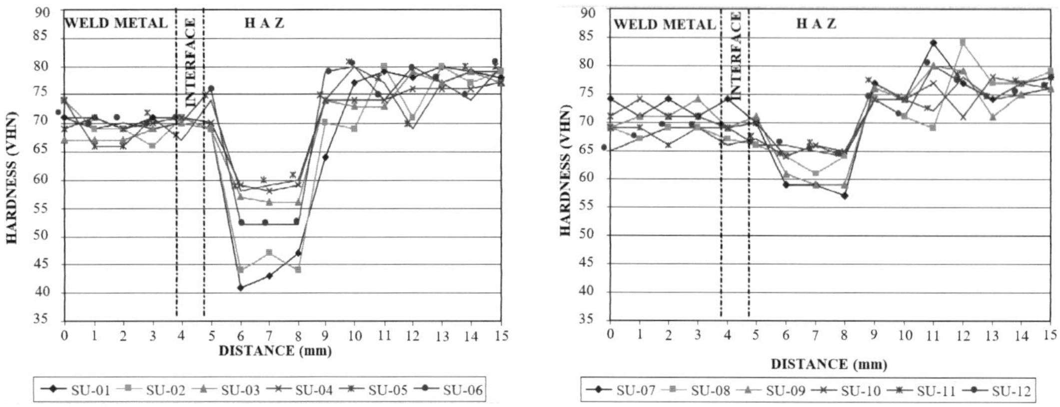

(a) (b)

Figure 3. Hardness distribution of weldments using filler wires: (a) ER-5556A and (b) ER-5087.

From Figure 3, it shows that the HAZ hardness distribution for welds using ER-5087 is more flat-distributed compared to the weldments using ER-5556A which is the HAZ hardness dropping very steeply to about 40 HV at the distance of 6 mm from the middle of weld metal. This is the minimum hardness for all welds produced.

3.4. Weld Porosity

The purpose of this calculation is to compare the use of filler metal at different welding AA5083 plates against the formation of porosity in the area of weld metal. Total percentage of porosity calculated on the weld metal for all weld parameters by using an optical microscope. Further images are recorded and the percentage using the image analysis software. Results can be seen in Table 4 and the macrostructure of porosity distribution, as shown in Figure 4. The porosity of weldment ER5087 containing zirconium is lower than the ER5556A containing no-zirconium. The rate of hydrogen absorption and solubility primarily determined by weldment composition. Pure aluminum is more prone to porosity formation, while Al-Mg alloys are much less due to the degassing effect from the evaporation of magnesium [13]. Furthermore, the addition of Mg, Ti, or Zr will tend to increase hydrogen solubility and have higher hydrogen concentration threshold values of porosity formation and thus more challenging to form porosity [14,15].

Table 4. Welding parameter of AA-5083 plate using two types of filler wires.

No.	Filler Metal	Linear Energy (J/mm)	Porosity (%)
1		669	3.36
2	ER 5556A	643	1.11
3		602	0.77
1		708	0.74
2	ER 5087	685	0.29
3		609	0.52

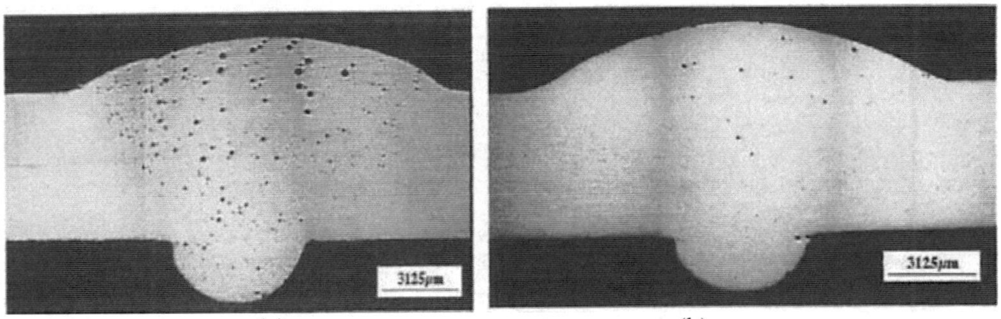

(a) (b)

Figure 4. Porosity distribution of weldments using filler wires: (a) ER-5556A and (b) ER-5087.

3.5. Relationship between the Grain Size and Zirconium Content in the Filler Wires

The level of grain refiner in the welding filler wire of ER5087 affects the solidification mode in the weld metal. Kou and co-workers [16] using the G/R ratio, showed that the effect of heat input and welding travel speed on the grain structure of weld metal. G denotes the temperature gradient in the solidification in the center of the weld pool, while R indicates the growth rate of the solidification. The stage of nucleation is influenced remarkably by the addition of grain refiner. Since ER5087 filler metal has a content of zirconium, some researchers investigated that low concentrations of zirconium aid grain refinement in Al alloys [5-11]. Their finding has shown that the second phase particles Al_3Zr may act as nucleating agents (sites) for aluminum alloy. Work by Kou et al. [16] who studied the parameters of welding and the weld metal grain has found that increased welding speed (TS) may enhance not only the G but also the S while increasing heat input (HI) is likely to lower the G. Thus, the reduction of G/R ratio when both the HI and the TS are increased. The decline of the G/R ratio can promote the constitutional supercooling [17,18]. The larger the constitutional supercooling, the higher the tendency for pore and the solid phase to nucleate and grow upon nucleating agent or site (Al_3Zr) in the super cooled liquid and a hence higher degree of grain refining. This super cooling, together with the presence of heterogeneous nucleation provided by zirconium in the filler wire, explains the reduction of grain size in weldment using ER5087 filler wire.

It is interesting to note that the grain size in the interface area (fusion line) and HAZ was smaller (finer) in the weldment of using 5087 filler wire than that of using ER-5556A. The possible explanation for this case is that the effects of grain refiner site of Al_3Zr may increase with decreasing the total of base metal melted (i.e., lower dilution) or high heat input (P) and high welding speed (S). This argument is supported by the research work on the effect of zirconium in aluminum [18] that zirconium was found to be a potential grain refiner of aluminum alloy.

4. Conclusion

This investigation has been primarily concerned with the influence of filler wire containing zirconium on MIG welded aluminum alloy 5083. The grain size of HAZ structure with the ER5087 containing zirconium was considerably finer than that of HAZ structure with the ER5556A containing no-zirconium. The grain size of the HAZ structure is coarser with increasing the linear energy (Q), ranging from 600 to 680 J/mm, this yields the decreasing in HAZ hardness. The hardness values in all welds vary with welding parameters employed. The HAZ hardness distribution of a weld produced using ER-5087 filler wire is more flat distributed compared to the weldments using ER-5556A, which is the HAZ hardness dropping very steeply to about 40 HV at a distance 6 mm from the middle of weld metal. Percentage porosity AA-5083 welds that use welding wire ER5087 has a smaller percentage than the weld using welding wire ER5556A.

Acknowledgment

The authors would like to acknowledge the support of research funding from the Directorate of Research and Community Service of Universitas Indonesia (DRPM UI) through "Hibah Penugasan Publikasi

Internasional Terindeks 9 (PIT 9)" in the fiscal year of 2019 with the contract number: NKB-0082/UN2.R3.1/HKP.05.00/2019.

References

[1] M. F. Ashby and D. R. H. Jones, Engineering Materials 2: An Introduction to Microstructures, Processing, and Design, 3rd Ed, Oxford: Butterworth-Heinemann, 2005.

[2] I. Polmear, D. StJohn, J.-F. Nie and M. Qian, Light Alloys: Metallurgy of the Light Metals, 5th Ed, London: Elsevier, 2017.

[3] J. E. Hatch, Aluminum: Properties and Physical Metallurgy, Aluminum Association Inc. and ASM International, 1984.

[4] P. Peasura and A. Watanapa, "Influence of Shielding Gas on Aluminum Alloy 5083 in Gas Tungsten Arc Welding," *Procedia Engineering 29,* pp. 2465-2469, 2011.

[5] H. A. Elhadari, H. A. Patel, D. L. Chen and W. Kasprzak, "Tensile and fatigue properties of a cast aluminum alloy with Ti, Zr and V additions," *Mater. Sci. Eng. A528,* p. 8128, 2011.

[6] W. T. Wang, X. M. Zhang, Z. G. Gao, Y. Z. Jia, and L. Y. Ye, "Influences of Ce addition on the microstructures and mechanical properties of 2519A aluminum alloy plate," *Journal of Alloys and Compound 491,* pp. 366-371, 2010.

[7] Y. L. Wu, F. H. Froes, C. G. Li, and A. Alvarez, *Metallurgical and Materials Transactions A,* vol. 30A, p. 1017, 1999.

[8] P. C. Seshagiri, B. S. Nair, G. M. Reddy, K. S. Rao, S. S. Bhattacharya, and K. P. Rao, "Improvement of mechanical properties of aluminium-copper-alloy (AA2219) GTA weld by Sc addition," *Science and Technology of Welding and Joining,* vol. 13, no. 2, p. 146, 2008.

[9] J. Zhao, F. Hang, H. G. Ban, K. Wen, L. Jiang and X. B. Chen, *Mater. Des.,* p. 306, 2010.

[10] Y. Dongxia, L. Xiaoyan, H. Dingyong, and H. Hui, "Effect of minor Er, and Zr on microstructure and mechanical properties of Al-Mg—Mn alloy (5083) welded joints," *Materials Science & Engineering: A,* vol. 561, pp. 226-231, 2013.

[11] Y. Deng, Z. Yin, K. Zhao, J. Duan, and Z. He, "Effects of Sc and Zr micro-alloying additions on the microstructure and mechanical properties of new Al—Zn—Mg alloys," *Journal of Alloys and Compounds 530,* pp. 71-80, 2012.

[12] B. S. Murty, S. A. Kori, and M. Chakraborty, "Grain refinement of aluminum and its alloys by heterogeneous nucleation and alloying," *International Mat. Reviews 47,* no. 1, pp. 3-29, 2002.

[13] R. Shore and R. McCauley, "Effects of porosity on high strength aluminum 7039," *Weld. J.* pp. 311s-321s, 1970.

[14] J. Charbonnier, "Gaz dans les alliages d'aluminium de fonderie," *techniques de l'ingénieur traité matériaux métalliques,* vol. section M218.

[15] R. Martukanitz and P. Michnuk, Sources of porosity in gas metal arc welding of aluminum, American society for materials, 1982.

[16] S. Kou and Y. Le, "Welding parameters and the grain structure of weld metal - a thermodynamic consideration," *Metallurgical Transactions A,* vol. 19, no. 4, pp. 1075-1082, 1988.

[17] C. E. Cross, O. Grong and M. Mousavi, "A model for equiaxed grain formation along the weld metal fusion line," *Scripta Materialia,* vol. 40, no. 10, pp. 1139-1144, 1999.

[18] A. A. Rao, B. S. Murty, and M. Chakraborty, "Role of zirconium and impurities in grain refinement of aluminum with Al—Ti—B," *Materials Science and Technology,* vol. 13, no. 9, pp. 769-777, 1997.

Preparation and Structures of Carbon-Copper Oxide Nanofibers from Electrospun Polyacrylonitrile Composite

Muhamad Nasir* and Rana Ida Sugatri
Research Unit for Clean Technology, Indonesian Institute of Sciences
Jl. Cisitu Sangkuriang, Bandung 40135, Indonesia

*mnasir71@yahoo.com

Abstract. Carbon-copper oxide nanofibers have been prepared by heat treatment of polyacrylonitrile-copper salt nanofibers. In this study, fabrication of PAN-copper salt using two different methods - firstly, synthesis of PAN-copper salt solution using electrospinning (PAN-CuAc NF), and secondly, immersion of electrospun PAN nanofiber in the copper salt solution (PAN-c-CuAc NF). Those ways had a distinct effect on the structure of carbon-copper oxide nanofibers. The morphologies and structures of carbon-copper oxide nanofibers were observed by scanning electron microscopy-energy dispersive x-ray spectroscopy (SEM-EDX), infrared spectroscopy (IR), X-ray diffraction (XRD), and Raman spectroscopy. The SEM image showed a randomly oriented nanofiber structure without bead formation, which average diameter of the nanofiber decreased with increasing process temperature. IR spectra and XRD patterns confirmed the formation of aromatic structures and graphite-like structures on carbon-copper oxide nanofiber. The Raman spectrum revealed that the intensity ratio (I_D/I_G) of carbonized nanofibers showed different values when the electrospun PAN composite nanofibers were synthesized using a different method, 1.59 for electrospun PAN nanofibers (CNF), 1.92 for carbonized PAN- CuAc nanofibers (C-copper oxide NF), and 0.60 for carbonized PAN-c-CuAc nanofibers (C-c-copper oxide NF).The presence of Cu element has been successfully detected in carbon-copper oxide nanofibers using EDX analysis.

***Keywords*:** Electrospinning, Carbon nanofibers, Copper oxide, Polyacrylonitrile, Structures

1. Introduction

Carbon-metal nanomaterials have been widely fabricated using based electrospun nanofibers through stabilization and carbonization processes. Their composites design potentially used as sensor, anodes and adsorbent application due to good biocompatibility, electrical conducting properties, and high surface area [1-3]. Incorporating metal salt in a precursor system before heat treatment supports the formation of graphite-like structure and metal oxide or pure metal in one step. The metal salt is able to enhance graphitization of the utilized polymer to form a graphite shell [4].

Carbon fibers generally have low densities, high thermal and chemical stabilities in the absence of oxidizing agents, good thermal and electrical conductivities [5]. PAN is well known as a precursor in producing carbon nanofibers due to the simplicity to be converted into carbon fibers in a high yield [6].

Some researchers have been reported advantages of carbon-metal nanofibers in some applications. Fu et al. [2] successfully synthesized carbon-copper composite nanofibers for catechol detection. Carbon-copper nanofibers showed better performance than carbon nanofibers as a laccase biosensor. The presence of Cu in the composite possessed rapid response, a wide linear range, good repeatability, reproducibility, long-term stability, and selectivity. The Carbon-SnO_2 nanofibers were fabricated by Bonino et al. [1] and obtained that tin oxide incorporated into carbon nanofibers was able to improve their performance when evaluated as lithium-ion battery anodes.

Most researchers synthesize carbon-metal nanofiber starting by dissolving metal salts and polymers in solvent, and after that the electrospinning process is carried out using that solution. After that the nanofiber is carbonized at a certain temperature to make carbon-metal nanofiber. [1-4, 7]. In this study, we observed the effect of methods in fabrication of electrospun PAN-copper acetate and their effect on stabilization and carbonization process to their chemical structures. Firstly, copper acetate monohydrate dissolved in NMP as solvent directly mixed with PAN before spun by electrospinning. Secondly, electrospun PAN nanofibers were immersed in a copper acetate solution and dried at room temperature. These composite nanofibers subsequently were treated by stabilization and carbonization process. The results showed those methods had a distinct effect on carbon nanofibers obtained.

2. Experimental method

Solution containing copper(II) acetate monohydrate (CuAc, Merck KGaA, Darmstadt, Germany) and polyacrylonitrile (PAN) (Mw= 150,000; Aldrich Chemistry, USA) was spun at a voltage of 12-13 kV for 4 h, the distance between the collector plate and needle was fixed at 18.5 cm and flow rate 1 mL/h. CuAc (1% w/v) solution was mixed with PAN (10% w/v) in 1-Methyl-2-pyrrolidone (NMP) (Merck Schuchardt, Hohenbrunn, Germany) at 80°C for 3 h and keep stirring at room temperature (PAN-CuAc). The second method, pristine PAN nanofibers that had been spun by electrospinning using the parameters as mentioned above was immersed in CuAc (1% w/v) solution (solvent: water) for 150 min and then dried at room temperature for overnight (PAN-c-CuAc). Furthermore, those electrospun nanofibers were heat-treated using a tube furnace (Carbolite Furnaces). Stabilization at 290°C for 2 h with a heating rate of 5°C/min under air atmosphere and proceeded to carbonization at 700°C for 1 h with 10°C/min under a pure Nitrogen atmosphere (150 mL/min).

The surface morphology was studied by scanning electron microscope (SEM) (JEOL IT-300 Japan Oxford Instrument X=MAX, UK) with energy-dispersive x-ray (EDX) analysis. Infrared spectrum analyzed by Attenuated Total Reflectance-Fourier Transform Infrared Spectroscopy (ATR-FTIR) Nicolet iS5 ATR iD5 Thermo Scientific (USA) for each sample using 64 scan rate in the range 500-4000 cm^{-1}. The phase and crystallinity structure were measured by using D8 Advance Bruker X-ray Diffractometer (XRD; Bruker, Germany) with Cu Kα (λ=1.540 Å) radiation over Bragg angle ranging from 10° to 80°. Raman spectroscopy (Bruker SENTERRA Raman Microscope) was used to characterize the structure of the carbonized electrospun nanofibers, using 532 nm laser as the excitation source and 10 mW power.

3. Results and discussion

3.1 Morphology analysis

Nanofibers that was spun using electrospinning and used as a precursor for producing carbon nanofibers had a randomly oriented structure nanofiber and without bead formation with the average diameter as shown in Table 1.

Table 1. Average diameter of electrospun nanofibers.

	Before heat treatment	**After Stabilization**	**After Carbonization**
PAN	470.40 ±50.16 nm	430.69 ±50.99 nm	357.73 ±50.72 nm
PAN-CuAc	854.35 ±71.40 nm	747.25 ±72.34 nm	633.03 ±57.94 nm
PAN-c-CuAc	480.39 ±56.83 nm	393.76 ±40.99 nm	363.61 ±33.48 nm

The presence of CuAc in PAN solution before fabrication in producing composite electrospun nanofibers affected enhancing the average diameter of composite nanofibers. It is probably due to the metal salt effect on electrospinning parameters used (voltage, a distance between needle and collector,

IOP Conf. Series: Materials Science and Engineering **924** (2020) 012019 doi:10.1088/1757-899X/924/1/012019

and flow rate). By the different methods, immersion technique, to obtain composite nanofibers we observed that increasing the average diameter of nanofibers took place as well. Dried copper salt coats part of the surface of the nanofibers. In that case, we hypothesize copper salt also adsorbed by pristine PAN nanofibers.

The fiber diameter is reduced along with increasing temperature process due to structural changes followed by removal of the non-carbon elements [5]. Moreover, dimension size of nanofiber sheets decreased into 50-53% for carbon nanofibers of pristine PAN (CNF) and carbon-copper oxide nanofibers that were obtained by immersion (C-c-copper oxide NF), respectively and about 67% for carbon-copper oxide nanofibers (C-copper oxide NF) after stabilization as well as about 40% for all nanofibers after carbonization compared with an initial size.

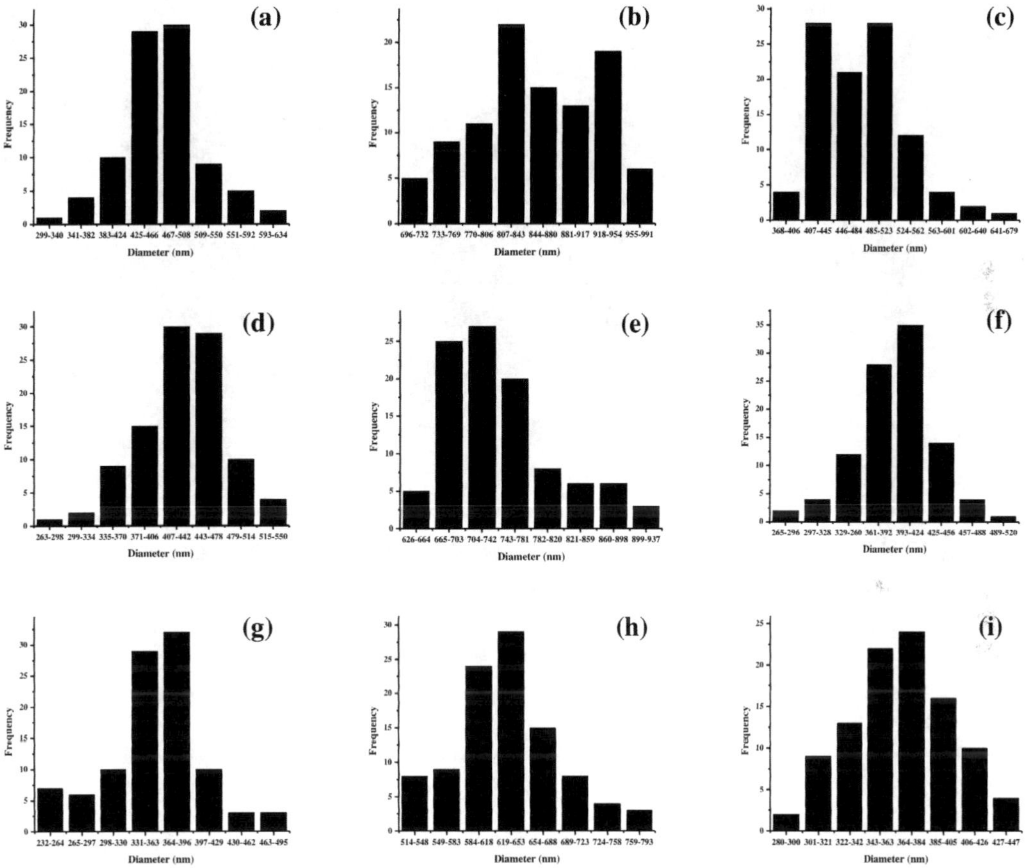

Figure 1. Distribution of diameter electrospun of (a) PAN, (b) PAN-CuAc, and (c) PAN-c-CuAc, stabilized electrospun of (d) CNF, (e) C-copper oxide NF, and (f) C-c-copper oxide NF, and carbonized electrospun of (g) CNF, (h) C-copper oxide NF, and (i) C-c-copper oxide NF.

Nanofibers colors also changed from white of PAN, the pale green of PAN-CuAc and white slightly covered blue of PAN-c-CuAc turned into brown after stabilization and black after carbonization. When the nanofibers were bent, stabilized and carbonized electrospun C-c-copper oxide NF was easily brittle compared to CNF and C-copper oxide NF.

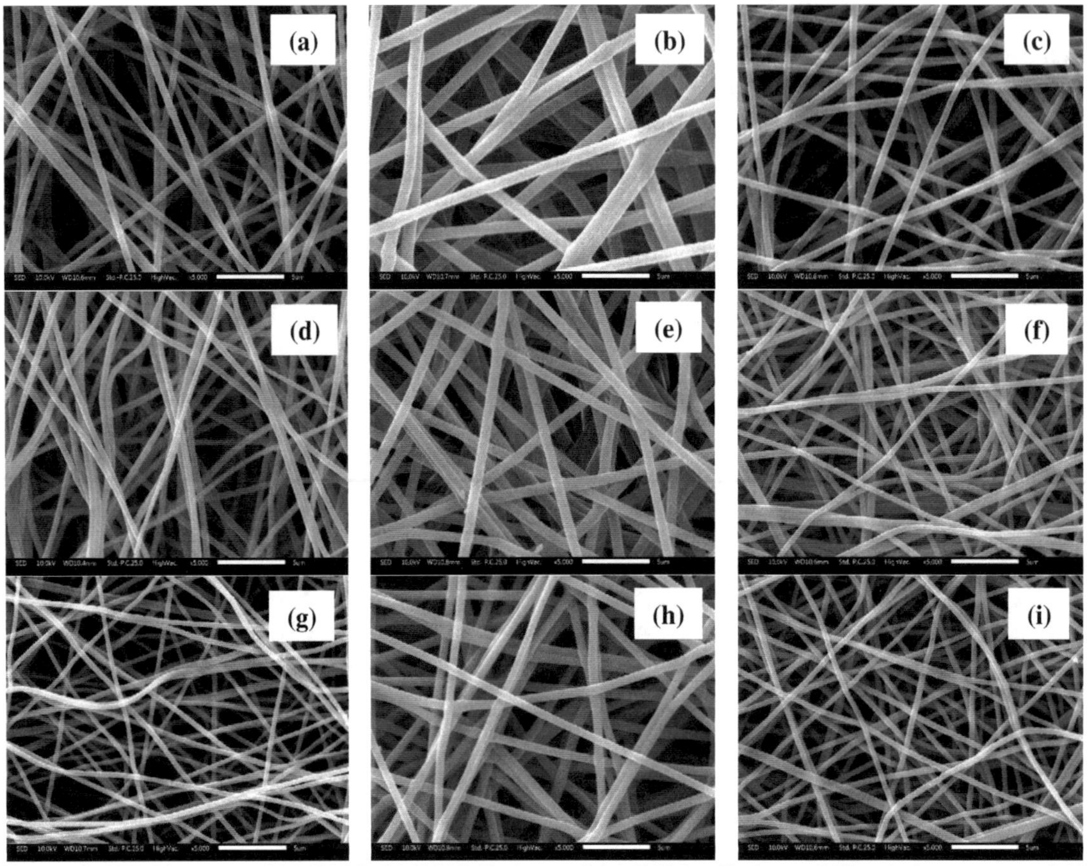

Figure 2. SEM images of (a) PAN, (b) PAN-CuAc, and (c) PAN-c-CuAc, stabilized electrospun of (d) CNF, (e) C-copper oxide NF, and (f) C-c-copper oxide NF, and carbonized electrospun of (g) CNF, (h) C-copper oxide NF, and (i) C-c-copper oxide NF.

3.2 IR spectra analysis

The IR spectra are shown in Figure 3-4. The characteristic peaks of electrospun PAN nanofibers were observed at around 2938 cm^{-1} attributed to CH_2 stretching, 2242 cm^{-1} of C≡N stretching and 1452 cm^{-1} of C-H bending in CH_2. In pristine PAN spectra, a strong band at 1693.19 cm^{-1} resulted from C=O group and medium intensities at 1282.42, 1110.80, and 1002.80 cm^{-1} attributed to C-N stretching from NMP as residual solvent after spinning. Besides, residual NMP also appeared at 1671.98 and 983.52 cm^{-1}, respectively for C=O stretching and C-N stretching in PAN-CuAc spectrum. The existence of CuAc in composite nanofibers was indicated by a medium intensity at 1070.30 cm^{-1} for C-O stretching of acetyl function groups. Differences between PAN-c-CuAc spectrum and others were a weak intensity at 1689.34 cm^{-1} and a medium intensity at 1558.20 cm^{-1} which also attributed to C=O and C-N of NMP. It probably was affected by the coating method that decreased the NMP content in electrospun PAN nanofibers structure.

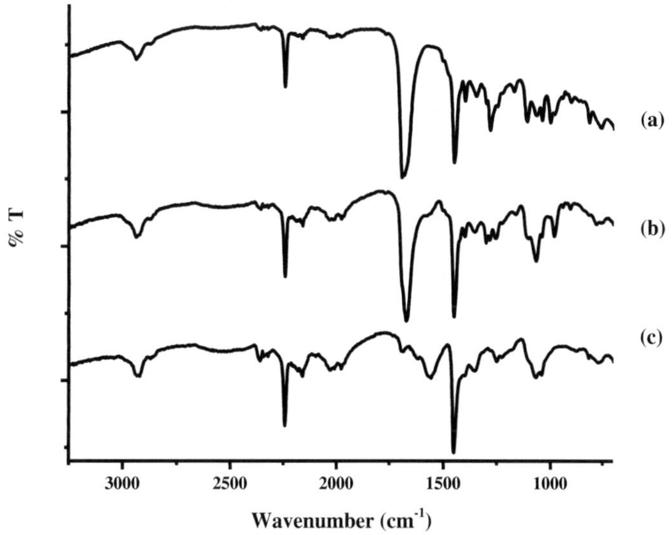

Figure 3. The IR spectra of electrospun of (a) PAN, (b) PAN-CuAc and (c) PAN-c-CuAc.

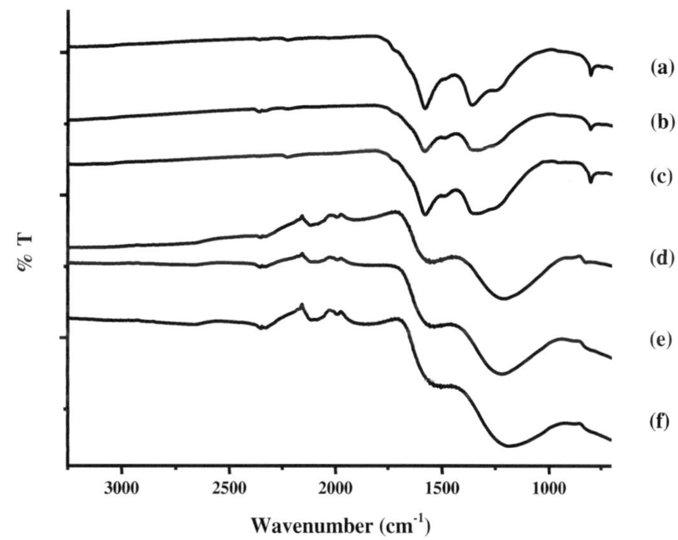

Figure 4. The IR spectra of stabilized electrospun of (a) CNF, (b) C-copper oxide NF, and (c) C-c-copper oxide NF, and carbonized electrospun of (d) CNF, (e) C-copper oxide NF, and (f) C-c-copper oxide NF.

On the stabilization of electrospun nanofibers at 290°C, the intensity of C≡N stretching of PAN at 2242.81 cm^{-1} decreased and shifted to 2229 cm^{-1}. In this process, a little nitrile groups change into amide groups [8] and initiated dehydrogenation of backbone and cyclization of nitrile groups which was proven by a strong peak at 1581 cm^{-1} for N-H bending in amide band II, C=N stretching and C=C stretching. As the result, a peak at 1452 cm^{-1} for C-H bending in CH_2 of PAN spectrum widened and shifted to 1361 cm^{-1} for C-H bending with a shoulder at 1253 cm^{-1} for mixed C-N stretching and C-C stretching. A strong and broad band at around 1300 cm^{-1} assigned to C-H in aromatic compound structure. A medium intensity band also appeared at 806 cm^{-1} attributed to C=C-H [8]. Carbonized CNF which was heated at 700°C had not shown characteristic peaks of PAN but appeared a peak at 1552 cm^{-1} attributed to C-C stretching and 1211 cm^{-1} attributed to C-H stretching of aromatic structure. The presence of copper oxide in PAN composition nanofibers was not detected in their IR spectra.

3.3 X-ray diffraction analysis

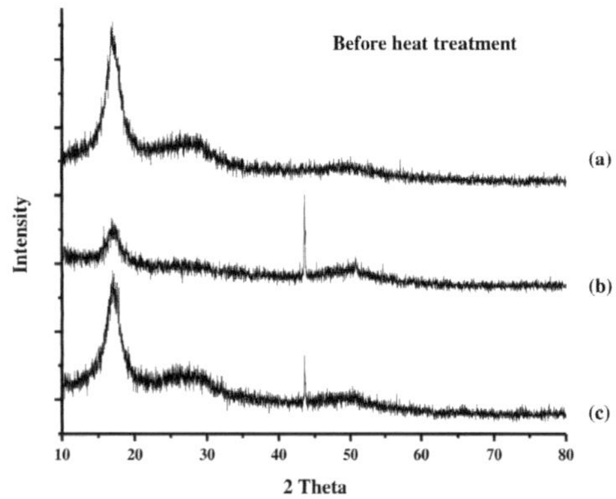

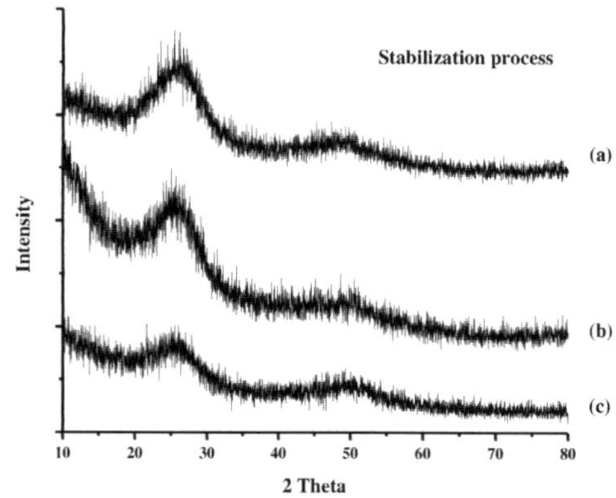

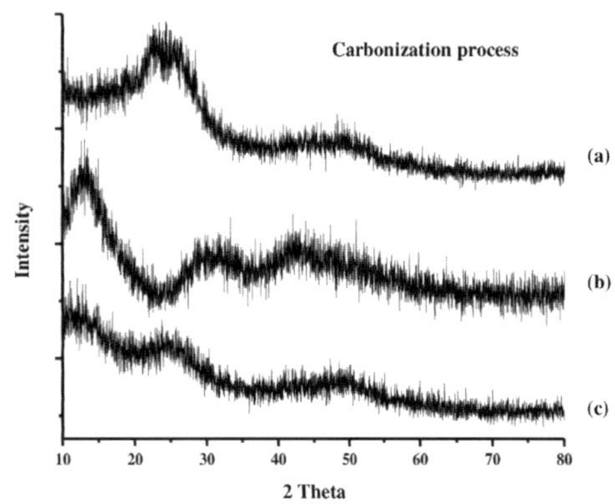

Figure 5. XRD pattern of electrospun (a) PAN, (b) PAN-CuAc and (c) PAN-c-CuAc nanofibers before and after heat treatment.

XRD patterns confirmed the structural properties of electrospun nanofibers before and after heat-treatment. As shown in Figure 5, electrospun PAN nanofibers have some typical peaks, a strong peak at 2θ=16.96° as well as two low-intensity peaks at 2θ= 27.71° and 49.90°. The existence of Cu in composite nanofibers was appeared at 2θ=43.38° and 43.58° corresponding to the (111) for FCC lattice of metallic Cu [2, 4, 9] beside of typical peak of PAN at 2θ=16.95°, weak intensity at 26.92° and 50.68° and also 2θ= 16.96°, 27.56° and 50.25°, respectively for PAN-CuAc and PAN-c-CuAc.

In the stabilization process, the linear structure of PAN is converted to a cyclic structure [3, 5] thus induced 2θ= ~16.9° was lost and ~25° attributed to graphite-like structure broaden followed by increasing the intensity. Subsequently, dehydrogenation and denitrogenation have been started during carbonization process. In 700°C and inert atmosphere (N_2) condition, CNF spectrum showed a broad and high intensity at 2θ=24.31° indicated graphite-like structure. However, C-copper oxide NF and C-c-copper oxide NF patterns showed distinct typical intensities. Three peaks at 2θ=13.05°, 31.64° and 42.05° of C-copper oxide NF strongly attributed to graphene-like structure [10] and partial structure of graphene and graphite observed in C-c-copper oxide NF. A weak intensity at 2θ=13.11° assigned to graphene-like structure and an intensity at 2θ=25.65° for graphite-like structure. A typical peak of PAN at 2θ=~50° shifted to ~48° occurred to stabilized and carbonized nanofibers. Here, no peak identified of copper was probably due to the low concentration of metal salt mixed in composition [2, 3].

3.4 Raman spectroscopy analysis

Figure 6 showed a typical Raman spectrum (800-2000 cm^{-1}) of carbonized electrospun nanofibers. The two prominent vibrational peaks at 1355 and 1578 cm^{-1} for CNF, 1346 and 1573 cm^{-1} for C-copper oxide NF, and 1392 and 1588 cm^{-1} for C-c-copper oxide NF attributed to D- and G- band vibrations, indicative of disordered and ordered graphitic structures, respectively [1, 7, 11-12]. Carbonized electrospun CNF and C-copper oxide NF had a more intense D band than the G band, which is related to the increase in the amount of the disordered phase. In contrast to carbonized electrospun C-c-copper oxide NF which had a lower D peak than the G band confirms that the material contains fewer defects [13]. That was also supported by the I_D/I_G value. The ratio of the intensity of the D to G bands (I_D/I_G), *R*, indicates the degree of the order in the graphitic material of the carbonized electrospun nanofibers [11, 12]. Those bands were fitted using a Gauss. The I_D/I_G values for the carbonized electrospun of CNF, C-copper oxide NF and C-c-copper oxide NF were calculated 1.59, 1.92 and 0.60, respectively. The higher I_D/I_G value of carbonized electrospun revealed more defects and lower I_D/I_G value for fewer defects [13].

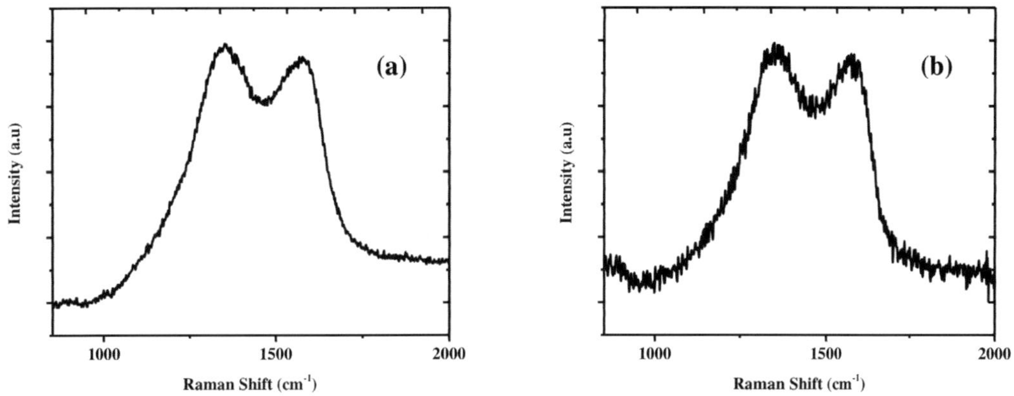

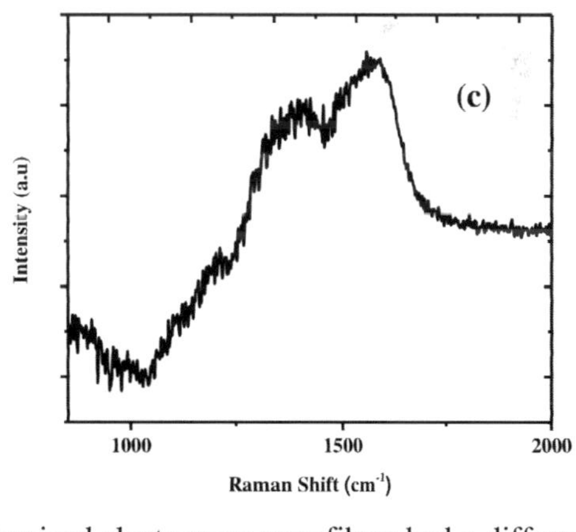

Figure 6. Raman spectrum of carbonized electrospun of (a) CNF, (b) C-copper oxide NF and (c) C-c-copper oxide NF.

The intensity ratio showed that the microstructure of carbonized electrospun nanofibers had a different effect when composite electrospun metal-nanofibers were synthesized using distinct methods.

3.5 Elemental analysis

Elemental content analyzed by EDX was showed in Table 2. Pristine PAN nanofibers have three main elementals which were detected namely C, N, and O. The existence of Oxygen content in the sample strongly caused by NMP as residual solvent after spinning. The detected Oxygen content of stabilized nanofibers was the highest that was led by the initiation of dehydrogenation and cyclization during the stabilization process [5]. In contrast to carbonized nanofibers, C content was higher than O due to aromatic structure had been formed with a little amount of N and O sourced from PAN structure and metal oxide. Cu content has been successfully detected in stabilized and carbonized C-copper oxide NF and C-c-copper oxide NF.

Table 2. Elemental content by EDX analysis.

Sample(s)	C	N	O	Cu
PAN	74.25	20.75	5.00	-
PAN-CuAc	71.43	20.37	5.10	3.10
PAN-c-Cu-Ac	70.77	20.77	3.93	4.53
Stabilized CNF	65.70	20.07	14.23	-
Stabilized C-copper oxide NF	64.79	19.35	11.88	3.98
Stabilized C-c-copper oxide NF	63.01	20.96	12.12	3.90
Carbonized CNF	84.25	10.69	5.06	-
Carbonized C-copper oxide NF	77.08	12.59	3.72	6.62
Carbonized C-c-copper oxide NF	73.04	13.85	3.76	9.35

4. Conclusion

Carbon-copper oxide nanofibers have been successfully synthesized by carbonizing of electrospun polyacrylonitrile-copper salt nanofibers. Polyacrylonitrile-copper salt nanofibers were prepared by two methods. Firstly, the electrospinning process used copper acetate/PAN solution. Secondly, immersing of PAN nanofibers in copper acetate solution. The heat treatment process has affected the reduction in the dimensions of the electrospun nanofibers sheet size and the average diameter of the nanofibers. Electrospun carbon-copper oxide nanofibers that were obtained by immersing method (C-c-copper oxide NF) was more brittle than CNF and C-copper oxide NF. IR spectrum showed the aromatic structure of carbon nanofibers formation and XRD pattern analyzed the graphite-like structure of all stabilized carbon nanofibers and partially structure of graphene-like for C-c-copper oxide NF. No peak identified of copper was probably due to the low concentration of metal salt in composition. Nevertheless, EDX analysis supported the presence of C, N, O and Cu elementals for carbon-copper oxide nanofibers. The intensity ratio (I_D/I_G) of the Raman spectra showed differences in the process of making polyacrylonitrile-copper salt nanofibers precursors affect the micro structure of carbon-metal nanofiber.

Acknowledgment

This research was supported by Indonesian Institute of Sciences (LIPI) for the laboratory facilities and Ministry of Research, Technology and Higher Education of the Republic of Indonesia for financial support in Program of National Innovation System Research Incentive (Insentif Riset Sistem Inovasi Nasional/INSINAS) 2019.

References

[1] C. A. Bonino, L. Ji, Z. Lin, O. Toprakci, X. Zhang, and S. A. Khan, "Electrospun carbon-tin oxide composite nanofibers for use as lithium ion battery anodes," *Applied Materials & Interfaces,* vol. 3, pp. 2534-2542, May. 2011

[2] J. Fu, H. Qiao, D. Li, L. Luo, K. Chen, and Q. Wei, "Laccase biosensor based on electrospun copper/carbon composite nanofibers for catechol detection," *Sensors,* vol. 14, pp. 3543-3556, Feb. 2014

[3] S. Kim, B. Bajaj, C. K. Byun, S. Kwon, H. Joh, K. B. Yi, and S. Lee, "Preparation of flexible zinc oxide/carbon nanofiber webs for mid-temperature desulfurization," *Applied Surface Science,* vol. 230, pp. 218-224, Sep. 2014

[4] N. A. M. Barakat, B. Kim, S. J. Park, Y. Jo, M. Jung, and H. Y. Kim, "Cobalt nanofibers encapsulated in a graphite shell by an electrospinning process," *Journal of Materials Chemistry,* vol. 19, pp. 7371-7378, Aug. 2009

[5] X. Huang, "Fabrication and properties of carbon fibers," *Materials,* vol. 2, pp. 2369-2403, Dec. 2009

[6] S. Kim, Y. Kuk, F. Jin, and S. Park, "Synthesis of polyacrylonitrile as precursor for high-performance ultrafine fibrids," *Bulletin Korean Chemical Society,* vol. 35, pp. 407-414, Nov. 13

[7] Y. Zhou, J. He, H. Wang, K. Qi, B. Ding, S. Cui, "Carbon nanofibers yarns fabricated from co-electrospun nanofibers," *Materials and Design,* vol. 95, pp. 591-598, Jan. 2016

[8] H. Kakida, and K. Tashiro, "Mechanism and kinetics of stabilization reaction of polyacrylonitrile and related copolymers II. Relationships between isothermal DSC thermograms and FT-IR spectral changes of polyacrylonitrile in comparison with the case of acrylonitrile/methacrylic acid copolymer," *Polymer Journal,* vol. 29, pp. 353-347, Oct. 1996

[9] A. Khan, A. Rashid, R. Younas, and R. Chong, "A chemical reduction approach to the synthesis of copper nanoparticles," International Nano Letters, vol. 6, pp. 21-26, Mar. 2016

[10] H. N. Lim, N. M. Huang, S. S. Lim, I. Harrison, and C. H. Chia, "Fabrication and characterization of graphene hydrogel via hydrothermal approach as a scaffold for preliminary study of cell growth," International Journal of Nanomedicine, vol. 6, pp. 1817-1823, Aug. 2011

[11] E. Zussman, X. Chen, W. Ding, L. Calabri, D. A. Dikin, J. P. Quintana, R. S. Ruoff, "Mechanical and structural characterization of electrospun PAN-derived carbon nanofibers," *Carbon,* vol. 43, pp. 2175-2185, May. 2005

[12] S. Lee, J. Kim, B. C. Ku, J. Kim, H. I. Joh, "Structural evolution of polyacrylonitrile fibers in stabilization and carbonization," *Advances in Chemical Engineering and Science,* vol. 2, pp. 275-282, Apr. 2012

[13] R. Muzyka, S. Drewniak, T. Pustelny, M. Chrubasik, G. Gryglewicz, "Characterization of graphite oxide and reduced graphene oxide obtained from different graphite precursors and oxidized by different methods using raman spectroscopy," *Materials,* vol. 11, pp. 1050, Jun. 2018

IOP Conf. Series: Materials Science and Engineering **924** (2020) 012020 doi:10.1088/1757-899X/924/1/012020

Characteristics of Magnesium/B_4C Reinforced Composite Fabricated by Stir Casting Method

Sutan Paulo Hang Jutanaiman and Anne Zulfia Syahrial*
Department of Metallurgy and Materials Engineering, University of Indonesia, Depok, 16436, Indonesia

*anne@metal.ui.ac.id

Abstract. At present the development of magnesium composites is increasingly being carried out, this is due to the main advantages of magnesium which has the lowest density when compared with other metals so that it can save fuel use when applied as a constituent of an automotive product yet still has great mechanical properties. In this research, magnesium composite is fabricated by mixing magnesium as composite matrix and micro B_4C as reinforcement particle with 2, 4, 6, and 8 Vf-% by stir casting method. The magnesium reinforced B_4C particles has improved the mechanical properties of the composite. The best mechanical properties of magnesium composite are shown by the addition of 8 Vf-% B_4C particles. These values increase up to 73 HRH for the hardness and 64 MPa for the UTS as well as, 0.11 J / mm^2 for the impact value. The wear rate also reduced to 0.0023 mm^3/m. The mechanical properties of magnesium composite are increased due to the increasing number of B_4C particles thatcaused new interfaces which will block dislocation movement and dispersed B_4C particles will also act as load restraint elements. Composite characterization was also carried out by using SEM-EDS and XRD test to find out the compounds and phases formed from the fabricated composites - then it was known that the possible compounds formed from the composites produced were MgB_2, MgO, B_4C, and $Mg_2B_2O_5$ respectively.

Keywords: Magnesium composite, B_4C particles, stir casting, mechanical properties

1. Introduction

At this time car is the main transportation mode that is often relied on by humans in meeting various needs every day. According to data from the central statistics agency, the number of four-wheeled vehicles that have been circulating throughout the Indonesian territory has reached up to more than 20 million[1]. However, the increasing number of circulation of four-wheeled vehicles is not supported by the availability of fuel oil which continues to soar every year. One solution to improve the efficiency of using fuel oil in a vehicle is by using lightweight metal as the material of the vehicle.

The use of lightweight metal as the basic material in vehicles can reduce the fuel oil needs of vehicles because with the mass of vehicles getting lighter, the vehicle engine only requires less fuel to drive the vehicle [2]. In this case, Magnesium is one of the most promising metals. Magnesium is one of the lightest metals on the face of the earth, with a specific gravity of only 1.74 g / cm3, where this density is 36% lighter than aluminum, 74% lighter than zinc, and 79% lighter than steel. However, there are limitations in the use of pure magnesium as an application of automotive parts because of the low mechanical properties of magnesium, so it is necessary to mix magnesium with other materials (ceramics) in order to form magnesium composites which have far better mechanical properties than pure magnesium.

The Composite material is a combination of two or more different components, this is done in order to obtain certain physical and mechanical properties that are better than the properties of each of the constituent components. In this study, the composite matrix used was pure magnesium metal while the amplifier used as boron carbide (B_4C) ceramic particles with micron-size 2, 4, 6, and 8 Vf-%. With the addition of macro-sized B_4C ceramic particles, magnesium composites that have low density, high rigidity, high hardness and high wear resistance [3] will be obtained. The fabrication method used is stir casting, this is done because the method is relatively simple and only requires a relatively low cost in the process.

2. Materials and Methods

2.1 Materials

In this research, the composite was fabricated by using magnesium as the matrix with addition of micro-sized B_4C particle with variation of 2, 4, 6 and 8 Vf-% supllied by Shanghai Xinglu Chemical Co., Ltd, China.

2.2 Experimental Methods

Magnesium ingot was sawn into small block form using bend saw machine, after that small block of magnesium was cleaned with soap and alcohol in order to remove the inclusion particle and oil that might be left on the surface. Before magnesium inserted into the furnace, crucible and mold were coated by zircon and thinner to prevent the chemical reactions that may happen during magnesium were melted. After that, before turn on the furnace, a vacuum pump was used to take out any gasses in the furnace and then furnace has flowed with argon gas until all of the magnesium fully melted. While waiting for magnesium to melt completely, micro-sized particle B_4C was pre-heated at 950°C in the muffle furnace to remove the moist of that reinforce. After the pre-heat process of reinforcing done, reinforce was inserted to molten magnesium and then stirred for 2 minutes with a metallic stirrer to get a homogenous mixture. Then after the stirring process complete, the molten composite was poured to mold and solidified completely in 3 minutes. Finally, a composite sample was ready to be used for further mechanic and chemical characterization.

2.3 Characterization

2.3.1 Chemical Characterization. Chemical characterization was carried out by using Optical Emission Spectroscopy (OES) to find out what chemical elements are present in magnesium composite with the addition of 2, 4, 6 & 8 Vf-% micro B_4C samples.

2.3.2 Mechanical Properties Characterization. Mechanical properties characterization was carried out by using tensile, impact, hardness, wear, and porosity & density testing. GoTech machine was used in tensile testing with JIS Z2241 standard, impact test was tested using Charphy method, hardness test was tested using Rockwell machine with HRH standard, pin-on-disk Ogoshi machine was used to wear test with 100m sliding distance, and density test was carried out using DahoMeter machine based on Archimedes Theory.

2.3.3 Microstructural Characterization. Microstructure characterization was carried out on each variable, the first step before testing was to mount all the sample variables and then grind it using abrasive paper with an elevated number of grit, then polishing using selvyt-cloth and kovac liquid. After that, an observation is made on all variables to view each microstructure that obtains on each variable. SEM-EDX testing was used too on one sample which had optimum mechanical properties value to know what elements are contained which can then be analyzed for any possible phase in this sample. XRD testing also carried out to prove the formed phase based on the results of the SEM-EDX test

3. Result and Discussion

3.1 OES Observation

The results of testing the chemical composition of pure magnesium and pure magnesium composite reinforced B_4C can be seen in Table 1.

Table 1. Comparison of OES Results for Magnesium Composite Materials.

Element	Percentage (Vf-%)				
(%)	*As-cast*	2 Vf-% B_4C	4 Vf-% B_4C	6 Vf-% B_4C	8 Vf-% B_4C
Mg	99,8	99,2	96,3	99,3	99,0
B	-	0,0026	0,0016	0,0017	0,0019
Al	< 0,005	0,527	0,599	0,570	0,574
Be	0,0001	< 0,0005	0,00007	0,00009	0,00052
Cu	< 0,001	< 0,00010	< 0,00010	< 0,00010	< 0,00010
Mn	0,0466	0,0022	0,0031	0,0031	0,0032
Zn	0,0377	0,0061	0,0059	0,0024	0,294
Ag	< 0,001	0,00011	0,00011	< 0,0010	< 0,00010
Ca	0,0023	0,00082	0,00056	0,00064	0,00051
Cd	< 0,001	< 0,00010	< 0,00010	< 0,00010	< 0,00010
Sn	0,0421	0,0055	0,0080	0,0113	0,0118
Sr	0,0007	0,00017	0,00012	< 0,00010	< 0,00010

Based on the data shown in Table 1, the percentage of magnesium in each variable is above 99%, ideally, the percentage of magnesium contained in each composite variable is at the number below 98%, the high percentage of magnesium content in each variable is due to a large amount of leftover magnesium remaining in the crucible which is used to melt magnesium where the remaining magnesium in the crucible wall dissolves into the molten metal liquid. The presence of magnesium remnants in the crucible is due to the oxidation of the molten metal from the previous melting, the oxidation of the molten metal then forms hard Magnesium Oxide (MgO) rocks [4] and is difficult to remove from the crucible.

3.2 Microstructure

Based on microstructural observations shown in Figure 1. Microstructure observations on reinforcement compositions of 2%, the presence of brownish-white base identified as pure magnesium phase and also blue-blackish grains scattered in the magnesium matrix, where the grains the blue-black colored grain is then identified as the B_4C phase, there are blue-black grains that clump in one area, clumping of the grains in one area is then called the agglomeration phenomenon[5].Then in the microstructure observation of the second composition, the number of blackish areas was identified as porosity. Furthermore, on the observation of the third composition, the B_4C particle agglomeration phenomenon was found to be greater than the other composition microstructure. Then, on the fourth composition observation, it was seen that B_4C particles were dispersed evenly throughout all parts of the pure magnesium matrix, and there was no visible porosity and agglomeration in the microstructure.

IOP Conf. Series: Materials Science and Engineering **924** (2020) 012020 doi:10.1088/1757-899X/924/1/012020

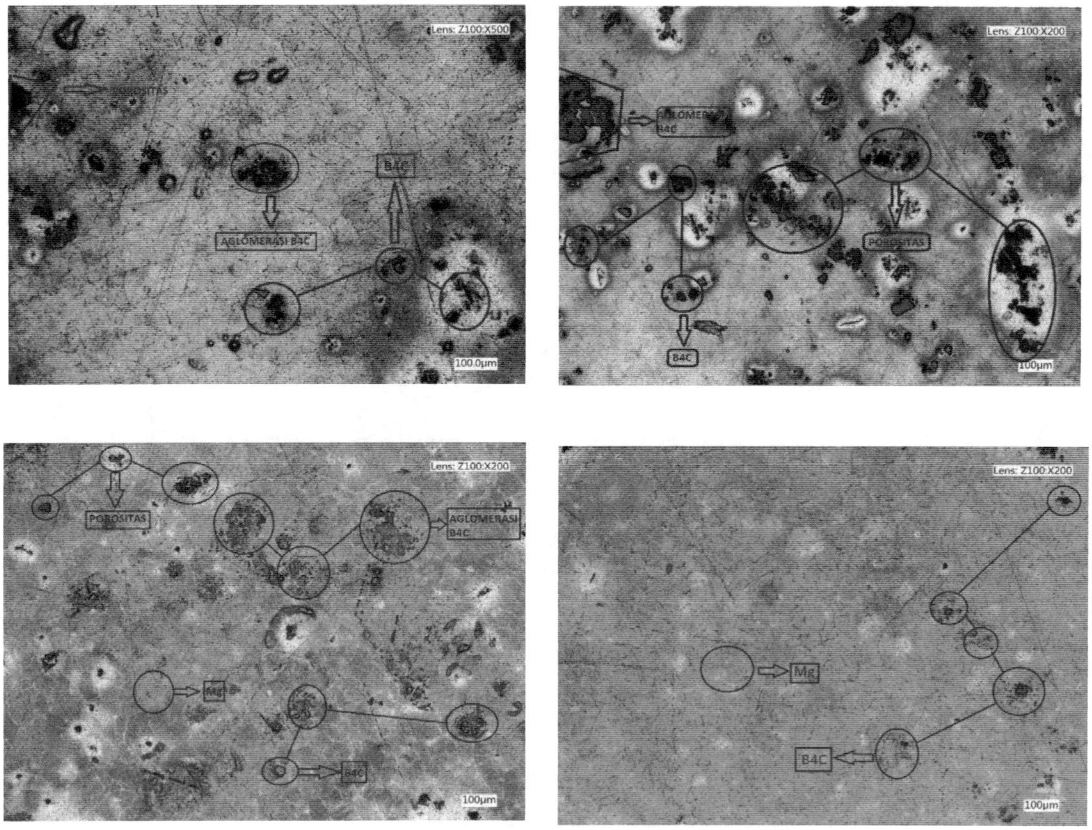

Figure 1. Microstructure of magnesium composite with addition of (a) 2 Vf-%, (b) 4 Vf-%, (c) 6 Vf-%, (d) 8 Vf-% B_4C.

3.3 XRD Observation

Observation of X-Ray Diffraction (XRD) on magnesium-reinforced B_4C particle composites was carried out on samples with a particle composition of 8 Vf-%B_4C, where this was done based on the results of testing mechanical properties, 8 Vf-% B_4C/magnesium composite had better mechanical properties than all other compositions.

In the graph of the obtained XRD testing in Figure 2, magnesium phases are seen at all peak points, this indicates that magnesium as the matrix is the most dominant phase in the composite sample tested. In addition, the other highest phase is MgO with a score of 16 and MgB_2O_5 with a score of 2. The number of MgO and $Mg_2B_2O_5$ phases formed is due to the fact that magnesium elements are very reactive and easily react with Oxygen to form MgO[6] as a compound that is very much found and is at each peak of the test results. In addition, there are also B_4C and Boron phases at the peak points of the test results, although in small amounts, this indicates that the distribution of reinforcing particles is quite good and evenly distributed on all parts of the pure magnesium matrix in pure magnesium composite B_4C particle reinforced.

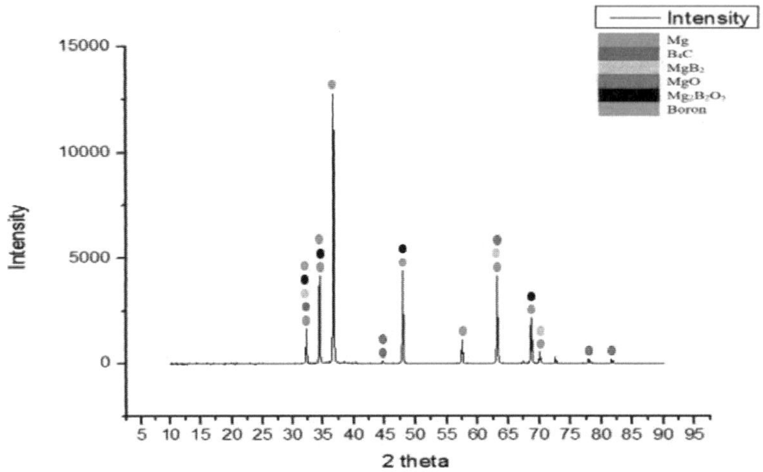

Figure 2. The XRD result of Mg/Micro B_4C8Vf-% Composite.

3.4 SEM-EDS Test

SEM-EDS testing is carried out on Magnesium composites with the addition of 8 Vf-% B_4C particles which have the best mechanical properties compared to each other variable.

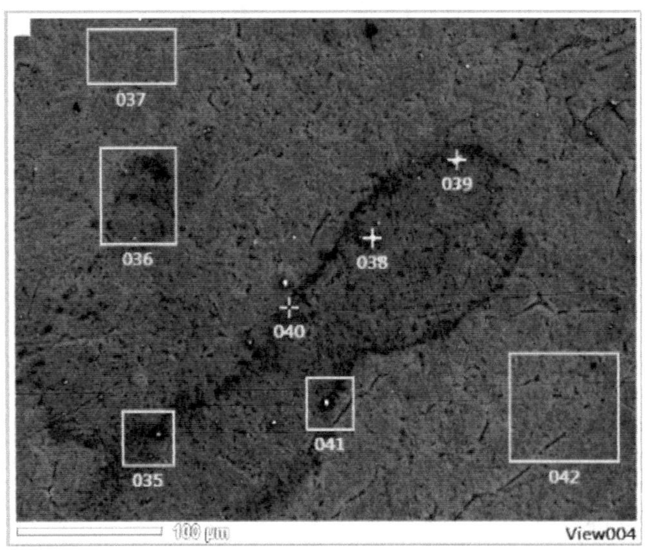

Figure 3. SEM Results of Mg/Micro B_4C8Vf-% Composite.

Based on the test results represented in Table 2, it can be seen that the elements contained are Mg, B, C, and O. Then based on the elements contained, it is predicted that the phases formed are MgO, MgB_2O_5, and MgB_2. Points 35, 36, 37 and 42 which are gray areas surrounded by black dots represent that the gray areas and black accents contain Mg and MgO phase. The formation of the MgO phase is caused by the element Mg which is very reactive with the surrounding environment, which in the surrounding environment contains oxide gases which then bind to molten metal molten magnesium and form a compound of Magnesium Oxide (MgO) [6].

Table 2. EDS Results of Mg/Micro B_4C8Vf-% Composite.

Point	Element (%)				Possible Formed Phase
	Mg	B	O	C	
35	94,71	2,72	2,46	0,12	Mg, MgO
36	90,53	6,76	2,23	0,47	Mg, MgO
37	91,35	5,80	2,36	0,49	Mg, MgO
38	79,19	12,78	6,44	1,58	MgO, B_4C
39	59,72	27,94	4,37	7,97	B_4C, $Mg_2B_2O_5$
40	78,28	17,04	1,74	2,94	MgO, MgB_2, MgB_2O_5
41	88,67	7,71	2,70	0,92	MgO, B_4C
42	90,49	6,30	2,47	0,74	Mg, MgO

Then at points 38, 39, 40 and, 41 there are bright white dots which indicate the presence of boron with a high enough percentage, the high level of boron added with the presence of Mg, C and O elements is predicted to form compounds B_4C, MgB_2O_5 and MgB_2 [7]. The formation of the MgB_2 and MgO phases is due to the presence of Mg and O bonds which then react with the B_4C amplifier [8].

3.5 Tensile and Elongation

Based on the results of the tests that have been done, the maximum tensile strength values are obtained in Figure 4(a).

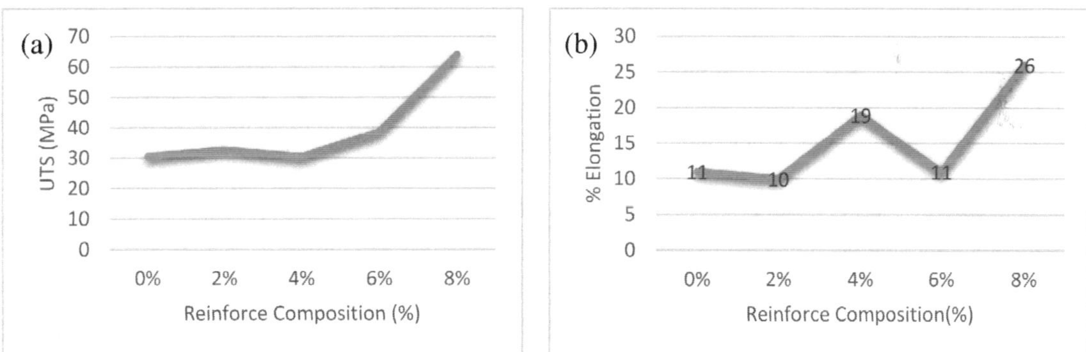

Figure 4. (a) Tensile, and (b) Elongation.

In general, there is a significant increase in maximum tensile strength along with the increasing number of B_4C reinforcing particles into the pure magnesium matrix. B_4C particles that are evenly dispersed in a pure magnesium matrix will increase the dislocation density, wherewith increasing dislocation density in the sample, greater energy is needed to deform the material the newly formed grains will block the movement of dislocations during deformation [9]. However, in a B_4C-reinforced pure magnesium composite sample of 4 Vf-%, the maximum tensile strength was decreased. This is caused by the high porosity percentage value in the composite sample. The highest elongation rate in this variable occurs because of good grain smoothing before it reaches the critical point of a material. This fine grain smoothing provides a more homogeneous strain distribution and decreases the stress concentration in the material so that the tenacity of the material becomes higher.

3.5.1 Hardness, Wear rate and Impact. Based on the graph shown in Figure 5(a), it can be seen that there is a consistent increase in hardness with the increasing number of added particles B4C. Increased hardness of the B4C reinforced composite is caused by dispersed B4C particle will make more and more interfaces formed between the pure magnesium matrix and B4C reinforce [10] then with the formation of a strong interface, the dislocation movement will be blocked so that the material becomes stronger [11].

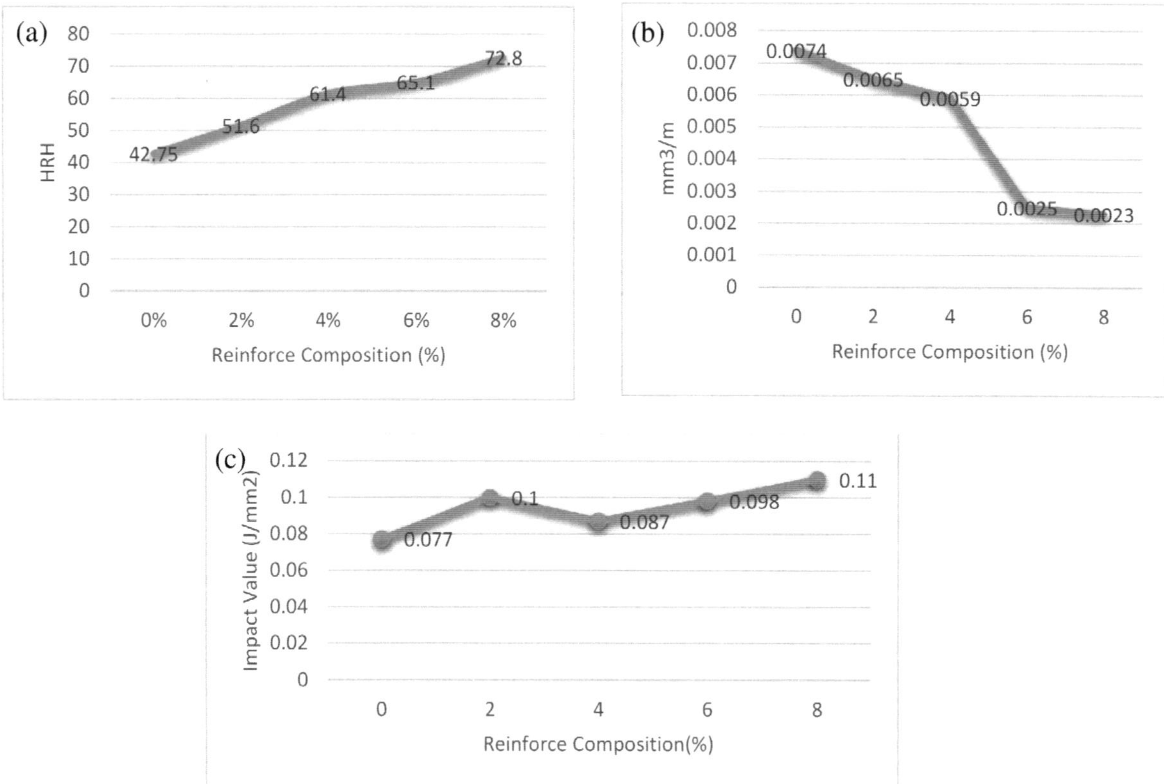

Figure 5. (a) Hardness, (b) Wear rate, and (c) Impact value of Mg/B_4C Composite.

There is an increase in wear resistance as the number of boron carbide particles is added, this is due to an increase in the hardness and strength of the composite as more reinforcement is added, this also corresponds to the Archard equation [12] where the wear rate of a material will be inversely proportional to the hardness of the material.

Based on the impact value graph in Figure 5(c), the addition of the B_4C reinforce tends to increase the impact value of the magnesium/B_4C composite. This is because the dispersed B_4C particles will act as a load restraining element which is the process of transferring the load first through the B_4C reinforce which has been dispersed in a pure magnesium matrix so that a larger load is needed to deform the material, hence the toughness of the composite will increase [7]. However, in the composition of 4% the value obtained tends to decrease this is due to the percentage of porosity in the composition is very high.

3.5.2. Density and Porosity. Based on Figure 6 (a), the actual density value obtained in all B_4C magnesium-reinforced composite compositions has a smaller value when compared to the theoretical density value, the smaller actual density value indicates the presence of porosity in the fabricated sample.

The larger difference between the theoretical density value and the actual density value, hence there is more porosity value obtained from the composite sample.

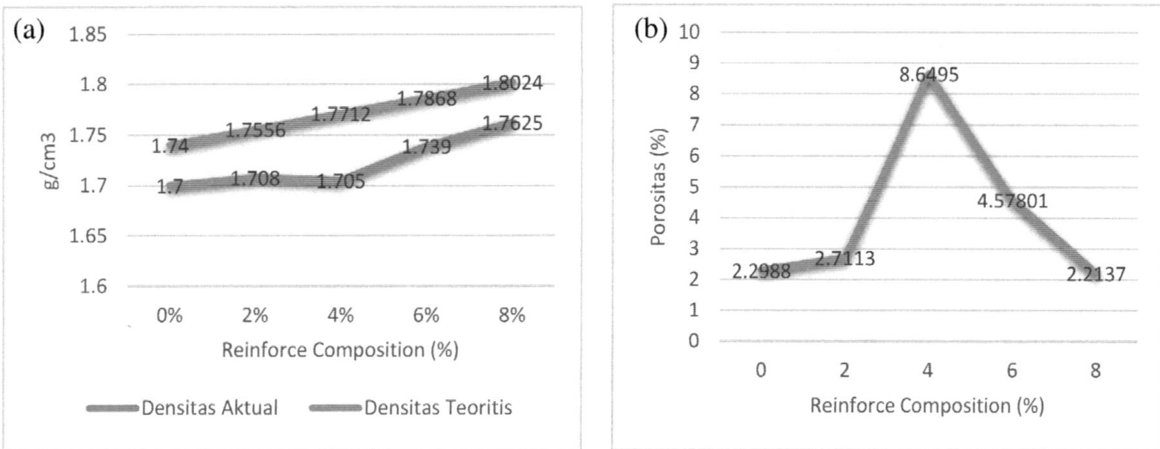

Figure 6. (a) Density and (b) Porosity of Mg/B_4C Composite.

Based on the graph in Figure 6 (b). It can be seen that the highest percentage porosity value occurs in samples with the addition of 4% B_4C reinforce. In this composition the porosity percentage value formed is 8.6495%, the highest porosity percentage value in the composite sample occurs due to the length of the stirring is longer than any other fabrication process so with this longer stirring process and dross extraction in the composite [13] sample resulted in contact or contamination of oxygen gas with molten magnesium metal that makes the porosity value is higher than any other composition [14].

4. Conclusion

Magnesium with the addition of B_4C reinforces particle composites has successfully fabricated using a stir casting method.The higher the difference between the actual density and the theoretical density in the tested sample, the higher the porosity percentage in the sample. Addition of reinforce particles B_4C to magnesium composites will increase composite hardness with a maximum value of the composition of 8 Vf-% B_4C which is equal to 72.8 HRH, Increase the composite impact value with the highest value in the composition of 8 Vf-% B_4C which is equal to 0.11 Joule / mm2, Reduce the composite wear rate with a minimum value of the composition of 8 Vf-% B_4C which is equal to 0.0023mm^3/m. The tensile test results show the optimal value of UTS and % elongation achieved in the composition of 8 Vf-% B_4C which is equal to 64.03 MPa and 26% elongation. XRD results of magnesium composite samples show there are phases of Mg, MgO, MgB_2, B_4C, and MgB_2O_5

Acknowledgments

This research supported by Directorate of Research and Community Services (DRPM), Universitas Indonesia for financial support through International Publications Index (PIT 9), 2019 with contract number: NKB-0042/UN2.R3.1/HKP.05.00/2019

References

[1]BadanPusatStatistik."PerkembanganJumlahKendaraanBermotorMenurutJenis," 1949-2016. Des 2017
[2]C. Blawert., N. Hort, K.U Kainer "Automotive applications of magnesium and its alloys," Center for Magnesium Technology, Institute for Materials Research, GKSS-Research Centre Geesthacht GmbH, Max-Planck-Str. 1, 21502Geesthacht, Germany. 2004

[3]S. Das, M. Chandrasekaran, and S. Samanta, "Science Direct Comparison of Mechanical Properties of AA6061 reinforced with (SiC / B 4 C) micro / nano-ceramic particle reinforcements," Mater. Today Proc., vol. 5, no. 9, pp. 18110–18119. 2018

[4] A. Dey, KM. Pandey. "Magnesium Metal Matrix Composites - A Review," Rev Adv Mater Sci. 2015;42(1):58-67. 2015

[5] A. Raturi, K. K. S. Mer, and P. Kumar Pant. "Synthesis and characterization of mechanical, tribological and micro-structural behavior of Al 7075 matrix reinforced with nano Al2O3 particles," Mater. Today Proc., vol. 4, no. 2, pp. 2645–2658. 2017

[6] Xi Y L, Chai DL, Zhang W X, and Zhou J E. "The interfacial reaction and microstructure in a ZK60-based hybrid composite,"Mater Lett 59 (2005) 1831. 2005

[7]H. Y. Wang, Q. C. Jiang, Y. Wang, B. X. Ma, and F. Zhao."Fabrication of TiB2participate reinforced magnesium matrix composites by powder metallurgy," Mater. Lett., vol. 58, no. 27–28, pp. 3509–3513. 2004

[8]Y.J. Liang, Y.C. Che."Data Handbook of Mineral Thermodynamics," edited by Northeastern University Press, (in China). 1993

[9]Materials H, Moradkhani A, Baharvandi H. International "Journal of Refractory Metals Mechanical properties and fracture behavior of B 4 Cnano / micro SiC composites," produced by pressureless sintering. Int J Refract Metals Hard Mater. 2018;70(September 2017):107-115. DOI:10.1016/j.ijrmhm.2017.10.001. 2017

[10] A. Mazaherydan M. O. Shabani. "Microstructural and abrasive wear properties of SiC reinforced aluminum-based composite produced by compocasting," Transactions of Nonferrous Metals Society of China, vol. 23, no. 7, pp. 1905-1914. 2013

[11] F. Aydin, Y. Sun, H. Ahlatci, and Y. Turen. "Investigation of Microstructure, Mechanical and Wear Behaviour of B 4 C Particulate Reinforced Magnesium Matrix Composites by Powder Metallurgy," Trans. Indian Inst. Met., vol. 71, no. 4, pp. 873–882. 2018

[12] Telliskivi, Tanel."Simulation of wear in a rolling-sliding contact by a semi-Winkler model and the Archard's wear law". Wear (0043-1648), 256 (7-8), p. 817. 2004

[13] R.P Cindy., Anne, Zulfia.S., Salahuddin, Y., Budi, W.Y. "Optimizing The Addition of TiB to Improve Mechanical Properties of the ADC 12/SiC Composite Through Stir Casting Process". E3S Web of Conferences 130, 01005. 2019

[14]J. Hashim, Looney, L., Hashmi, M.S.J. "Particle distribution in cast metal matrix composites—Part I,".Journal of Material Processing Technology.Volume 123 pp. 251-257. 2008

The Effect of Al_2O_3 Nano Particles on Characteristics of Mg-Al-5TiB Grain Refiner Composites Produced by Stir Casting Method

Alintiya Zhafara and Anne Zulfia S*
Department of Metallurgy and Materials Engineering, Faculty of Engineering, University of Indonesia, Depok, 16436, Indonesia

*anne@metal.ui.ac.id

Abstract. High needs for light material with good mechanical characteristics made magnesium and its alloys as an alternative choice because of magnesium's low density, 1.74 gr/cm^3. However, its use is limited by its low strength and ductility. In this experiment, a fabrication process of the composite which is Mg-Al-5TiB as Matrix with the addition of Nano Al_2O_3 as a reinforced with volume fraction varied from 0.10 Vf-% to 0.25 Vf-% by stir casting method. The composites then were characterized by both microstructure and mechanical properties. The microstructure was analyzed by an optical microscope and scanning electron microscope link to energy-dispersive spectra to analyze the phases present in composites and confirmed by X-Ray Diffraction. The XRD shows some phases such as Mg, MgO, $MgAl_2O_4$, and B_2O_3. The optimum mechanical properties of composites were obtained at the addition of 0.2 Vf-% Al_2O_3, where the tensile strength, elongation, wear rate and impact strength were 46.1 MPa, 14%, 0.3x10-^{5} mm^3/s, and 0.06 J/mm^2. The hardness of composites was value, 40.1 HRH. The porosity maximum was found at the addition of 0.25 Vf-% with the value of 9.37% caused by nano Al_2O_3 particle forming clustering (porosity is related to clustered particles) and inhibiting liquid metal moving rate. The addition of grain refiner makes solidification went faster and refined the grains.

***Keywords*:** Magnesium Composite, Mg-Al-5TiB, Al_2O_3 Nano Particles, Stir Casting Method

1. Introduction

Indonesia is an active country in developing various industrial fields, such as the automotive industry, arms industry, manufacture, and infrastructure industry, etc. These developments reflect the development progress of a country, including the automotive industry. Along with the growth of the needs of the automotive market in Indonesia, the demands for more effective and efficient fossil for fuels have also increased. By increasing fuel efficiency, the use of fuel for a vehicle can be saved and the pollution produced can be minimized. However, world oil conditions that are declining in capacity and its fluctuating prices and tight controls and efforts to reduce emissions in order to reduce environmental impacts are triggers the current industries to incessantly search for new materials as an alternative over conventional materials. Since the significant influence, the automotive industry gives to a country, one alternative is the choice of lightweight material as the main material. One of the light materials that can be used is Magnesium.

Magnesium is the lightest of all structural metals. Magnesium has a specific gravity of 1.74 g/cm^3

IOP Conf. Series: Materials Science and Engineering **924** (2020) 012021 doi:10.1088/1757-899X/924/1/012021

[1], which is approximately one quarter of the weight of steel and two-thirds of aluminium. Moreover, the availability of magnesium in nature is very high since magnesium is the most abundant element of the sixth order on earth, 2.7% of the earth's crust composition is magnesium [2].

It has an HCP (Hexagonal Close Packed) structure so that it brings Magnesium to have highest specific strength. Magnesium as based metal matrix composites (MMCs) has attracted considerable interest because of their unique mechanical properties over monolithic alloy. This is due to the fact that monolithic alloys possess low elastic modulus. This limitation can be circumvented by incorporation of harder and stiffer ceramic particulates and alloying elements in the matrix. However, the selection of the type, size and volume fraction of particulates are essential to give the optimum properties to these composite materials. One of the reinforcements that can be used is the nano-sized aluminium oxide ceramic particle (Al_2O_3), also the addition of Al-5Ti-B as a grain refiner. Ceramic reinforcement has isotropic properties, excellent mechanical properties, chemical inertness properties, is stable at high temperatures, and its expansion coefficient can be controlled [3], thereby ceramic particles are considered fair for metal matrix composite fabrication. Al-5Ti-B can improve the mechanical property, reduce porosity, increase resistant of hot cracking, and change the structure [4]. Thus, Mg-Al-5Ti-B is used as the main matrix then reinforced by nano-sized Al_2O_3 particle.

Stir casting is the fabrication method selected in this research because of its advantages which are economical, easy, flexible, and can be used for production in large quantities that is very suitable to be applied in the automotive industry. However, this method has some difficulties such as the highly chemical activity of Mg, hardly uniform distribution of particles caused by agglomeration of the nano- Al_2O_3 if it does not stir well, the precipitating or floating reinforcing particle (depending on the specific gravity of the reinforcement phase and the metal matrix), the breaking of the reinforcing particles during the agitation process, the moisture between the particles and the matrix, and the porosity in the composite matrix. Therefore, a composite material with magnesium base matrix is developed with variation addition of nano-Al_2O_3 particles as reinforcement particle using a stir casting method to investigate the influence and mechanism of the number and shape of nano-Al_2O_3 reinforcing particles in improving the mechanical properties of magnesium matrix composites.

2. Experimental Method

2.1 Materials

In this research, pure magnesium 90% wt and Al-5Ti-B 10% wt was used as the matrix and nano-sized Al_2O_3 particle with 0.10, 0.15, 0.20, and 0.25 Vf-% was used.

2.2 Sample Preparation

The magnesium ingot and Al-5Ti-B (matrix) were sawn into small block form using miniband saw. After sawn, magnesium and Al-5-Ti-B are weighted as the same as the mass balance. Magnesium was cleaned with soap and 96% alcohol in order to remove the inclusion particle and oil that might be left on the surface. Al_2O_3 (reinforce) was agglomerated by ultrasonic vibrator within 380 seconds in order to remove inclusions while the furnace and the mold were coated with mixing thinner and zircon so that the furnace and the molt have high thermal resistance to avoid cracking. After that, the magnesium and Al-5Ti-B were inserted into the furnace, before turning on the heater, they were fluxed in 6 minutes and drained with argon gas in 20 seconds. The matrix was melted at 650°C in the furnace and before being poured into the matrix, the reinforce were pre-heated at 900°C in the muffle furnace to remove the moist. After the preheat process, nano Al_2O_3 particle was inserted into molten Mg-Al-5Ti-B and then stirred for 1 minute to obtain a homogenous mixture. Then, they were poured into the mold and the mold was opened after the composite was solidified. Finally, this composite sample was being used for further characterization.

2.3 Characterization

2.3.1 Chemical Characterization. Chemical characterization was performed using Optical Emission

Spectroscopy (OES) to determine what elements were present in magnesium ingot samples and Mg-Al-5Ti-B composite with 0.10, 0.15, 0.20, and 0.25 Vf-% nano-Al_2O_3. Chemical characterization using OES is generated in elemental form and the amount of a particular area is large enough. Elements obtained are further analyzed in relation to variations in the addition of nano-Al_2O_3 and the process of the stirring casting method.

2.3.2 Mechanical Properties Characterization. Each sample was subjected to destructive testing (tensile, impact, hardness, and wear testing). Prior to the destructive test, each specimen was tested to find its actual density using a tool based on Archimedes Theory. The tensile test was conducted using the GoTech machine based on JIS Z2241 at room temperature. The impact test was carried out using the Charpy method at room temperature. The hardness test was completed using the Brinell test with HRH type. The wear test was executed using a pin-on-disk machine named Ogoshi by sliding a specimen into a harder material for a 100-meter sliding distance.

2.3.3 Microstructural Characterization. One sample which had optimum value was prepared by grinding using an abrasive paper with the machine automatically then polished with micron alumina paste on velvet cloth also automatically. After sample preparation was done, the microstructure characterization was conducted using optical microscope to observe microstructure and perform grain calculation, SEM with EDX integration to see topography of specimen in 3D, surface fracture of impact testing, element mapping, and analysis, XRD ad for composite specimen with value hard composite, composite reinforcement, and composite with the most reinforcing volume fraction underwent an XRD test to know the formed phase.

3. Result and Discussion

3.1 Chemical Composition

OES (Optical Emission Spectrometry) is a test to confirm the presence of elements in magnesium as-cast or non-reinforced magnesium as well as magnesium composite quantitatively. Thus, one will know the effect of the casting process on metal characterization.

Table 1. Chemical composition of Mg as-cast and Composite Material.

Element (%)	Material				
	Mg *as-cast*	**Mg-Al5TiB/ 0.10 Vf-% nano Al2O3**	**Mg-Al5TiB/ 0.15 Vf-% nano Al2O3**	**Mg-Al5TiB/ 0.20 Vf-% nano Al2O3**	**Mg-Al5TiB/ 0.25 Vf-% nano Al2O3**
Mg	>96,6	95,5	93,8	>96,0	95
Al	1,96	3,88	5,45	3,37	4,29
Ti	0,00087	0,0536	0,0782	0,0565	0,0692
B	0	>0,0264	>0,0264	>0,0264	>0,0264
Be	0,00007	0,00028	0,00041	0,00023	0,00022
Cu	0,0007	<0,00010	<0,00010	<0,00010	<0,00010
Mn	0,0073	0,0134	0,0167	0,0113	0,0125
Zn	0,0152	0,0484	0,0336	0,0218	0,0224
Ag	0	0,0003	0,00048	0,00036	0,00038
Ca	0	0,0012	0,0018	0,0012	0,0021
Cd	0,00001	<0,00010	<0,00010	<0,00010	<0,00010
Sn	0,0333	0,0398	0,0548	0,035	0,34

Sr	0	0,00015	0,00015	0,00013	0,00011

Table 1 above shows the chemical composition of the materials. According to characterization, magnesium as-cast has 96.6% Mg, which indicates that there is still a possibility of composition level up to 99% in accordance with the mass balance at the beginning. Instead, there are other elements that were not supposed to be in the composite, which are Al, Ti, Be, Cu, Mn, etc in a tiny amount. Nonetheless, these elements would not give a significant effect on the characteristics itself. For the composite, the result shows that the increase of reinforcing particles reduces the content of magnesium, while it decreases on the 0.20 and 0.25 Vf-% nano Al_2O_3. It was caused by the agglomeration, also while pouring the liquid into the mold was not fast enough so that the liquid composite reacted too long with the air.

3.2 SEM and EDS

SEM and EDS tests were performed using magnesium as-cast and magnesium composite with 0.20 Vf-% to represent all of the composite samples.

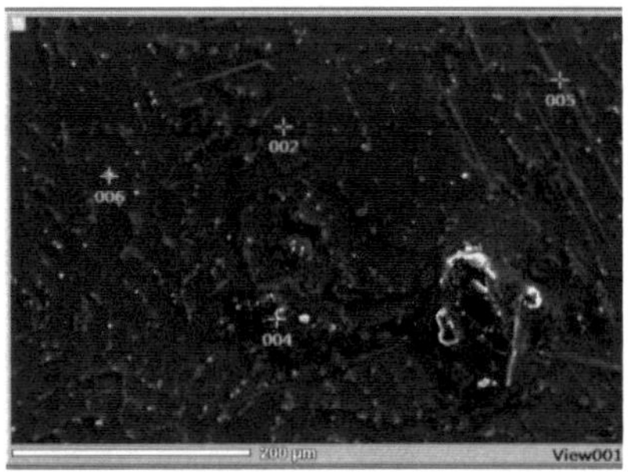

Figure 1. The SEM result of Mg as-cast.

Table 2. The EDS result of Mg as-cast.

Points	Elements (%)				Possible formed phase
	Mg	Al	Si	O	
002	100	-	-	-	Mg
004	79,44	7,9	-	12,66	MgO, Al_2O_3
005	100	-	-	-	Mg
006	85	-	2,83	12,16	MgO

Based on Figure 1 and Table 2 above, they show some elements like Mg, Al, Si, and O which probably formed some compounds; MgO an Al_2O_3. MgO was formed due to the magnesium property which is very easy to react with air. Furthermore, Al_2O_3 was formed at point 004, it was caused by the presence of inclusions. Also based on the OES result, it shows that Mg as cast indeed has 1.96% Al.

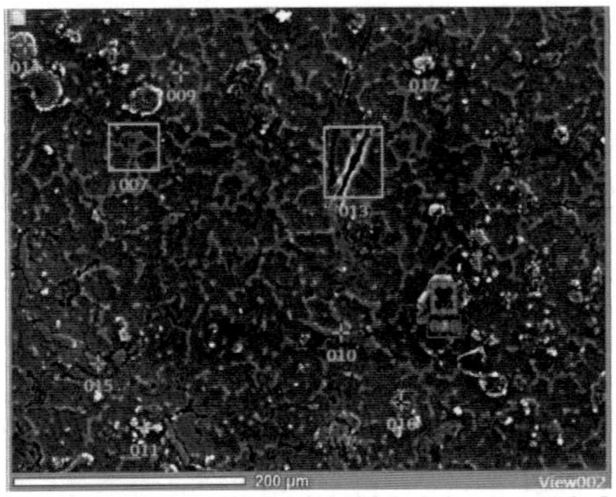

Figure 2. The SEM result of Mg-Al-5Ti-B/ nano Al_2O_3 0.20 Vf-%.

Table 3. The EDS result of Mg-Al-5Ti-B/ nano Al_2O_3 0.20 Vf-%.

Point	Elements (%)						Possible formed phase
	Mg	Al	Ti	B	O	C	
007	36,57	3,69	0,18	9,54	48,7	1,23	MgO, Al_2O_3, B_2O_3, $MgAl_2O_4$, $Al_{12}Mg_{17}$
008	14,65	1,6	5,89	16,82	53,45	6,44	MgO, Al_2O_3, $MgAl_2O_4$, $Al_{12}Mg_{17}$, B_2O_3, TiO_2,
009	57,68	1,08	-	-	39,14	1,4	MgO, Al_2O_3
010	24,81	1,11	0,05	16,84	54,74	2,39	MgO, Al_2O_3, $MgAl_2O_4$, $Al_{12}Mg_{17}$, B_2O_3, TiO_2
011	47,79	3,22	0,33	3,85	43,2	1,5	MgO, Al_2O_3, $MgAl_2O_4$, $Al_{12}Mg_{17}$, B_2O_3, TiO_2
013	30,89	2,94	0,09	12,87	51,7	1,31	MgO, Al_2O_3, $Mg_2Al_2O_4$, $Al_{12}Mg_{17}B_2O_3$, TiO_2
014	35,72	2,98	10,15	4,41	43,09	3,34	MgO, Al_2O_3, $MgAl_2O_4$, $Al_{12}Mg_{17}$, TiO_2, B_2O_3
015	100	-	-	-	-	-	Mg
016	35,15	1,05	0,03	11,91	50,59	1,2	MgO, Al_2O_3, $MgAl_2O_4$, $Al_{12}Mg_{17}$, TiO_2, B_2O_3,
017	21,88	2,18	0,22	16,91	54,17	4,47	MgO, Al_2O_3, $MgAl_2O_4$, $Al_{12}Mg_{17}$, TiO_2, B_2O_3

According to Figure 2 and Table 3 above, they identified 10 locations. Almost at all points, they had Mg, Al, O, Ti, and B which many compounds formed such as MgO, Al_2O_3, $MgAl_2O_4$, $Al_{12}Mg_{17}$, B_2O_3, and TiO_2. MgO as the rust product and showed at all points, this is because magnesium is a highly reactive metal and very easily oxidized. At point 013, TiO2 formed is based on its morphology that looks like a straight line. At location 007 was a little bit different, because it might show $Al_{12}Mg_{17}$as dendrite. In addition, there was C element at almost all points, C was indicated as an impurity which can only form from a stirrer that might have rust.

3.3 XRD

XRD is a test to find out what phases are formed in the composite microstructure. In this research, magnesium composite with Al_2O_3 0.20 Vf-% was observed to represent all the composite samples.

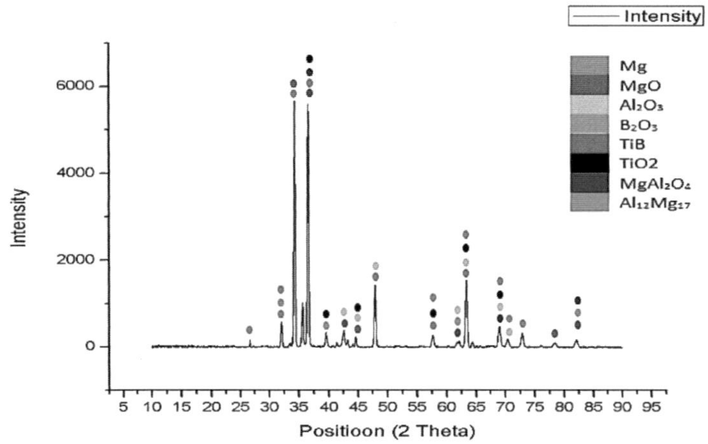

Figure 3. The X Ray Diffraction of Mg-Al-5Ti-B reinforced 0.20 Vf-% nano Al_2O_3 composites.

Based on the Figure 3, there were 16 peaks at °2Θ, which are 13,2612; 19,5809; 26,68; 34,3753; 35,6481; 36,5975; 40,7468; 44,1643; 44,6572; 47,9456; 57,7107; 63,3855; 64,4163; 69,0999; 82,1061; 70,4862, dan 84,6962. The highest scores were Mg; MgO; $MgAl_2O_4$; B_2O_3. Mg phase had the highest peak amongst all with a score of 41. This indicates magnesium as a matrix has the dominant phase. After the Mg phase, there was a MgO phase with a score of 28 which signified magnesium properties that is very easy to react with air. There was a B_2O_3 phase, where boron will react with oxygen easily at a very high temperature and at the end form B_2O_3. Also, there was a $MgAl_2O_4$ phase which is a reaction between Mg and nano Al_2O_3 with a score of 7. The presence of reinforcement particles in magnesium composites will make magnesium composites much better than as-cast magnesium.

3.4 Microstructure

The observation used a digital optical microscope with 200x, 500x, and 1000x magnification. All samples were taken from the tensile test specimens and were prepared metallographically.

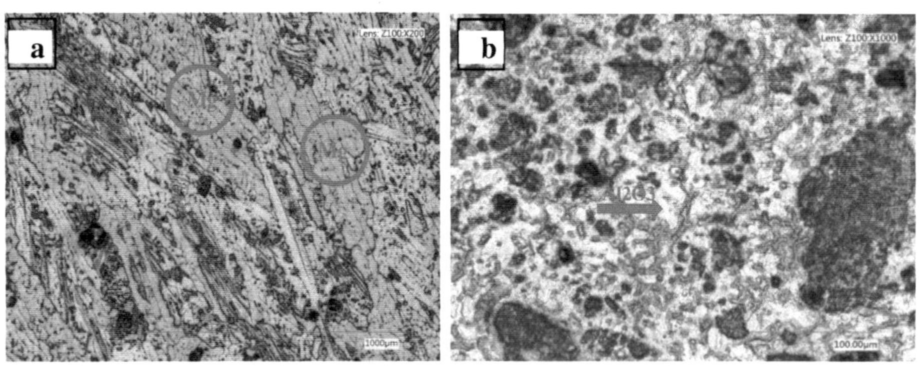

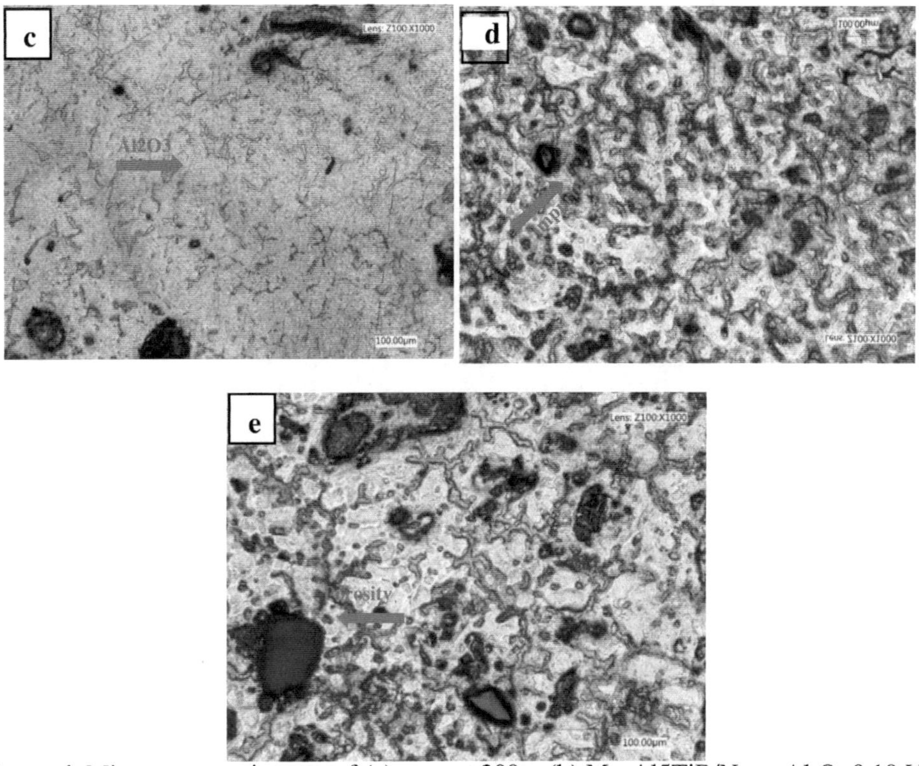

Figure 4. Microstructure images of (a) as-cast 200x; (b) Mg-Al5TiB/Nano Al_2O_3 0,10 Vf-%; (c) Mg-Al5TiB/Nano Al_2O_3 0,15 Vf-%; (d) Mg-Al5TiB/Nano Al_2O_3 0,20 Vf-%; (e) Mg-Al5TiB/Nano Al_2O_3 0,25 Vf-%1000x.

Based on Figure 4, all samples appeared to have block spots which indicate the presence of impurity as well as porosity. The highest amount of corrosion and impurities were found in composite nano Al_2O_3 0.25 Vf-%. The impurity and porosity were caused by the presence of metal oxides formed during the stirring process [5]. Magnesium as-cast has a different morphology than the composite, it was white and oval-shaped. The oval shape was produced when the metal was poured and filled the mold. The wall part of the mold was the first area touched by the liquid so that the cooling rate was fast and became the initial freezing area. Magnesium composite has an equiaxed grain shaped [6]. It was caused by the addition of Al-5Ti-B as grain refiner which would affect its microstructure [4].

As we can see from the microstructure, there was no twinning phenomenon. It was caused by the addition of grain refinement agent Al-5Ti-B as much as 10 wt% which was able to prevent the formation of twinning in the grain boundary and form finer and random grains, thereby the material required larger force to form twinning [7]. There was dendrite in almost all microstructure. It was formed often in cast metal products. The rate of cooling is the most important factor for dendrite formation. The mechanical properties will increase in parallel with dendrite reduction [8].

3.5 Mechanical Properties

All samples were prepared and tested to measure mechanical properties (tensile strength, hardness, wear rate, impact strength) and density. Before performing destructive test, each specimen were tested to find its density using Archimedes' Theory.

3.5.1 Tensile and Elongation

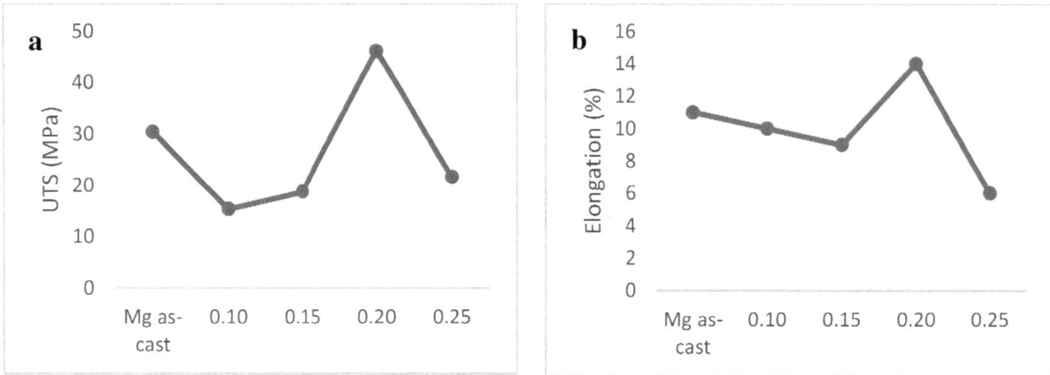

Figure 5. Mechanical properties of Mg-Al-5Ti-B reinforced 0.20 Vf-% nano Al_2O_3 composites (a) Tensile, and (b) Elongation.

Based on Figure 5, the highest tensile strength achieved in the addition of nano Al_2O_3 0.20 Vf-% and the lowest is at nano Al_2O_3 0.10 Vf-%. The increase of tensile strength is due to the addition of Al_2O_3 and Al-5Ti-B which act as a barrier from dislocation. The nano-size of Al_2O_3 makes the particle has a greater surface area than the micron size and it is effective to inhibit dislocation [9]. The tensile strength affects the elongation of the composite. The optimum level is shown at 0.20 Vf-% which is 14%. However, the nano-size particle of reinforcing will not give significant value for elongation, yet for tensile strength without decreasing the elongation itself [9].

3.5.2 Hardness, Wear rate and Impact

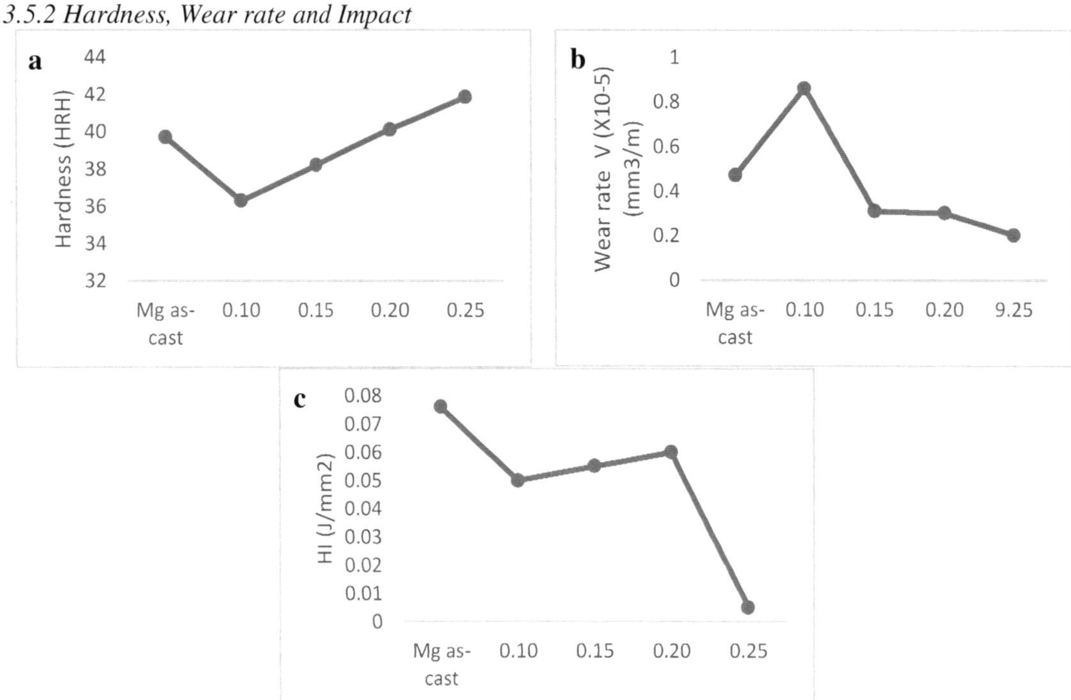

Figure 6. Mechanical properties of Mg-Al-5Ti-B reinforced 0.20 Vf-% nano Al_2O_3 composites (a) Hardness, (b) Wear rate, and (c) Impact Strength.

According to Figure 6, as the Al_2O_3 and Al-5Ti-B added increased, the hardness of cast increased as well. The maximum hardness is obtained by adding nano Al_2O_3 0.25 Vf-%. The addition of nano-sized particles would trigger the formation of fine interphase so that it eases the transfer loading process. The hardness will influence the composite's wear rate. The harder the material, the higher the wear rate, as shown that nano Al_2O_3 0.25 Vf-% had the lowest wear rate. The hard particle will minimize the matrix from the shear stresses [10]. Moreover, when the hardness increases, it tends to change the material property from ductile to brittle. If the material is brittle, it will reduce the absorbed energy during fracture [11].

3.5.3 Density and Porosity

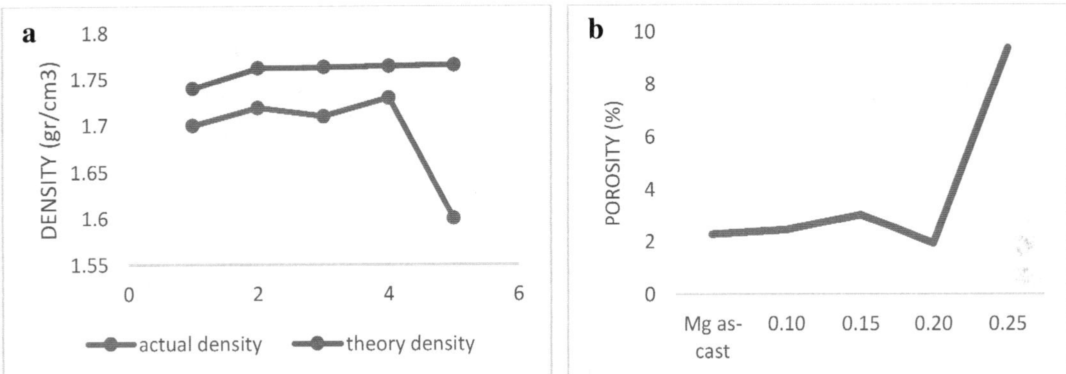

Figure 7. (a) Density and (b) Porosity of Mechanical properties of Mg-Al-5Ti-B reinforced 0.20 Vf-% nano Al_2O_3 composites.

Density test results using the Archimedes principle obtained the actual density. The results of the actual density of the entire samples are compared with the theoretical density obtained from the calculation. From the results of density testing, the porosity percentage can be obtained. According to Figure 7, all of the stirred casting samples showed an actual density value much lower than the theoretical density and pure magnesium density. The addition of nano Al_2O_3 0.25 Vf-% has the largest gap value between actual density and theoretical density, as well as the porosity had the largest number on that composition. It can be caused by the decrease of wetting effect as the increase of reinforcing particle's amount, in addition to the tendency of agglomeration that causes the higher value of porosity [12].

4. Conclusion

The stir casting method may be used to fabricate Mg-Al-5Ti-B composite of the nano-Al2O3 particle. The addition of nano-Al2O3 can increase the UTS and elongation value at 0.20 Vf-% specifically, reduce the impact value, increase the hardness, decrease the wear rate, and increase the density of the material. The addition of 10%wt Al-5Ti-B's aim to refine the material's grain by forming initial nucleation, so that solidification occurs faster and is able to increase the material's strength. The result of the microstructure shows the presence of MgO, $MgAl_2O_4$, $Al_{12}Mg_{17}$, and B_2O_3 phase.

Acknowledgments

Authors would like to thank Directorate of Research and Community Services (DRPM), Universitas Indonesia for financial support through International Publications Index (PIT 9), 2019 with contract number: NKB-0042/UN2.R3.1/HKP.05.00/2019.

IOP Conf. Series: Materials Science and Engineering **924** (2020) 012021 doi:10.1088/1757-899X/924/1/012021

References

[1] M. Gupta and Sharon 2011 N.M.L. Magnesium, Magnesium Alloys, and Magnesium Composites. John Wiley & Sons, Inc
[2] J. B. Clark *et al* 1988 Phase Diagrams of Binary Magnesium Alloys, ASM International, Metals Park, Ohio, USA
[3] Paramsothy, M., Hassan, S. F., Srikanth, N., Gupta, M. Simultaneous Enhancement of Tensile/Compressive Strength and Ductility of Magnesium Alloy AZ31 Using Carbon Nanotubes, J. Nanosci. Nanotechnol. 10 pp. 956-964. 2010
[4] D.P, Mondal and N. Jha 2014 "Effect of Grain Refinement on Mechanical Properties and Sliding Wear Resistance of Extruded Sc-free 7042 Aluminum Alloy," Mater. Des., vol. 22, no. 5, pp. 1001– 1011
[5] A. Safitiri Yani 2018 "Development of Magnesium Matrix Composite with Volume Fraction of Nano-Al2O3 Particles Produced by Stir Casting", FTUI pp 1678-1685. 2018
[6] S.H. Emamy, Seyed Ebrahimi, M. J. Mater Des. 31 2010, 200–209
[7] Y. C. Lee, Dahle, A. K. and StJohn, D. H 2000 "The role of solute in grain refinement of magnesium," Metall. Mater. Trans. A, vol. 31, no. 11, pp. 2895–2906
[8] M., Vasheghani Farahania, Emadoddina, E, Emamyb, M. Honarbakhsh Raoufa 2013 A. Effect of grain refinement on mechanical properties and sliding wear resistance of extruded Sc-free 7042 aluminum alloy
[9] A. Raturi, Mer, K. K. S. and Pant, P. K 2017 "Science Direct Synthesis and characterization of mechanical , tribological and micro structural behaviour of Al 7075 matrix reinforced with nano Al 2 O 3," *Mater. Today Proc.*, vol. 4, no. 2, pp. 2645–2658
[10] V. Bharath, Nagaral, M., Auradi, V., and Kori, S. A 2014 "Preparation of 6061Al-Al 2 O 3 MMC ' s by Stir Casting and Evaluation of Mechanical and Wear Properties," *Procedia Mater. Sci.*, vol. 6, no. Icmpc, pp. 1658–1667
[11] A. Pramanik 2016 "Effects of reinforcement on wear resistance of aluminum matrix composites," *Trans. Nonferrous Met. Soc. China (English Ed.*, vol. 26, no. 2, pp. 348–358
[12] A. Mazahery, Abdizadeh, H. and Baharvandi, H. R 2009 "Materials Science and Engineering A," vol. 518, pp. 61–64.

Study on the Effect of Volume Fraction (Vf%) SiC Addition to The Characteristics of Mg-Al-Sr / SiC Composites Using Stir Casting Method

Ronan Ramadhan and Anne Zulfia*
Department of Metallurgy and Materials Engineering, University of Indonesia, Depok, 16436, Indonesia

*anne@metal.ui.ac.id

Abstract. In this study a magnesium metal matrix composite fabrication process was carried out combined with Al-15Sr alloy which will create intermetallic phases of $Mg_{17}Al_{12}$ and $Mg_{58}Al_{38}Sr_4$. Reinforce material that is used for this study is SiC with an amount of 2, 4, 6, 8 vf%. Stir casting is used as the fabrication method as it doesn't require much maintenance and relatively cheap. Argon is used as a shielding gas so oxidation does not occur to magnesium and prevent combustion. The composite is then casted into a SKD 61 metal mold and then cooled so it can be characterized by the various tests it will go through. SiC is added as the reinforcement material in hopes to increase the mechanical properties of the composite. This can be seen when the composite go through a number of tests, including microstructure analysis, SEM, XRD testing, and various mechanical tests. Through this procedure is it then concluded that composite with 6 vf% SiC is the optimum amount of reinforce material which results in UTS of 51,09 MPa, hardness of 73 HRH, impact strength of 0.0367 J/mm^2 and abrasion rate of 5.68 x 10^{-3} mm^3/m.

Keywords **:** Magnesium based composite, Mg-Al-Sr, SiC, stir casting

1. Introduction

With the increasing use of energy resources, the industry continues to look for substitutes for conventional materials to increase efficiency and the use of lightweight materials to be one solution that cannot be avoided. Magnesium is a promising material for this problem.

Magnesium is one type of metal used to create blocks of machines with low mass. Magnesium is the metal that has the lowest density which is 1.738 g/cm3, only two-thirds of the value of light metal density is usually used, namely aluminum so that in terms of lightness, magnesium is a metal that is suitable for making high-efficiency fuel engine blocks of vehicles. In addition, the availability of magnesium in nature is very abundant [1].

However, for application to the engine block it cannot use pure magnesium because the mechanical properties of pure magnesium are not good enough to use. Magnesium is difficult to be cold worked due to the crystal structure of HCP, the process of increasing the strength of magnesium is by adding a mixture of other elements. The most common element used as a magnesium alloy for industrial applications is aluminum because it can increase strength, hardness and castability of magnesium. However, the heat resistance of the Mg-Al alloy is still low, so it is very susceptible to creep deformation [2]. One way to increase magnesium resistance at high temperatures is by refining grains [3]. Addition of strontium metal in Mg-Al alloys is able to soften microstructure and increase the strength and resistance to the mechanical load of the alloy in high temperature applications. Addition of Sr can also be contributed to the increased creep strength of the alloy [4].

The addition of various types of reinforcement on magnesium alloys can improve mechanical properties such as fatigue and wear resistance [5]. In addition, monolithic metal alloys have a low modulus of elasticity if no reinforcement is added. This limitation can be overcome by combining harder ceramic particles into the metal matrix. When compared with aluminum matrix composites, research on magnesium matrix composites is still limited due to the nature of magnesium which is very reactive and easily oxidized when melted. The method of protecting molten magnesium usually uses fluxes and protective gases in the form of inert gases [6]. In this study a magnesium metal matrix combined with Al-15Sr alloy which is further reinforced by the SiC particles was fabricated.

2. Experimental method

2.1. Materials
In this research, pure magnesium 95% wt and 4,25% Al and 0,75% Sr was used as the matrix and micro sized SiC particle with 2, 4, 6, and 8% volume fraction was used as the reinforcement material.

2.2. Sample preparation
The magnesium ingot and 4,25Al-Sr (matrix) were cut into small blocks using the band saw. Magnesium and Al-Sr rod are weighted according to the mass balance calculation. Magnesium is cleaned with 96% alcohol in order to remove inclusion and oil that might be left on the surface from the cutting procedure. After that, the magnesium and Al-Sr are inserted into the furnace. Before the heater is turned on, the furnace is fluxed for 6 minutes with argon gas. The matrix was melted at 650°C in the furnace and before being poured into the matrix, the reinforce were pre-heated at 900°C in the muffle furnace to remove the moisture. After the pre heating process, SiC particle was inserted into molten Mg-Al-Sr and then stirred for 1 minutes to get a homogenous mixture. Then, they were poured into the mold and the mold will be opened after the composite has solidified. Finally, this composite sample will be used for further characterization [7].

2.3 Characterization
2.3.1 Chemical Characterization. Chemical characterization is performed using Optical Emission Spectroscopy (OES) to determine what elements were present in magnesium ingot samples and Mg-Al-Sr composite with with 2, 4, 6, and 8% Vf SiC. Chemical characterization using OES is generated the form of elements. Elements obtained are further discussed in relation to variations in the addition of SiC and the process of the stirring casting method.

2.3.2 Mechanical Characterization. Each sample is subjected to destructive testing (tensile, impact, hardness, and wear testing). Prior to destructive test, each specimen were tested to find its actual density using Archimedes Theory. The tensile test was done using GoTech machine based on JIS Z2241 at room temperature. The impact test was tested using the Charpy method at room temperature. The hardness test was done using Brinell test with HRH type. The wear was done by Ogoshi by sliding a specimen into harder material for 100 meter sliding distance.

2.3.3 Microstructural Characterization. Firstly all of the samples are grinded and polished. After sample preparation is done, the microstructure characterization is done using optical microscope to view microstructure, SEM with EDX integration to see topography of specimen in 3D and analysis, and XRD to confirm the presence of substances in the material.

3. Results and discussion

3.1 Chemical Composition

OES (Optical Emission Spectrometry) is a test to confirm the presence of elements in magnesium as-cast or non reinforced magnesium also magnesium composite quantitatively. So that, we will know the effect of the casting process on the metal characterization.

Table 1. Chemical composition of Mg as-cast and Composite Material.

Element	**wt%**			
	Composite 1 (2% SiC)	**Composite 2 (4% SiC)**	**Composite 3 (6% SiC)**	**Composite 4 (8% SiC)**
Mg	94,6	95,2	94,8	91,7
Al	3,99	3,18	2,83	4,17
Be	0,00025	0,00022	0,00016	0,00019
Cu	<0,0001	<0,0001	<0,0001	<0,0001
Mn	0,0106	0,0089	0,0069	0,0092
Zn	0,0263	0,0195	0,0170	0,0222
Ag	0,00058	0,00045	0,00041	0,00061
Ca	0,0062	0,0065	0,0076	0,0164
Cd	<0,0001	<0,0001	<0,0001	<0,0001
Sn	0,0330	0,0286	0,0220	0,0221
Sr	0,0904	0,08	0,08	0,0888
Si	1,11	1,29	2,07	3,7

Table 1 above show the chemical composition of the material. There are a difference in the amount of added elements which strays from the mass balance, instead there are another elements that were not supposed to be in the composite, which is Ti, Be, Cu, Mn, etc but with tiny amount. Even though, these elements would not give significant effect on the characteristics itself. According to calculation, as you increase the number of reinforcement particles it will decrease the wt% of magnesium.

3.2 Microstructure

The observation used digital optical microscope with 200x, 500x, and 1000x magnification. Observations using optical microscopes show the microstructure of MgAlSr / SiC composite materials. The main phase observed in the microstructure is α-Mg in the grain which is flat and brightly coloured. Observations using optical microscopes did not clearly show the presence of the intermetallic phases of $Mg_{17}Al_{12}$ and $Mg_{58}Al_{38}Sr_4$ in the generated microscope

In Figure 1 there are arrows of different colors, where the blue arrow shows the α-Mg phase, the orange arrow shows dendritic which is estimated to contain the Mg17Al12 phase and the Mg58Al38Sr4 phase and the white arrow shows the amplifier namely SiC [8].

In pure Magnesium as seen in figure 3, grain boundaries can be seen and their size is much greater than the grain boundary size of Mg-Al-Sr composite. This is due to the Aluminum content which will refine the Magnesium grain due to the presence of other phases namely the intermetallic phase of $Mg_{17}Al_{12}$ and $Mg_{58}Al_{38}Sr_4$ [9].

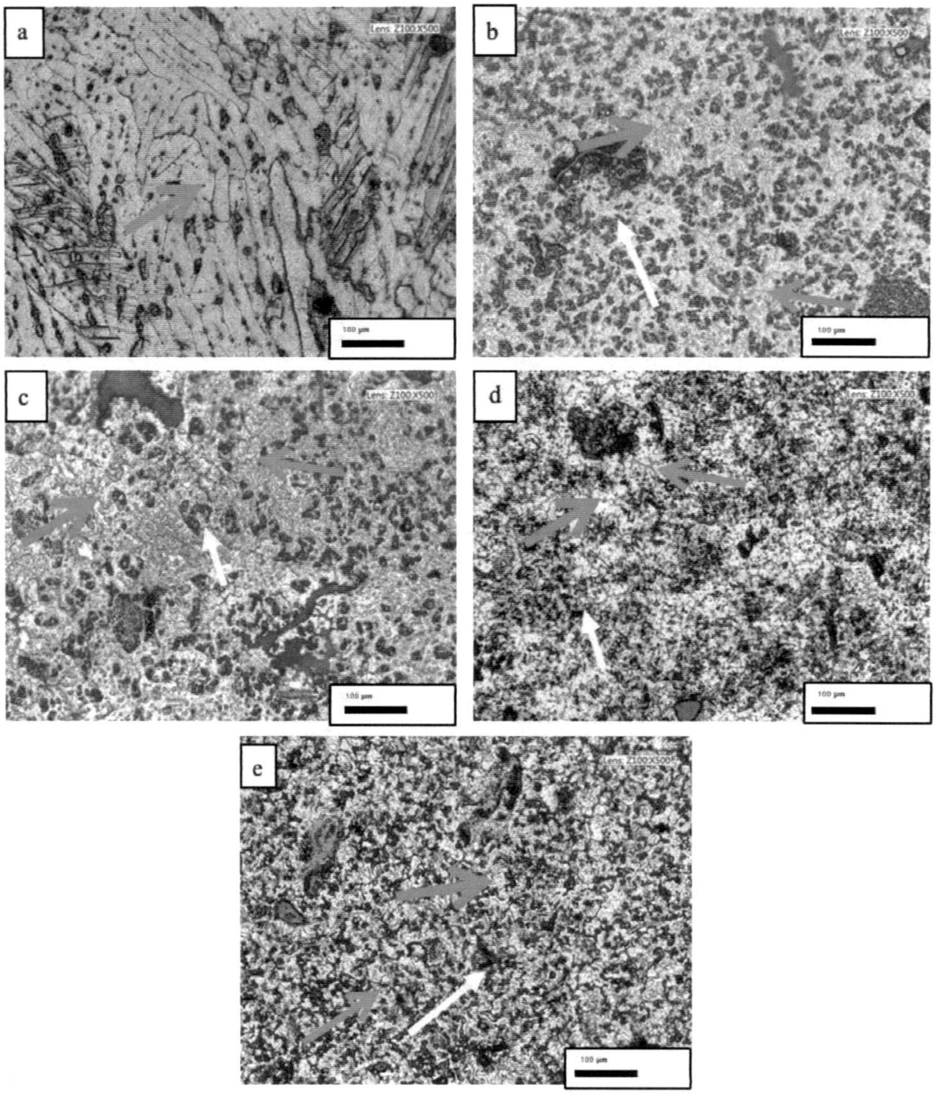

Figure 1. Microstructure images of (a) pure Mg (b) MgAlSr 2% SiC ; (c) MgAlSr 5% SiC; (d) MgAlSr 6% SiC; (e) MgAlSr 8% SiC [500x].

3.3 SEM and EDS

SEM and EDS tests were performed using magnesium composite with 6% Vf to represent all of the composite sample.

Table 2. The EDS result of Mg-Al-Sr.

Point	Wt (%)				
	Mg	Al	Sr	Si	O
A	95,67	4,33	-	-	-
B	89,85	9,97	-	0,18	-
C	68,12	6,04	-	19,91	5,78
D	84,79	6,06	1,28	3,64	4,23
E	78,52	7,15	-	8,55	5,78
F	83,45	5,41	-	6,4	4,74

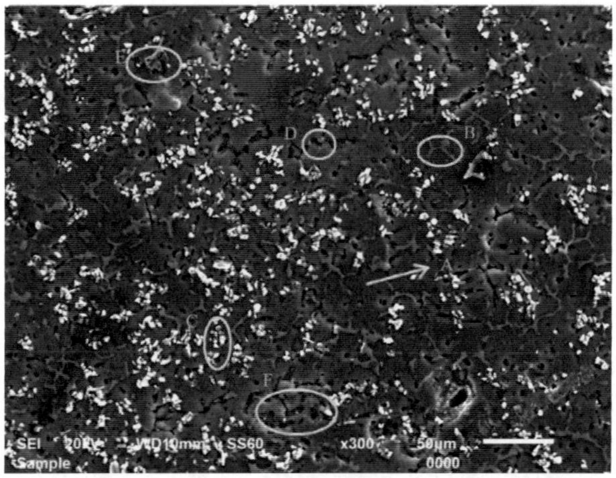

Figure 2. The SEM result of Mg as-cast.

Based on the Figure 2 and Table 2 above, it shows some elements like Mg, Al, Si, and O which probably formed some compounds; MgO an Al_2O_3. MgO was formed due to the magnesium properties which is very easy to react with air. In point A it shows α-Mg. In point B, according to the elements that are present shows that it forms $Mg_{17}Al_{12}$. Point C shows the reinforce particle SiC. Point D shows MgAlSr alloy. Point E and F shows MgAlSi composition.

3.4 XRD

XRD is a test to find out what phases are formed in composite microstructure. In this research, magnesium composite with SiC 6% Vf was observed to represent all the composite samples.

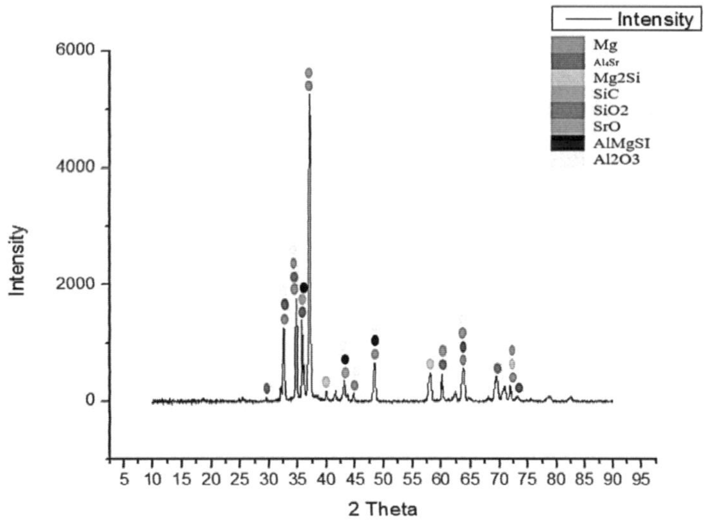

Figure 3. The XRD result of Mg-Al-5Ti-B/ nano Al_2O_3 0.20% Vf.

Based on the data obtained, there are seven identified compounds, namely Mg_2Si, SiC, SiO_2, AlMgSi, Al_2O_3, Al_4Sr and SrO. Compounds that have high diffraction peaks are SrO and Al_2O_3. This is different from the results of SEM observations and micro-observations showing precipitates formed in the Mg-Al-Sr / 6 Vf% SiC composites are estimated to be $Mg_{17}Al_{12}$ and $Mg_{58}Al_{38}Sr_4$. From the XRD results above, it can be seen that the intermetallic phases formed are Al_4Sr and AlMgSi. This difference is caused by non-equilibrium cooling during the metal solidification process which results in a difference from the results of the calculation rule. AlMgSi is formed because of the high amount of Si content. SiO2 is formed from the pre-oxidation of the SiC particles [10]. The main phase of Mg, and SiC have higher intensity than that of the secondary phases and they make up 55% of the composite. Which is 25% higher than the secondary phases which makes up 45% of the composite.

3.5 Mechanical Properties

3.5.1 Tensile and Elongation

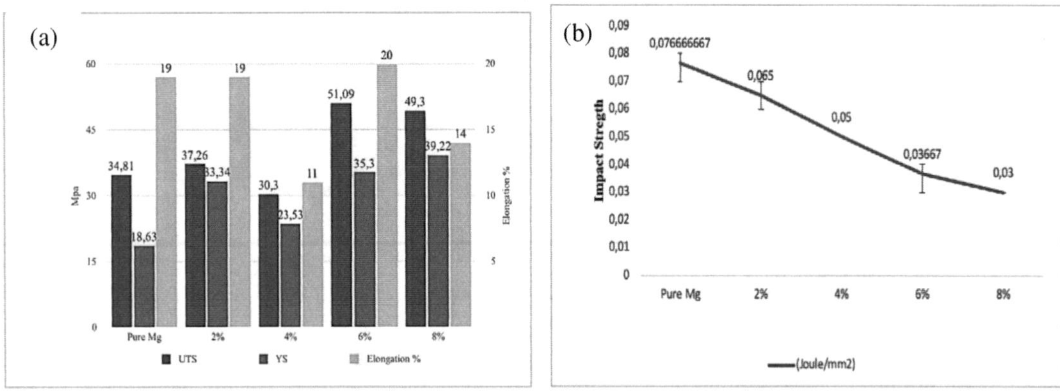

Figure 4. (a) Tensile strength and Elongation (b) Impact value of composite material.

Based on the Figure 4, the highest tensile strength achieved in the addition of 6% Vf SiC and the lowest is 2% Vf SiC. The increase of tensile strength is due to the addition of SiC and AlSr which act as a barrier from dislocation. The increased amount of SiC will increase surface area and create more interface between matrix and reinforcement material. The increased tensile strength between pure magnesium and magnesium alloy is caused by the formation of the phase $Mg_{17}Al_{12}$ which will cause the grain size to reduce as seen in the microstructure [11].
According to the data acquired on figure 4, the impact value of the matrix will decrease with the addition of SiC particles, this is caused by the increased hardness of the material [12].

3.5.2 Hardness, Wear rate, and Impact

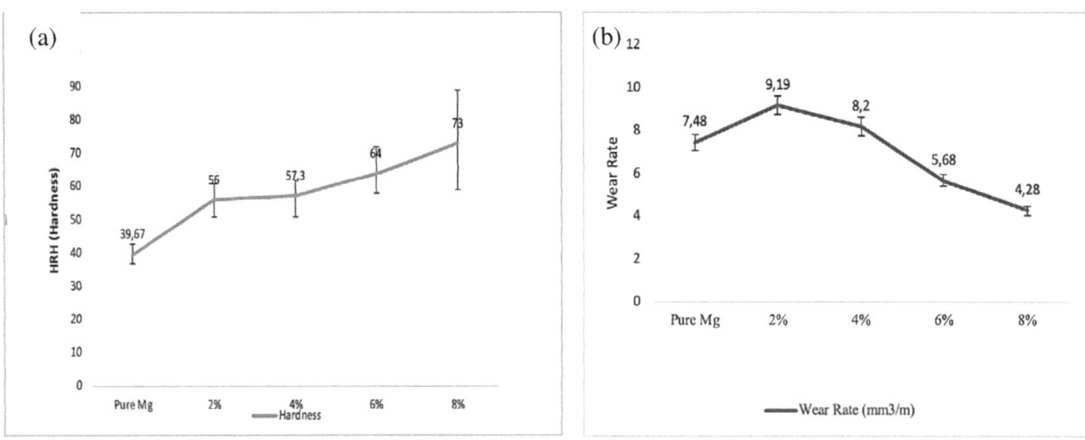

Figure 5. (a) Hardness, (b) Wear rate.

According to the Figure 5, as the SiC particles is increased, the hardness of composite increase as well. The maximum hardness is obtained by adding 8% Vf SiC. The hardness will influence wear rate of composite, the harder material the lower its wear rate, and it shows that 8% Vf SiC had the lowest wear rate. The hard particle will minimize the matrix from the shear stresses [13]. This happens because of the presence of a reinforcer, where SiC particles and intermetallic such as Mg_2Si will make dislocations difficult [14].

3.5.3 Density and Porosity

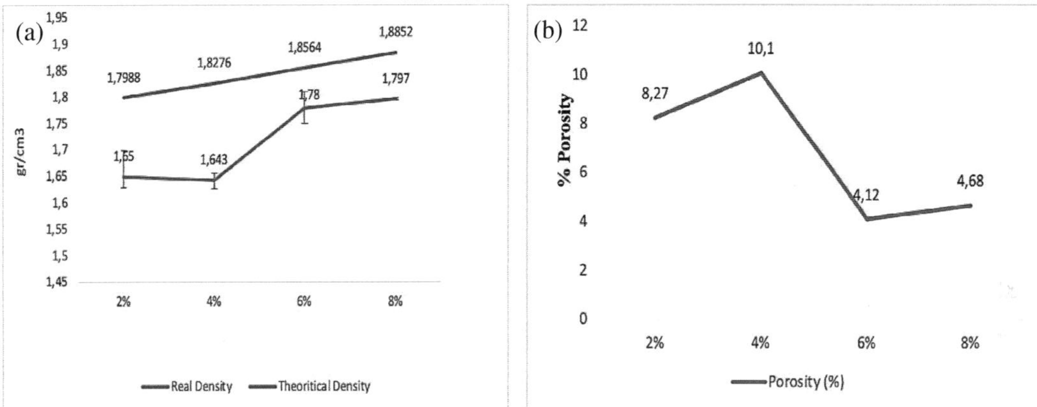

Figure 6. (a) Density and (b) Porosity of composite material.

Density test using Archimedes principle results in the actual density of the composite. The result of the actual density are compared with the theoretical density, which is obtained from calculation. From the comparation of density percentage of porosity can be obtained. According to the Figure 6, all of the stirred casting samples showed an actual density value much lower than the theoretical. The addition of 4% vf SiC has the largest amount of porosity. It can be caused by gases that are trapped in the molten metal, which can come from turbulence during casting [15].

4. Conclusion

Composite MgAlSr with SiC amplifier can be fabricated using the Stir Casting method. The least amount of porosity and agglomeration in this study was obtained in variable 3 with the use of SiC as much as 6% vf. Increasing the amount of SiC amplifier will improve the mechanical properties of the MgAlSr composite. The optimum mechanical properties obtained have a UTS value of 51.09 MPa, hardness value of 73 HRH, impact value of 0.0367 J/ mm2, and wear rate of 5.68 x 10-3 mm3/ m. The phase formed in the MgAlSr composite according to micro-observations is the α-Mg phase, β ($Mg_{17}Al_{12}$) and $Mg_{58}Al_{38}Sr_4$ phases. The XRD results show that the phase formed in the MgAlSr / SiC composites is α-Mg, AlMgSi, and Al_4Sr. This difference is caused by non-equilibrium cooling.

Acknowledgment

Authors would like to thank Directorate of Research and Community Services (DRPM), Universitas Indonesia for financial support through International Publications Index (PIT 9), 2019 with contract number: NKB-0042/UN2.R3.1/HKP.05.00/2019

References

[1]. M. Gupta and N. Sharon, "Magnesium, Magnesium Alloys, and Magnesium Composites". Hoboken, N.J.: Wiley, pp. 1-2, 2013.

[2]. T. Nayeri, M. Yari and S. Sadreddini, "Effect of Sr on the Microstructure and Properties of Mg-6Al Alloy", Protection of Metals and Physical Chemistry of Surfaces, vol. 52, no. 2, pp. 273-278, 2016.

[3]. J. Du, Y. Shi, M. Zhou, and W. Li, "Effect of Sr on Grain Refinement of Mg– 3%Al Alloy Containing Trace Fe by Carbon-Inoculation", Journal of Materials Science & Technology, vol. 32, no. 12, pp. 1297-1302, 2016.

[4]. C. Zhao, C., F. Pan, L. Zhang, H. Pan, K. Song, A. Tang, "Microstructure, mechanical properties, bio-corrosion properties and cytotoxicity of as-extruded Mg-Sr alloys". Materials Science and Engineering: C 70, 1081–1088, 2017.

[5]. K. Deng, K. Wu, Y. Wu, K. Nie, and Zheng, M., "Effect of Submicron Size SiC Particulates on Microstructure and Mechanical Properties of AZ91 Magnesium Matrix Composites", Journal of Alloys and Compounds, vol. 504, no. 2, pp. 542-547, 2010.

[6]. A. Luo, "Magnesium Casting Technology for Structural Applications", Journal of Magnesium and Alloys, vol. 1, no. 1, pp. 2-22, 2013.

[7]. A. Inegbenebor, C. Bolu, P. Babalola, and O. Fayomi, "Aluminum Silicon Carbide Particulate Metal Matrix Composite Development Via Stir Casting Processing", Silicon, 2016.

[8]. M. Mezbahul-Islam, A. Mostafa and M. Medraj, "Essential Magnesium Alloys Binary Phase Diagrams and Their Thermochemical Data", Journal of Materials, vol. 2014, pp. 1-33, 2014.

[9]. A. Tang, F. Pan, M. Yang, and R. Cheng, "Mechanical Properties and Microstructure of Magnesium-Aluminum Based Alloys Containing Strontium", vol 6, no 49, 1203–1211, 2008.

[10]. C. Li, Z. Wang, M. Zha, C. Wang, H. Yu, H. Wang, and Q. Jiang, "Effect of Pre-Oxidation Treatment of Nano-SiC Particulates on Microstructure and Mechanical Properties of SiC/Mg-8Al-1Sn Composites Fabricated by Powder Metallurgy Combined with Hot Extrusion", Materials, vol. 9, no. 12, p. 964, 2016.

[11]. B.S. Pugazhendhi, A. Kar, K. Sinnaeruvadi, and S. Suwas, "Effect of Aluminium on Microstructure, Mechanical Property and Texture Evolution of Dual Phase Mg-8Li Alloy in Different Processing Conditions", Archives of Civil and Mechanical Engineering 18, no. 4, pp. 1332–44. 2018.

[12]. K.S. Prakash, P. Balasundar, S. Nagaraja, and P.M. Gopal, "Mechanical and wear behaviour of Mg – SiC – Gr hybrid composites", Journal of Magnesium and Alloys, vol 3, no. 4, 197–206, 2016.

[13]. K.B. Nie, X.J. Wang, K. Wu, X.S. Hu, M.Y. Zheng, L. Xu, "Microstructure and tensile properties of micro-SiC particles reinforced magnesium matrix composites produced by semisolid stirring assisted ultrasonic vibration", Materials Science and Engineering: A 528, 8709–8714, 2011.

[14]. A. Asgari, M. Sedighi, P. Krajnik, "Magnesium alloy-silicon carbide composite fabrication using chips waste", Chalmers University of Technology, Journal of Cleaner Production, 2019.

[15]. B.C. Kandpal, J. Kumar, H. Singh, "Manufacturing and technological challenges in Stir casting of metal matrix composites– A Review", Materials Today: Proceedings 5, 5–10, 2018.

Preliminary Study of Zirconium Carbide Ceramic Deposition on Austenitic Stainless Steel by Pulsed Laser Deposition

N. Shabrina*, Sumaryo, B. Sugeng, D.N. Haerani, Mardiyanto, A.K. Rivai
Center For Science and Technology of Advanced Materials- National Nuclear Energy Agency of Indonesia – BATAN, Kawasan Puspiptek Gd.71, Tangerang Selatan, Banten

*shabrina_n@batan.go.id

Abstract. Coating Zirconium Carbide (ZrC) on the stainless steel (SS) will increase the corrosion resistance of SS which suffer from the intergranular attacks during operation as construction materials in reactors. ZrC coating have been partially deposited on Austenitic 316L SS by Pulsed Laser Deposition (PLD) at 800°C temperature substrate with constant oxygen flow injection of 40 sccm (Standard Cubic Centimeters per Minute), chamber pressure around 229-244 mTorr and 105 the numbers of laser shots. The sample was analyzed using X-Ray Diffraction (XRD), Microscope Optic (MO), Scanning Electron Microscope – Energy Dispersive X-Ray Spectroscope (SEM-EDS), and Atomic Force Microscope (AFM) to characterize the phase, morphology, thickness of the films and surface topography. The results showed that the thickness of the thin film around 4.93-8.10 µm with 49.6 nm surface roughness. The film surface was very smooth. The present of Zr and C was detected in SEM & EDS results but not appeared in XRD pattern, possibly due to diminish tension of the thin films on the substrates caused by low lattice mismatch between ZrC and Austenitic 316L SS substrate. Furthermore, ZrO_2 phase has been found on the substrate which is predicted because of the interaction between ZrC target and oxygen background gas.

***Keywords*:** Zirconium Carbide, Stainless Steel, Coating, Pulsed Laser Deposition

1. Introduction

Austenitic stainless steels are materials that very important in both conventional and advanced reactor technologies, especially in their application as cladding material. Austenitic stainless steels 316L used reactors construction materials in worldwide [1]. Beside that, because of its good corrosion resistance and good compatibility with hot sodium, 316L SS used in the core of sodium cooled fast reactors [2] .

During operation in stainless steel, several problems was found. Stainless steel suffers from intergranular attacks which result in the loss a plasticity and strength because of crystal structure deformation caused by a localized attack along the grain [3]. Beside that, irradiation can modify the dislocation contributions, which can harden the SS. These problems may impact the life extension in the reactor [3].

Coating method is one of the methods that will increase the corrosion resistance of SS. By employing the coating can be protect the SS substrate against several problem above in high temperatures. Coating based Al_2O_3 and NiAl cement showed the excellent performance to protect on the high temperature corrosion. However, they are very bad as thermal insulator than ceramics [4].

IOP Conf. Series: Materials Science and Engineering **924** (2020) 012023 doi:10.1088/1757-899X/924/1/012023

Ceramic ZrC is being considered for increase the corrosion resistance of SS. ZrC has high melting point (3532°C), high hardness number (25GPa), good wear resistance and high thermochemical stability [5]. Xin et al stated that ZrC have attached much attention due to low cost, their relatively low density, excellent ablation resistance and good strength at high temperatures [6].

Several deposition techniques have been applied to obtain the thin film such as Chemical Vapor Deposition (CVD) and Physical Vapor Deposition (PVD). ZrC coating have been done by CVD, but the result showed that CVD consume much time due to the low deposition rate, thus increasing the manufacturing process and limited application [6]. Some report said the deposition was not being deposited well. Sunil A used radio frequency cold plasma assisted chemical vapor deposition (CVD) to coated ZrC on a mild steel substrate [7]. But the result showed that ZrC layer was not deposited on the substrates which caused by inadequate vacuum and or the leak of oxygen to the reactor [7]. Another deposition technique, PVD have been employed to obtain ZrC thin film for various application [4,7-9]. One of the PVD techniques is PLD which a very sustainable technique for produce the protective coating. Craciun et al deposited the ZrC thin films on (100) Si substrate of 500°C by PLD and have confirmed the elemental composition of the deposited films by AES and XPS. Several problems of non-uniform thickness of coating and wettability have been found on the ZrC films [5].

In this study, ZrC ceramic was used as coating materials on austenitic stainless steel (316L) using PLD and the purpose is to develop the thin film ZrC with smooth roughness of surface and uniform thickness coating.

2. Experimental method

2.1. Materials

316L SS was used as substrate with chemical composition (in wt%) of steel as following: 17.2 Chromium, 10.9 Nickel, 2.1 Molybdenum, 1.6 Manganese, 0.02 Carbon and balance Iron. The surface of substrates was prepared by grinding, polishing to maintain the clean surface, and sandblasted with 0.05 microns alumina particle. Sandblasted with alumina particle can enhancing the adhesion strength between substrate and coating [3]. For coating, round shape of ZrC target was used with 6,73 g/cc density and 99.9% purity by Beijing Loyal Target company.

2.2. Sample preparation

Thin films of ZrC have been deposited on austenitic steel 316L SS using Pulsed Laser Deposition (PLD) at laboratory facilities of Center For Science and Technology of Advanced Materials – National Nuclear Agency of Indonesia. This PLD use Nd:YAG solid state laser with the 266 nm wavelength and ~ 100 mJ energy. The ZrC thin film was deposited in 800°C temperature substrate with the constant oxygen flow injection of 40 sccm (Standard Cubic Centimeters per Minute) that produce a chamber pressure around 229-244 mTorr and the numbers of laser shots were 10^5.

After deposition, the samples were analyzed using X-Ray Diffraction (XRD), Microscope Optic (MO), Scanning Electron Microscope – Energy Dispersive X-Ray Spectroscope (SEM-EDS), and Atomic Force Microscope to characterize the phase, morphology, thickness of the films and surface topography.

3. Results and discussion

The thickness of the samples was analyzed from the cross section of the samples with MO and SEM. Figures 1 show that thickness of the ZrC thin film deposited in 316L SS was 4.93-8.10 μm.

Figure 1. Cross Section of ZrC film deposited in 316L SS (a) Optic result and (b) SEM result.

Figure 2. AFM topography analysis of ZrC deposited in 316L SS.

Figure 2 show the AFM topography analysis of ZrC deposited in 316L SS. The results show that the roughness layer of ZrC deposited in both of 316L SS have a nano-mater scale. The Ra value of sample has 49.6 nm. The roughness value of the sample is smaller if we compare to the Craciun's et al works [4], which the roughness value around 4-6 micrometer. The films surface was very smooth. It's found that deposition using PLD could minimize the formation of micro porous and could develop very smooth layer [10].

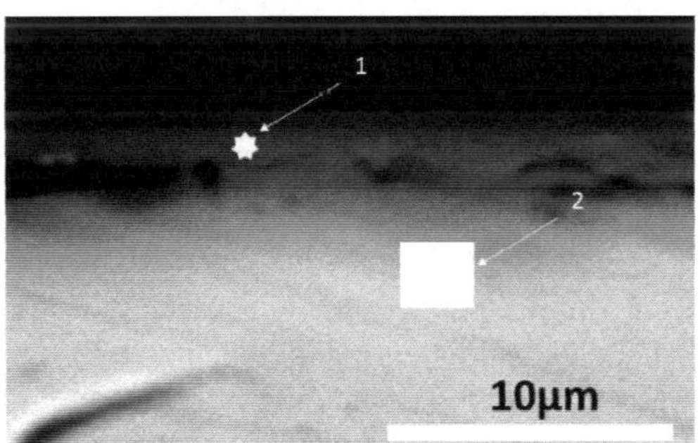

Figure 3. SEM result of cross section ZrC film deposited in 316L SS.

Table 1. EDS result with the position indicated in Figure 3

	Cr (wt%)	Ni (wt%)	Mo (wt%)	Zr (wt%)	Fe (wt%)
1	15.77	10.93	-	1.92	Balance
2	15.11	9	1.69	-	Balance

EDS result showed element on the position 1 point area as thin film and position 2 square area as substrate. The result showed that the substrate element composition was Chromium iron nickel with composition above confirmed as austenitic stainless steel. Position 1 showed the Zr element. However, the wt% of Zr is relatively small, it can be obtained that the number of Zr deposited on substrate was very small.

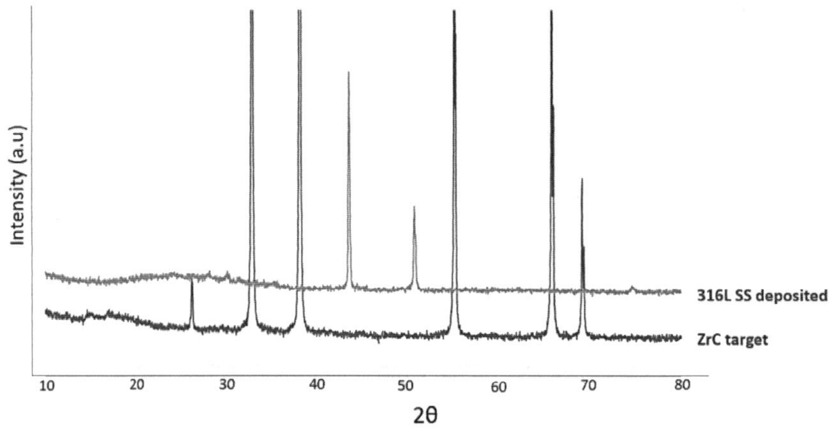

Figure 4. XRD Pattern of ZrC film deposited in 316L SS.

The phase of the samples was investigated by XRD (PANalytical) using Cu Kα radiation (λ= 1.5406 Å). Figure 4 shows the XRD pattern of ZrC film deposited in 316L SS. The XRD pattern showed that ZrC peaks was not observed in the 316L SS deposited. This result contributed to the small number of ZrC deposited on the substrate as shown in EDS result. So, ZrC partially deposited in the 316L SS. This happen because of the diminish tension of the ZrC thin films on the substrate. The lattice parameter of ZrC is 9.386 Å (#00-032-1489) while the lattice parameter of Austenitic SS is 3.5911 Å (#00-033-0397). This low lattice mismatch contributed to the low-quality epitaxial growth so ZrC can't be fully deposited to the austenitic SS.

The XRD pattern of 316L SS deposited showed small peak of ZrO_2 at 2θ=30.19 which identified as main peak of ZrO_2 tetragonal (#00-080-0784). The formation of small ZrO_2 phase was possible because Zr^- ions due to plasma ablation from ZrC target interacted with O^{2+} from the environment. The formation of ZrO_2 phase can be made because the film was deposited in oxygen flow injection. This result confirmed with Sunit et al works, they deposited ZrC on Zr-alloy substrate using the CVD technique [6]. Zr and C presented in EDS result but not presented in XRD pattern, means the ZrC layer partially deposited on the substrate. This attributed to inadequate vacuum and or the leak of oxygen [6].

4. Conclusion

The thin layer of ZrC have been partially deposited in austenitic 316L SS using PLD with temperature substrate was 800° C. The thickness of the thin film around 4.93-8.10 μm with 49.6 nm surface roughness. The films surface was very smooth. The present of Zr and C was detected in SEM & EDS results but not appeared in XRD pattern, possibly due to diminish tension of the thin films on the substrates. It can be established that ZrC was partially coated on the surface of the substrate. Small portion of ZrO_2 have been found on the substrate because the film was deposited in oxygen flow injection. Therefore, to improve the quality of thin film the background gas should be replaced with non-oxygen containing gas.

Acknowledgment

Authors would like to express gratitude to Bandriyana, M.Si and Dra. Deswita, for their help in discussing and characterization of the samples using SEM-EDS, and Head for Center For Science and Technology of Advanced Materials (PSTBM) – BATAN for his support and encouragement. This research supported by Ministry of Research, Technology and Higher Education of the Republic of Indonesia.

References

[1] R. B. Rebak, "Alloy Selection for Accident Tolerant Fuel Cladding in Commercial Light Water Reactors," *Metall. Mater. Trans. E*, vol. 2, no. 4, pp. 197–207, 2015.

[2] M. Saravanan, A. Devaraju, N. Venkateshwaran, A. Krishnakumari, and J. Saarvesh, "A review on recent progress in coatings on AISI austenitic stainless steel," *Mater. Today Proc.*, vol. 5, no. 6, pp. 14392–14396, 2018.

[3] D. de Souza Gomes, A. Abe, A. T. e Silva, C. Giovedi, and M. R. Martins, "Evaluation of corrosion on the fuel performance of stainless steel cladding," *EPJ Nucl. Sci. Technol.*, vol. 2, p. 40, 2016.

[4] N. Abu-warda, M. D. López, and M. V. Utrilla, "High temperature corrosion and wear behavior of HVOF-sprayed coating of Al2O3-NiAl on AISI 304 stainless steel," *Surf. Coatings Technol.*, vol. 359, no. December 2018, pp. 35–46, 2019.

[5] D. Craciun *et al.*, "Optical and mechanical properties of nanocrystalline ZrC thin films grown by pulsed laser deposition," *Appl. Surf. Sci.*, vol. 352, pp. 28–32, 2015.

[6] X. Yang, Z. A. Su, Q. Z. Huang, L. Y. Chai, P. Zhong, and L. Xue, "A zirconium carbide coating on graphite prepared by reactive melt infiltration," *J. Cent. South Univ.*, vol. 21, no. 2, pp. 472–476, 2014.

[7] A. Sunil, "The application of plasma assisted chemical vapour deposition of ZrC onto a Zr-alloy substrate," North-West University, 2017.

[8] V. Craciun, E. J. McCumiskey, M. Hanna, and C. R. Taylor, "Very hard ZrC thin films grown by pulsed laser deposition," *J. Eur. Ceram. Soc.*, 2013.

[9] D. Craciun, G. Socol, G. Dorcioman, N. Stefan, G. Bourne, and V. Craciun, "High quality ZrC, ZrC/ZrN and ZrC/TiN thin films grown by pulsed laser deposition," *J. Optoelectron. Adv. Mater.*, 2010.

[10] M. hui Ding, H. Sen Zhang, C. Zhang, and X. Jin, "Characterization of ZrC coatings deposited on biomedical 316L stainless steel by magnetron sputtering method," *Surf. Coatings Technol.*, vol. 224, pp. 34–41, Jun. 2013.

Investigation on Performance of Cu/Ni Film as Low Temperature Sensor

Moh. Toifur[1*], Nurmalia Agustin[1], Azmi Khusnani[1], and Okimustava[2]
[1]Magister of Physics Education Department, Pascasarjana Program, University of Ahmad Dahlan.
[2]Physics Education Department Faculty of Teaching and Training, University of Ahmad Dahlan Jl. Pramuka No. 42 Sidikan 55161, Indonesia.

*toifur@mpfis.uad.ac.id

Abstract. This research has investigated the effect of deposition time on the performance of Cu/Ni thin film as low temperature sensor. The sample was carried out by using electroplating technique on various deposition times from 1–5 minute. The liquid nitrogen was used as low temperature medium which the temperature was regulated from 0°C down to -200°C and vice versa. All Cu/Ni sensors perform a good response on the temperature change with variation of the output voltage range, time constants, sensitivity, and hysteresis losses. The Cu/Ni film produced during 1 minute deposition is the most sensitive sensor with sensitivity of (1.25 ± 0.01) mV/°C, but rather slow in responding temperature change and rather less stable voltage compared with the Cu/Ni film produced in 4 minute deposition. Unfortunately, the 4 minute sample is less sensitive in responding temperature change, that is (1.00 ± 0.01) mV/°C. We also found that all sensors have various hysteresis losses. The Cu/Ni film produced in 3 minute deposition has the minimum hysteresis area that is (6.640 ± 0.005)V°C.

Keywords : Cu/Ni thin film, low temperature sensor, time response, sensitivity, hysteresis losses

1. Introduction

Currently the low temperature sensor is urgently needed in various fields such as in agriculture for food preservation, in medical field for organ preservation, and in animal husbandry field for preservation of animal sperm which will be used for artificial insemination [1-3]. All of that, requires a low temperature storage medium below -150°C. Therefore, a thermometer that can monitor the temperature of the medium at any time is needed. The most often used thermometers to measure low temperatures are gas thermometer, helium vapor pressure thermometer, Coulomb blockade, RTD, and p-n junction [4].

Nevertheless, making a temperature sensor below -150°C (cryogenic temperature) is not easy because the electrical and thermal properties of the material at this temperature are most non linear [4], so that the response of sensor acting at this temperature is also non linear and individual [5]. One of sensor materials that has the potential to be used as a low temperature sensor is CuNi alloy and Cu/Ni film [6]. The thin film shape is the main consideration because the electrical properties are quite different from the bulk. This is something that many researchers called as the size effect. The use of Ni combined with Cu or coated on Cu is to increase the sensitivity of the sensor, because Ni has a higher resistivity (7.3 μΩcm) than Cu (1.7 μΩcm) [4].

In the previous work, we have done the study on the quality of Cu/Ni film as low temperature sensor materials prepared by electroplating on the various deposition time [7]. From the microstructure characterization with XRD it was found that all samples had a crystalline structure. Deposition time can affect the diffraction angle, shifted left or right, so that the interplanar spacing (*d*-spacing) changes. In the direction of (111) the smallest *d*-spacing corresponds to a sample of 3 minute deposition of 0.855Å. From the SEM photo it appears that the deposition time influences the size of Ni particle, the longer the deposition time the greater the particle size. The largest Ni particle size is in accordance with the sample produced in 5 minute deposition, that is 1.38 Å. This value is almost 4 times the grain size of the sample produced in a 2 minute deposition. From the sheet resistivity test, it is obtained that sheet resistivity of all samples are above sheet resistivity of Cu and the longer the deposition time the sheet resistivity decreases.

From these results, the study is continued on the calibration of the samples as low temperature sensors ranging from 0 down to -200°C. To calibrate the Cu/Ni sensor we use the thermocouple TCA-BTA wich has range of temperature measurement from -200°C to 140°C. From the calibration we can obtain information regarding sensor sensitivity. Besides this, we also investigate the response times [8], stability of output voltage, and hysteresis losses. These four parameters are important parameters for the sensors.

2. Experimental method

2.1. Materials

The temperature sensor used as object in this research is the Cu/Ni films resulted from electroplating technique with deposition parameters as specified in Table 1. The experimental devices are arranged according to design shown in Fig.1. In the sensor calibration, 3 liters liquid nitrogen as low temperature medium is inserted into the flask that has a volume of 5 liters. Temperature of medium is measured using Cu/Ni sensor by immersing the sensor into the flask slowly at an average speed of 0.17 cm/s to the bottom of flask about 50 cm depth. With this speed, the decreasing temperature data from time to time gradually can be recorded properly.

Table 1. Specifications of Deposition Parameters in Making Cu/Ni Sensors.

Cathode	: Cu
Anode	: Ni
Electrodes distance	: 4 cm
Electrolyte	: H_3BO_3 (7.5 gr), $NiSO_4$(75 gr), $NiCl_2$(10 gr), and H_2O (250 ml).
Electrolyte temperature	: 60°C
Current density	: 0.07 A/mm^2
Deposition time	: 1-5 minute
Estimated Thickness of Ni film	: (10.3 – 81.5) nm

Temperature measurement using Cu/Ni sensors cannot directly obtain the temperature quantity, but in the potential difference between the two ends of the sensor. Potential differences occur because of changes in resistivity in the Cu/Ni sensor as a result of changes in medium temperature [9-12]. Therefore, it is important to ensure that the measured voltage is pure from the sensor. To ensure the potential difference at the ends of the sensor is really from the sensor not included the voltage from the wire connected at the two ends of the sensors, the sensor is connected to the 4-WCB (Four Wire Configuration Bridge) transducer. To enlarge the output voltage (i.e. in order of millivolts), the voltage is amplified using Op-Amp up to 100 times more so that the output voltage is able to be acquisited with a VP-BTA voltage probe sensor that has range 0 – 5 volts.

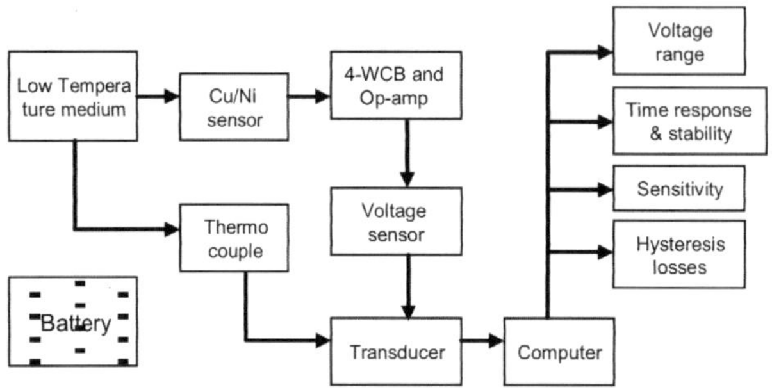

Figure 1. Block diagram of research.

Furthermore, the voltage probe from the Cu/Ni sensor and the thermocouple probe are connected to the LabQuest Mini transducer and then from the transducer to the computer. The process of voltage and temperature data logging can be done coincident with the help of software loggerpro 3.8. Sampling rate is set to 2 data/s for 5 minute.

The next step is inserting Cu/Ni sensors into a flask together with the TCA-BTA thermocouple. Here, thermocouple is used to calibrate Cu/Ni sensors. According to data catalog this thermocouple has the ability to measure temperatures from -200°C to 1400°C, and for specific range temperature measurement, i.e. -200°C to 0°C it has a precision of ±5°C. When the sensor is still outside the flask, the sensor measures the room temperature. Since the sensor starts to enter the flask, the sensor measures the temperature of the medium which gradually drops from 0°C down to -200°C. The temperature difference between the top flask and the bottom one is due to the top of flask containing nitrogen gas so that the temperature is higher than that at the bottom containing liquid nitrogen. As revealed by Kar and Sharma, nitrogen starts to melt at -196°C [8-9], while the bottom of the flask contains liquid nitrogen, the temperature reaches -200°C.

After the temperature of -200°C is reached, the sensor is pulled up slowly until it is outside the flask. The logged data is displayed in two versions that are numeric and graphic. The numerical data contains the arrays of time data (t_i), sensor voltage (V_i) and temperature of thermocouple (T_i), while the graphical data contains the curve of V-t from Cu/Ni sensor and T-t from thermocouple. Data is then analyzed to obtain a range of sensor voltages, response times, stability, sensitivities, and hysteresis losses.

2.2. *Procedure of Data Analysis*

1) *Voltage range*. The temperature to be measured has the range of 0°C down to -200°C. The measured temperature is expressed in voltage quantities. In fact the magnitude of the output voltage of each sensor varies in the range of temperatures between 20°C to 200°C. Voltage range of each sensor is obtained from the output voltage difference when the sensor will be used to measure the two temperatures that are 20°C and -200°C. Range voltage is a parameter that contributes to the resolution of sensor. The greater the size of the voltage range leads to the greater resolution.
2) *Accuracy of Sensor*.The accuracy of sensor is shown through a clear and consistent relationship between output voltage and temperature. Some theories revealed that the relationship between resistivity and temperature is linear, polynomial order-3, or even any order polynomial. However, the simplest form is linear [13-14]. Shortly speaking, the most important thing is the consistency of the relationship between temperature and resistivity. Because of the resistivity here is

represented by the voltage, the relationship between resistivity and temperature is substituted by the voltage and temperature $V(T)$.

3) *Sensitivity and stability of sensor.* The description of the sensitivity level of sensor is done by matching the sensor voltage data (V) to temperature (T) in the temperature range -140°C to 0°C. The voltage variation of the sensor describes the variation in sensor resistance because the temperature of the medium changes when it is electrified.
As shown in the Matthiessen formula for pure metals [15-16], the total resistivity of the sensor follows the PTC (Positive Temperature Coefficient) pattern. This resistivity is contributed by two factors, i.e.: first, from electron scattering when collides to the lattice vibration, and second, by scattering electrons due to collision with the residue, that consists of crystal defects and impurities in the material. The first type is dependent temperature, while the second type is independent one.
From the V-T curve, the most sensitive sensors have the largest slope. Furthermore, the stability of the sensor is shown by the precision level of the sensor in measuring temperature. Precision level can be seen from the magnitude of the data deviation (V_i, T_i) to the V-T curve. Overall, it can be represented by the standard deviation. Then, the sensor that has the smallest standard deviation will indicate that the sensor is the most accurate in appointing the measurement temperature, or the most stable.
4) *Response time of sensor.* To determine the response time of sensor, the activity is divided into two parts, the first is when the temperature decreases from -20°C down to -200°C and the second is when temperature rises from -200°C to -20°C. This division is important by considering that the response time of sensor to the decreasing temperature is different from the response time for increasing one, so it needs an explanation separately. The upper limit of temperature is limited to -20°C because of a consideration that in the future the application of sensor will be dedicated more specifically for cryogenic sensors. As an example in the animal husbandry, the upper temperature limit for storaging cow cement is -100°C. So that temperatures above -40°C are less relevant. Then, response time of sensor is obtained by drawing a graph of voltage (V_i) with respect to time (t_i). The most suitable equation to express the relationship between V_i data and t_i is exponential, i.e.:

$$V = m_1 e^{-m_2 t} + m_3 \quad (1)$$

where V is voltage, m_1, m_2 and m_3 are fitting constants. The m_2 is a time constant which is a measure of the sensor's speed in response to the temperature. The increasing m_2 leads to the sensor faster in responding the medium temperature.
5) *Hysteresis losses.* An error in the sensor is ordinary thing. There is no perfect sensor even for commercial equipment produced by the same factory even was set to the same parameters. The results of the reading can be slightly different from the true value. May be this is relating with selection of the quality of the dc voltage source, the quality of the circuit, properly connection between electronic components, and the compatibility between the current from DC power and the ability of sensor to accept and flowing the current [17–21].

A good temperature sensor is a sensor that can exhibit the same temperature even used at other times. Hysteresis loss in here is a loss arising due to the difference voltage, if a sensor is used to measure the same temperature. If the voltage difference occurs when the medium temperature is decreased from -20°C down to -200°C and vice versa the hysteresis lop will be formed.
The hysteresis curve should be like a leaf, which is balanced between the left and right. But in this research-due to technical reasoning- the thermocouple is only able to measure the low temperatures down to -140°C, meanwhile the Cu/Ni sensor is able to continue its measuresment up to the temperature of liquid nitrogen (-200°C). Then a part of the hysteresis loop from -140°C to -200°C is not appeared (see Fig. 2).

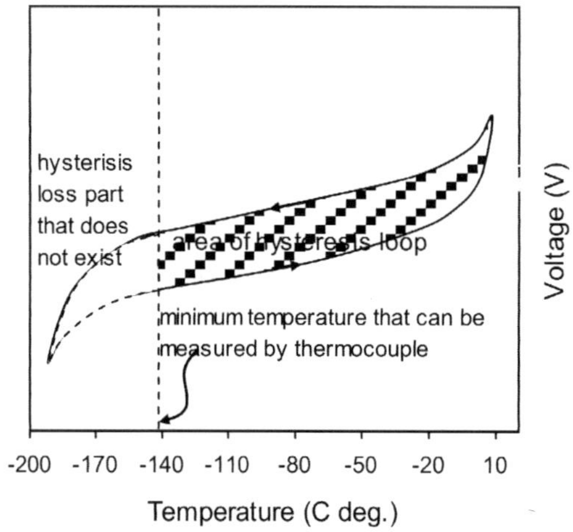

Figure 2. Determination of the area of hysteresis loop.

The hysteresis loss is expressed by the shaded area. The area is computationally determined by following the trapezoidal formula i.e.

$$A = \frac{h}{2}\left(y_0 + y_n + 2\sum_{1}^{n-1} y_i \right) \tag{2}$$

where h is the distance between temperature T_i that is about 0.55 min, and y is voltage V.

3. Results and discussion

3.1. Output Voltage in Measuring Liquid Nitrogen

In Fig. 3, as an example, the measurement of liquid nitrogen temperature done by thermocouples and by Cu/Ni sensor is displayed together. The Cu/Ni sensor was produced from 5 minute deposition. The data logging was set for 5 minute with the sampling rate of 2 data/second. On the left side, the time, temperature and potential data are displayed numerically, while on the right side output voltage and temperature against the time is displayed graphically.

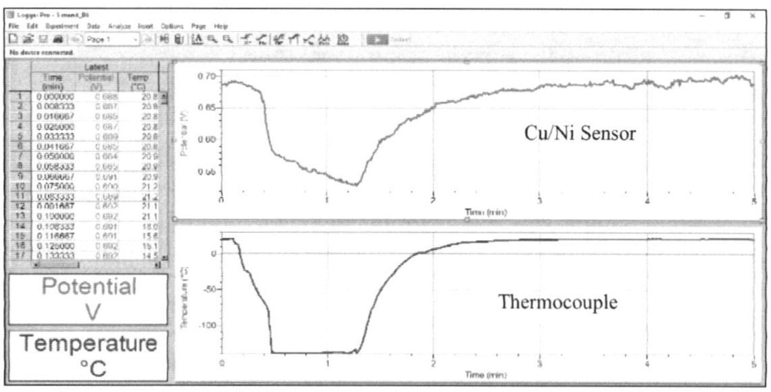

Figure 3. Time, voltage and temperature data arrays (left), and the voltage and temperature with time curves (right) displayed by loggerpro software. the data was obtained from measurements by thermocouples and Cu/Ni sensor produced from 5 minute Ni deposition on Cu.

From the figure, one can see the minimum temperature from thermocouples that can be reached that is -140°C, which is shown by horizontal lines (the botttom fig.), while Cu/Ni sensors still continue to decrease until they reach the temperature of -200°C. Then, with the same way, the data collection of liquid nitrogen temperature is carried out for all sensors together with the thermocouple. After processing data, the results are displayed together in the V-t curve as shown in Figure 4.

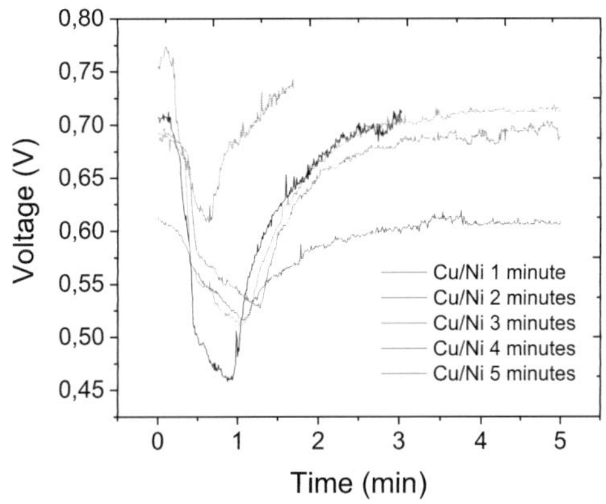

Figure 4. The curves of voltage from Cu/Ni sensors resulted from deposition Ni at variations of deposition time 1–5 minute in measuring temperature of liquid nitrogen.

All sensors exhibit their ability to respond the changes on temperature of liquid nitrogen (Fig. 4). The minimum temperature that can be measured by thermocouples is -140°C while Cu/Ni sensors are still able to continue to measure temperature to -200°C.

Unfortunately, the ripple is found to still accompanies the output voltage with different amplitude and position. On relatively high temperature, above -40°C the ripple looks more fluctuating than on below -40°C. This ripple has strong relation to the fluency of electric current flowing on the surface of sensor. The ripple occurs due to the poor quality on connection between sensors and cables, cables and electronic components, and the incopatibility between the electric current flowing in sensor and the size of sensor.

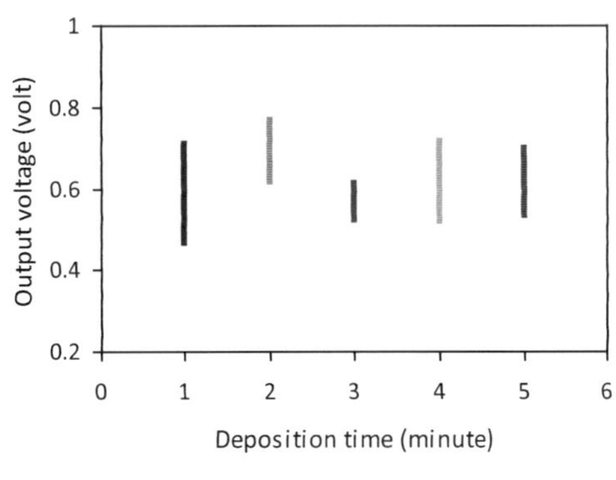

Figure 5. The range of output voltage of Cu/Ni sensors when used to measure temperatures from 20°C – -200°C. (a) for decreasing temperature, (b) for increasng temperature.

From the investigation on the range of output voltage of Cu/Ni sensors as shown in Fig. 5, the values vary from 0.10 V to 0.25 V. The largest range is in accordance to Cu/Ni sensors produced from 1 minute deposition that is 0.25 volt. The range of voltage in temperature sensor has important role due to the relation to sensitivity of sensor. Sensor that has a greater voltage range can respond a wider temperature range so it is more accurate in appointment the temperature because the little change in temperature has been able to produce the change of voltage. Thus, the Cu/Ni sensor obtained from deposition for 1 minute has the largest sensitivity compared to the sensor resulted from another deposition time.

3.2. Response Time of Cu/Ni sensors

In Fig. 6 it can be reported the time response of sensor in sensing liquid nitrogen from 20°C to -200°C and vice versa. For the temperature decrease from 20°C to -200°C the curve (Ti, ti) of the thermocouple almost coincides with the curve (Vi, ti) of Cu/Ni sensor.

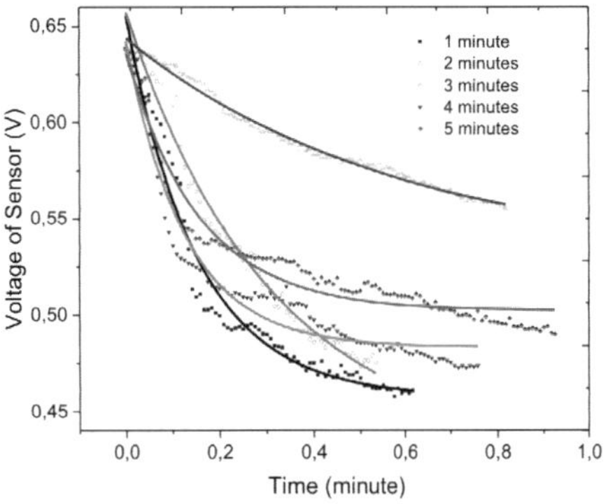

(a)

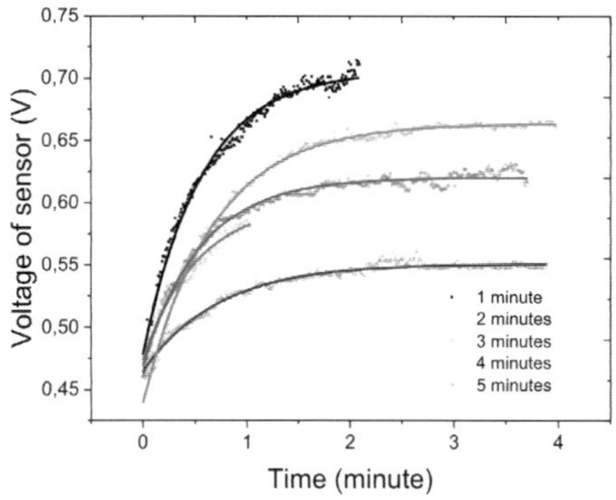

(b)

Figure 6. Data fitting of voltage-time to determine the time constant. (a) decreeasing temperature of N_2, (b) rising temperature of N_2.

This shows that Cu/Ni sensors can respond to temperature changes as well as thermocouples. However, this becomes different for the rising temperature from -200°C to 20°C, the position of the Cu/Ni sensor curve is rather more right than the position of thermocouple curve. This indicates that Cu/Ni sensors experience slight delays in responding temperature changes if it is compared to thermocouples. Indeed, for thin layer-based temperature sensors - as said by Anand - the average response time is slower than that for thermocouples [22]. However, the advantage of Cu/Ni sensor is on the ability to measure the low temperature up to -200°C lower than the low temperature able to be reached by the thermocouples. To determine the response time of each sensor the fitting of data (Vi, ti) is performed for the decreasing temperature and also for the rising temperature according to the exponential function according to (1). The result is displayed on Fig. 6. The value of the time constant (m2) is tabulated in Table 2 column 2 for decreasing nitrogen temperature and column 4 for increasing nitrogen temperature.

Table 2. Time Constant (m_2) for Increasing and Decreasing Temperatures of N_2, and Time to Reach Temperature -200°C.

Deposition Time (min)	Time constant for decreasing temperature of N_2 (min.$^{-1}$)	Relative error	Time constant for increasing temperatureof N_2 (min.$^{-1}$)	Relative error	Time to achieve Temp. of -200°C (min.)
1	6.63 ± 0.37	5.6%	1.77 ± 0.03	1.7%	0.93
2	3.28 ± 0.30	9.1%	1.93 ± 0.15	7.8%	0.66
3	1.77 ± 0.11	6.2%	1.39 ± 0.02	1.4%	1.08
4	7.94 ± 0.43	5.4%	1.50 ± 0.01	0.7%	1.08
5	6.42 ± 0.34	5.3%	1.80 ± 0.02	1.1%	1.08

Due to the differences in the time constants, the time needed for the sensor to reach a temperature of -200°C also varies. For decreasing temperature N_2, sensor Cu/Ni resulted from 4 minute deposition time is the fastest sensor in responding to the changes of temperature. This sensor has the time constant of (7.94 ± 0.43)/minute. Meanwhile for increasing temperature of N_2, the sensor produced from 2 minute deposition time is the fastest in response to temperature changes and also the fastest to reach a temperature of -200°C that are 1.93/minute and 0.66 minute respectively.

From previous work [7], regarding to XRD analysis it is obtained that the sensor resulted from 2 minute of Ni deposition has the highest diffracted beam intensity compared to other samples so that it is the most regular crystal structure. But from analysis of crystal size by using Scherrer formula [23,24], that sensor has the smallest particle size i.e. 0.38Å. The small particle size is thought due to the thickness of the Ni layer that relatively thin so that the deposit is still dominated by diffusion layer. From the formula of the relationship between the thickness of the layer d (μm) and the deposition time t (minute) as has been obtained, i.e. $d = 1.28\times10^{-5}t - 4.43\times10^{-7}$ (m), the thickness of Ni produced in 2 minute deposition is 0.25 μm while the thickness for that in 4 minute deposition is 0.51 μm [7].

As known, a good low temperature sensor is an alloy between Cu and Ni to form CuNi [25]. In electroplating this can not be carried out because the Ni layer attached to the surface of Cu to produce a Cu/Ni layer even though at the interface there are Ni atoms which diffuse into the Cu surface to form diffusion layer. This part play an important role in determining the quality of temperature sensor. The diffusion process can also move atoms, molecules, and clusters of atoms on the surface of the material so that mutual diffusion or interdiffusion is possible to be occured. Therefore, diffusion clearly affects the morphology of the layer.

The diffusion of Ni atoms into the Cu layer is relatively easy because the size of both atoms is almost the same, which is 0.1278Å for Cu and 0.1246Å for Ni. The diffusion process only occurs in a few moments until the crystal defects in Cu are filled, and after that Ni atoms continue the coating process to produce a continuous layer to form Stranski Kastranov type [26]. The thickness of effective diffusion layer formation according to Fellner is about 3 μm [27]. When the layer is more continuous, the current density that passes through it increases [28]. Similarly, by considering that Cu is a good conductor and low resistivity (1.7×10-8 Ωm) while Ni has a high resistivity (7.0×10-8 Ωm), if Cu and Ni are combined the resistivity of Cu will increase.

As in previous explaination, the presence of a ripple in the output voltage signal indicates that the electric current on the sensor is not yet stable. The level of stability of the output voltage signal can be seen in the relative error of the time constant (column 3 and 5, Table 2). The more stable output voltage signal is indicated by the small relative error. From this, for N2 temperature down Cu/Ni sensor ressulted from 5 minute deposition is the most stable output voltage with a 5.3% relative error. However, for rising N2 temperature, Cu/Ni sensor resulted from 4 minute deposition is the most stable with relative error of 0.7%. Conversely, the samples produced from 2 minute deposition had the largest relative error both for decreasing and rising N_2 temperature, that are 9.1% and 7.8%. This sensor is the most unstable voltage signal and not good if it will be used as temperature sensor.

3.3. Sensitivity of Sensor

Sensitivity is the most important quantity of sensor quality because the essence of study about sensors is to find materials which sensitive to the change in other quantities. The other quantities are complementary, although it is also very important. The relationship between sensor output voltage (V) and medium temperature (T) is displayed in Fig.7. To explain the sensitivity of the sensor, an analysis of the graphs of data fitting (Vi, Ti) was carried out according to the linear regression. The result is a linear equation as shown in Table 2. From the value of the determination index R2 approaching 1 as shown in column 3, it indicates that all samples have a strong linear relationship between the output voltage and the temperature of N2. This is an advantage of this Cu/Ni sensors, by considering the difficulty of finding the cryogenic sensors that have a linear relation in both electrical and thermal properties to the temperature, as revealed by Lebioda and Rymaszewsky [4] or Yeager and Courts [2].

Information about sensitivity levels of sensor can be investigated through the slope of graph. The slope of graph illustrates the change of the sensor voltage due to the changes in temperature of the medium. The larger slope of the curve the more sensitive the sensor. Sensitivity level of each sensor is shown in Table 2 column 4.

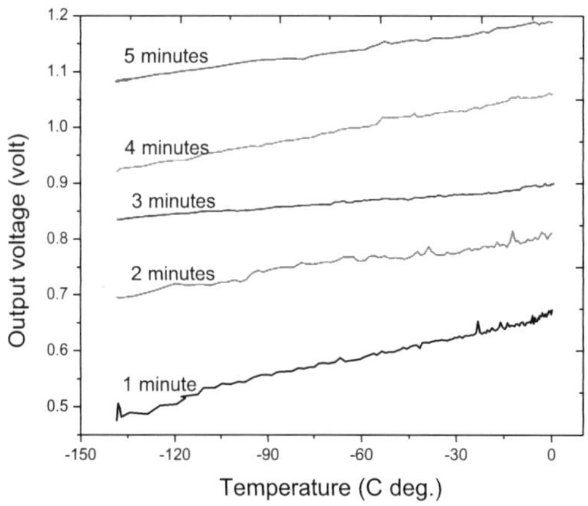

Figure 7. Sensitivity of the sensor.

Table 3. Linear Equation between Sensor Output Voltage and Medium Temperature.

Deposition Time (minute)	Relationshio between voltage and temperature (volt)	Determination Index, R^2	Sensitivity (mV/°C)
1	V=0.0012 T + 0.6634	0.99	1.26 ± 0.01
2	V= 0.0007 T + 0.7213	0.95	0.74 ± 0.02
3	V=0.0004 T + 0.5803	0.98	0.43 ± 0.01
4	V=0.0010 T + 0.6672	0.99	1.00 ± 0.01
5	V=0.0008 T + 0.6340	0.99	0.76 ± 0.01

In general, all samples have shown a positive response to temperature changes of N_2 from -140°C to 0°C. Table 2 column 4 show that the sensitivity level is influenced by the deposition time. The sensitivity level of sensor ranges from 0.43 mV/°C to 1.25 mV/°C. This sensitivity is higher than that has been resulted by Yang for Cu/CuNi sensor to measure temperatures from 30°C to 200°C. The sensitivity is about 43.94 μV/°C to 46.49 μV/°C [29]. Furthermore, the highest sensitivity of sensor that Yang obtained is 46.49 μV/°C concerning to the sample as thick as of 0.5 μm. In this study, the deposition of Cu/Ni sensors within 1 minute is according to the thickness of 12.40 μm and has the highest sensitivity compared to other sensors, that is (1.25 ± 0.01) mV/°C.

This sample is interesting as the sensitive sensor and quite fast in responding to temperature changes especially for the decreasing medium temperature, with a time constant of (6.63 ± 0.37) /minute which is number 2 after the time constant for sensor produced in 4 minute deposition. Similarly, the time needed to reach a temperature of -200°C from 20°C is quite short, that is 0.66 minute. The next level of sensitivity is (1.00 ± 0.01) mV/°C that is in accordance with the sample produced from 4 minute deposition. This sample also has the highest time constant the decreasing medium temperature that is (7.94 ± 0.43)/minute for.

3.4. Hysteresis Losses

In completing information about the characteristics of Cu/Ni sensors, information about hysteresis loss was displayed in the form of hysteresis area, as pointed on Fig. 5. One of the characteristics of a good temperature sensor is having the smallest hysteresis area. For the ideal sensor the measured temperature of the medium in the decreasing temperature process is same to the temperature in increasing process. The hysteresis loss describes the loss due to inequality of sensor to measure the temperature of the medium at different times that are for decreasing and increasing temperatures. Hysteresis loss is also related to reproducibility.

According to Aydemir [30] and dan Fank [28] the existence of hysteresis is influenced by the structure of film. They also suggested for reducing hysteresis losses 1/2 up to 1/3 times that should be done by annealing the film at the certain temperature. In this study the measurement of temperature is done by Cu/Ni sensors, and the output in a voltage quantity.

Hysteresis loop is formed from depict (*Vi, Ti*) data for decreasing temperature from 20°C to -140°C connected continuously by the data for increasing temperature from -140°C to 20°C. In Fig. 8 the hysteresis curve of each sample is displayed. Here, a part of hysteresis loop on the left side is not present because the minimum temperature that is possible be measured by thermocouple is -140°C, although the Cu/Ni sensor is still able continue measuring the medium temperature lower that is -200°C. So the curve V-T below -140°C doesn't exist. Therefore, the hysteresis loss is presented by the hysteresis area according to (2) not in percent hysteresis. From the figure it appears that all samples have a different hysteresis loss.

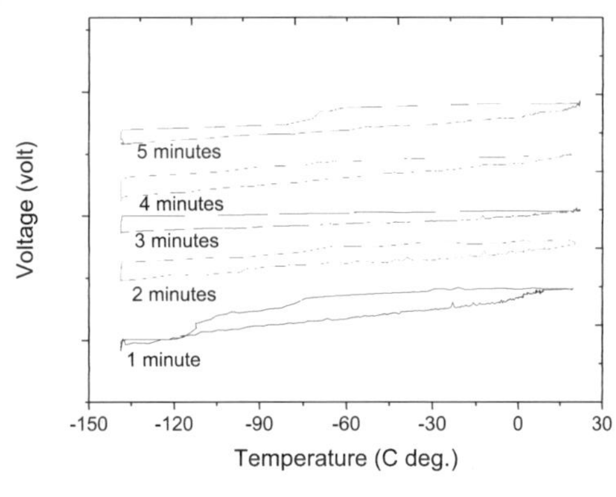

Figure 8. Hysteresis loop of the voltagetemperature curve of the cu/ni sensor, resulted from deposition in a time variation.

The smallest hysteresis is (6.640 ± 0.005) V°C corresponds to the Cu/Ni sensor from deposition of Ni within 3 minute. Meanwhile, the largest hysteresis loss is in accordance with the sensor resulted from 4 minute Ni deposition, that is (20.653 ± 0.005) V°C.

Sensors that have a small hysteresis area will be more accurate if used to measure temperature at any time. Unfortunately, the hysteresis loop curve of all these sensors is less complete, that is for the section at temperature below -140 °C so that in this study we cannot conclude exactly which sensor that has the smallest hysteresis loss. At least the existence of these results has been able to provide an overview of the limitations of sensors in providing precision temperature measurements. As a consideration, from the previous work on study about microstructure of samples it was found that there was a tendency that the increase in deposition time made the arrangement of Ni crystals became irregular even though their thickness increased. The irregularity of the crystal arrangement is characterized by the decreasing intensity of the XRD diffraction peak. But d-spacing is getting closer even this sample has the smallest that is 1.2166Å. With the small d-spacing, it is expected that electric conduction will take easier both the decreasing and rising medium temperatures. It leads to the hysteresis area smaller, and then reducing the hysteresis losses.

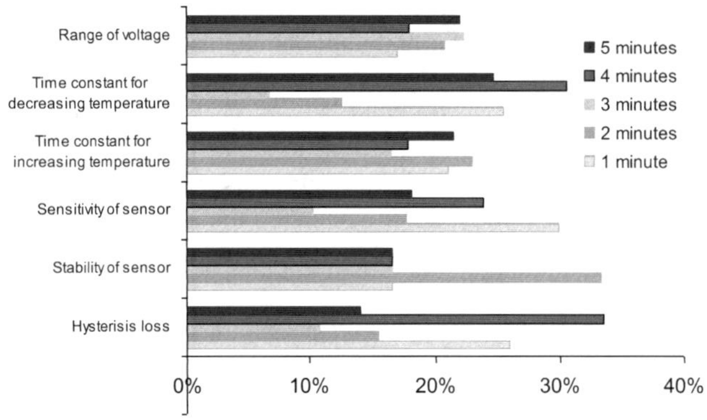

Figure 9. Characteristic of Cu/Ni sensors on the various deposition time.

As a resume from the discussion and analysis that have been done, in Fig. 9, an illustration of the advantages and disadvantages of each sensor which consists of a range of voltage, time constant for decreasing and increasing temperatures, sensitivity of sensors, stability of sensor, and hysteresis losses is given. It is not easy to obtain a sensor produced by electroplating in a variation of the deposition time which has advantages on all sensor quality parameters. But the most important parameter here is sensor sensitivity, while other parameters role as contributing factors in complementing the main parameter. The Cu/Ni layer produced from 1 minute deposition is the most sensitive one compared to the other sensors but has the smallest voltage range, while the response time if used to measure the medium temperature which goes both decrease and increase is quite well. The second one is Cu/Ni sensor produced on 4 minute of Ni deposition. This sensor is most responsive to the decreasing medium temperature, sensitive, has stable voltage signal in both decreasing and increasing medium temperatures but the time constant is not so large for the increasing temperature so the time to reach the -200°C is rather slow compared with the sensor produced in 1 minute of Ni deposition. Unfortunately, both 1 minute and 4 minute sensors of Ni depositions have the large hysteresis losses. Furthermore, another sensor is not considered because of the weakness in the main quality parameter, that is sensitivity.

The sensor is stable enough compared to other sensors axcept the sensor resulted within 2 minute of Ni deposition. Similarly, the loss of hysteresis is still quite large. The Cu/Ni sensor resulted from 2 minute of Ni deposition has a sensitivity level below the Cu/Ni film sensor resulted from 1 minute deposition. The range of the output voltage of this sensor is not so large, and also the time constant for the decreasing medium temperature is less good, while the time constant for the rising medium temperature is quite good, the output voltage signal in response to temperature changes is stable enough, but this sensor has a large enough hysteresis loss.

Therefore, in this preleminary study, eventhough we have obtained deposition time which is capable to produce Cu/Ni sensors with high sensitivity, it still contains a further tasking in finding solutions to some deficiencies that accompany on these parameters of sensitivity. It is expected that the results of further research can overcome these deficiencies so that the Cu/Ni low temperature sensor which has more quality parameters can be found.

4. Conclusion

The plating of Ni on Cu substrate by varying the mass fraction of Ni in solution influences the thickness of Ni film, microstructure, and sensitivity of Cu/Ni as a low-temperature sensor. The

IOP Conf. Series: Materials Science and Engineering **924** (2020) 012024 doi:10.1088/1757-899X/924/1/012024

thickness of Ni is proportional to the electrolyte concentration. Analysis on resistivity obtained information that resistivity of Cu/Ni depends not only on the thickness of the Ni film but also by the regularity of the crystal and grain size. All samples can be used as low-temperature sensors to measure temperatures from 0°C to -140°C. Sensitivity is a linear relation to ambient temperature where S5 has the highest sensitivity that is 1.05×10^{-3} V/°C.

Acknowledgment
The researcher expresses his deepest gratitude to the ministry of Research, Technology and Higher Education of the Republic of Indonesia which has provided research funding through the 2019 Higher Education Advanced Applied Research (PTUPT) scheme, contract number PTUPT-005/SKPP.TJ/ LPPM UAD/III/2019.

References

[1] V. M. Soto, D. Leandro, A. L. Aldaba, J. J. B. López, J.I.P. Landazábal, J. L. Auguste, R. Jamier, P. Roy and M. L. Kar Amo, "Study of Optical Fiber Sensors for Cryogenic Temperature Measurements", Sensors, Vol. 17, pp. 1-12, Nov. 2017.
[2] C.J. Yeager, S.S. Courts, "A review of Cryogenic Thermometry And Common Temperature Sensors". IEEE Sens. J., Vol.1, no. 4, pp. Dec. 2001.
[3] A. Ukil, H. Braendle, P. Krippner, "Distributed Temperature Sensing: Review Of Technology And Applications", IEEE Sens. J., pp. 1-8, May 2012.
[4] Lebioda, M., Rymaszewsk, J., Dynamic Properties of Cryogenic Temperature Sensors, Przegląd Elektrotechniczny, ISSN 00332097, R. 91 NR 2/2015.
[5] T.J. Quinn, "The International Temperature Scale of 1990 (ITS-90)", Physica Scripta. VO. 41, pp. 730-732, Oct. 2017.
[6] B.W. An, J.H. Shin, S.Y. Kim, J. Kim, S. Ji, J. Park, Y. Lee, J. Jang, Y.G. Park, E. Cho, S. Jo and J.U. Park, "Smart Sensor Systems for Wearable Electronic Devices", Polymers, Vol. 9, No.8, pp. 2-40, July 2017.
[7] M. Toifur, A. Nurmalia, Okimustava, I. Sukarelawan, "The Efeect of Deposition Time on The Film Thickness, Microstructure, Sheet Resistivity of Cu/Ni Thin Film Resulted from Deposition by Electroplating Technique (Pengaruh Waktu Deposisi Pada Tebal Lapisan, Struktur Mikro, Resistivitas Keping Lapisan Tipis Cu/Ni Hasil Deposisi Dengan Teknik Elektroplating)", Jurnal Material Dan Energi Indonesia, 2017.
[8] S. Kar and R.G., Sharma, "Cryogenic Temperature Sensors", DEF SCI J, Vol. 57, No.3, pp. 195-208,May 2007.
[9] M.A. AlJalali, "Temperature Dependence Behavior of Electrical Resistivity in Noble Metals at Low Temperatures", Journal Of Advances in Physics, Vol. 5, No. 3, pp. 982-992, Oct. 2014.
[10] S. Pal, "Electrical Resistivity of Noble Metals", Canadian Journal of Physics, Vol. 51, No. 21, 2225-2232, Nov. 1973.
[11] A. Alam and M. Tomak, "Electrical Resistivity of Liquid Noble Metal Alloys", Physica Status Solidi (B), Vol. 126, No. 729, pp. 729-732, Sept. 1984.
[12] A. Kaushal, A. Vardhan, A.C. Tiwari, S.K. Saluja, "Study The Effect Of Cryogenic Cooling On Orthogonal Machining Process", American Journal of Engineering Research (AJER), Vol. 5, No.3, pp.96-104, March. 2016.
[13] Acherman, S.O. "Studies on The Temperature Dependence of Electric Conductivity for Metals in the Nineteenth Century: A Neglected Chapter in The History of Superconductivity", Revista Brasileira de Ensino de F´ısica, Vol. 3, No. 4, pp. 1-13, Dec. 2011.
[14] A.A.A. Elfaki, A.A. Elamin, M.D.A. Alla, R.A. Elgani, A.S. Mohammed, A.A. Mohammed, S.A. Elhouri, "The Effect of Temperature on Conductivity of Conductors and Superconductors", American Journal of Physics and Applications, Vol. 5, No.1, pp1-5, Feb. 2017.
[15] E. Borchi, S. De Gennaro, P.L. Tasselli, "On the LowTemperature Electrical Resistivity of

IOP Conf. Series: Materials Science and Engineering **924** (2020) 012024 doi:10.1088/1757-899X/924/1/012024

Copper and Gold Borchi", Journal of Physics F Metal Physics, Vol. 10, No.10, pp.270-274, Oct. 2015.

[16] E. Borchi, S. De Gennaro, and P. L. Tasselli, "Low Temperature Electrical Resistivity of Noble Metals", Phys. Rev. B 12, Vol. 12, No. 12, pp. 5479-5487, Dec. 1975.

[17] L.P. Madr, J. Fonseca, A. Matute, E.I.G. Velásquez and C.A. Palacio, "SelfCompensated Driving Circuit for Reducing Drift and Hysteresis in Force Sensing Resistors", Electronics, Vol. 7, No. 146, pp. 1-20, Aug. 2018.

[18] A. Matute, L.P. Madrid, G. Moreno, F. Cárdenas, and C.A. Palacio, "A Novel and Inexpensive Approach for Force Sensing Based on FSR Piezocapacitance Aimed at Hysteresis Error Reduction", Hindawi Journal of Sensors, pp. 1-6, April, 2018.

[19] H. Hilber and B. Jakoby, "Hysteresis and Material Effects of Printed Strain Gauges Embedded in Organic Coatings", in Proceedings Euro Sensors Paris, 2017, pp. 2-4.

[20] C. Gui, R. Legtenberg, H.A.C. Tilmans, J. H. J. Fluitman, and M. Elwenspoek, "Nonlinearity and Hysteresis of Resonant Strain Gauges", Journal of Microelectromechanical Systems, Vol.7, No.1, pp.122-127, March 1998.

[21] M. Rakotondrabe, C. Clevy and P. Lutz. (2008, Sep). Hysteresis and Vibration Compensation in a NonLinear Unimorph Piezocantilever IEEE/RSJ International Conference on Intelligent Robots and Systems, IROS'08. [online] pp.558-563. Available: http://hal.archives-ouvertes.fr/hal-00325162.

[22] A. Anand, A. Aggarwal, J. Kumar, "Resistance Temperature Detector (RTD)", International Journal of Innovative Research In Technology, Vol.1, Issue 12, pp.15-19, 2015.

[23] A. Monshi, M. R. Foroughi, M. R. Monshi, "Modified Scherrer Equation to Estimate More Accurately NanoCrystallite Size Using XRD", World Journal of Nano Science and Engineering, Vol.2, No.3, pp. 154-160, Sep. 2012.

[24] S. Bykkam, M. Ahmadipour, S. Narisngam, V. R. Kalagadda, S. C. Chidurala. (2015, Feb.). Extensive Studies on XRay Diffraction of Green Synthesized Silver Nanoparticles, Advances in Nanoparticles [Online], 4, pp.1-10. Available: http://www.scirp.org/journal/anp.

[25] R. G. Delatorre, M. L. Sartorelli, A. Q. Schervenski, and A. A. Pasa, Thermoelectric Properties of Electrodeposited CuNi Alloys on Si, Journal of Applied Physics, Vol. 93, No. 6154, pp. 6154-6157, May 2003.

[26] A.L. Vázquez de Parga, J.J. Hinarejos, F. Calleja, J. Camarero, R. Otero, R. Miranda, "Quantum oscillations in surface properties", Surface Science 603, Vol. 603, pp. 1389-1396, January 2009.

[27] P. Fellner and M.C. Paučírová. "Kinetics of Formation of Diffusion Layer at The Electrolytic Deposition of Aluminium on Copper Base from Molten Electrolyte", Chem. Zvesti, Vol. 36, No.3, pp. 322-325, Sep. 1982.

[28] N.D.M. Sin, M.H. Mamat, and M.R. Mahmood, "Effect of Deposition Time On Properties Of Nanostructured ZnO Thin Films Deposited by RF Magnetron Sputtering", Advanced Materials Research, Vol. 832, pp. 460-465, Nov. 2014.

[29] L. Yang, Y. Zhao, C. Feng, H. Zhou, "The Influence of Size Effect on Sensitivity of Cu/CuNi Thinfilm Thermocouple", Physics Procedia, Vol. 22, pp.95-100, Nov. 2011.

[30] B. Aydemir, L. Yagmur, S. Fank, "Hysteresis Errors of Commonly Used Sensor Materials", Measurement, Vol. 43, pp.792-796, Feb. 2010.

Hydrochloric Acid Leaching Approaches for Tantalum and Niobium Pentoxides Enrichment in 1st Tin Slag from Bangka Island

Sulaksana Permana[1,2*], Rafdi Abdul Majid[3], and Johny Wahyuadi Soedarsono[1]
[1]Centre of Mineral Processing and Corrosion Research, Department of Metallurgy and Materials, Universitas Indonesia, 16424 Depok, Jawa Barat, Indonesia
[2] Institut Sains dan Teknologi Al-Kamal Jl. Raya Al Kamal No.2 Kedoya Selatan, Kebon Jeruk Jakarta Barat, DKI Jakarta 11520, Indonesia
[3] Department of Chemical Engineering, Faculty of Industrial Technology, Universitas Muslim Indonesia, Panakukang Makassar 90231

*sulaksana@yahoo.com

Abstract. Indonesia is the second largest tin producer in the world that resulting abound of by-product. This by-product contains some of valuable metal oxides, such as 0.30% tantalum (Ta_2O_5) and 0.58% niobium pentoxide (Nb_2O_5). The supply deficit of these minerals induces researchers worldwide to conduct investigations. This study explored the effect of roasting at 900°C, NaOH quenching, and HCl leaching with 45, 65, and 85°C temperature variables and 4M, 6M, 8M concentration variables. These measures were employed to upgrade the contents of tantalum and niobium pentoxide (TNP). The filtration products of quenching and leaching were characterized by X-Ray Fluorescence (XRF). On the one hand, 900°C roasting and NaOH quenching decreased the 100 mesh-measuring TNP contents, 0.14% Ta_2O_5 and 0.2% Nb_2O_5. On the other hand, HCl leaching with a 65°C temperature variable showed an optimal yield ratio of 3.75 Ta_2O_5 and 3.57 Nb_2O_5. The total yield ratio of TNP is 3.63.

Keywords **:** 1st tin slag, tantalum and niobium pentoxide, roasting, quenching, hydrochloric acid leaching

1. Introduction

Tantalum and niobium are elements which are not found freely in nature, but are always associated with several types of metals to form oxide minerals. Tantalum and niobium are included in the category of critical metal in the world, this is related to the assessment results of 78 types of metal produced by the European Commissioning in 2017 [1]. On the other hand, both of two metals have many uses in high-tech applications such as the aircraft industry, electronics, super alloys, catalysts, capacitors, energy and superconductors [2]. A survey on tantalum supply and demand in 2013 indicated that in 2015, this chemical element would undergo a deficit of 79,000 lbs. and in 2016, of 212,000 lbs. Roskill estimated that the total demand for tantalum would have been growing by 3.3% CAGR for the next nine years, 2017-2026 [3]. Research on a tantalum secondary source from tin slag has been conducted for several decades, between 1964 and now [4-24].

The general process of tantalum extraction involves smelting raw materials comprised of tin concentrates, flux additives, and reducing agent [24]. This process manufactures a by-product called 1st tin slag (1TS) which holds approximately 20 % tin. The next process is a tin recovery which also employs smelting. This process then produces 2nd tin slag (also called as Bangka Tin slag (BTS) [23]) by-product.

After 2011, research on tantalum from tin slag relatively increased. The rise of exploration was driven by a tantalum deficit. Supply of tantalum is predicted to exist until 2020 [25]. Studies of metal recovery mostly involved roasting and quenching since these two processes could create more pores and extend the wet touched areas. These wet touched areas would improve the effectiveness of dissolution or leaching. Roasting and quenching were applied to research on rich husk [26] and tin slag [23] to develop the quality of the substance which would be recovered.

Both dissolution and leaching are some of the methods used for metal extraction. In this investigation, leaching was employed to dissolve oxides. Several prior explorations also exploited this method for processing REE and scandium elements [27-31]. A phase diagram thermodynamic analysis of Ta_2O_5, Nb_2O_5 and rare earth was done using HSC Chemistry [10], [32]. This device was also involved in research focusing on (1) lithium extraction from lepidolite [33-35], (2) roasting of nickeliferous and smelting of ferronickel, both of which were extracted from laterite [36,37], and (5) simulation of cesium reaction during a high temperature voloxidation [38]. In additional, the use of HSC Chemistry was for training simulator for processing minerals [39] and conversion of alternative hydrocarbon-based feedstock to hydrogen [40].

Some previous studies shows that the use of alkaline and acid solvents can increase contents of tantalum and niobium oxide in tin slag through the dissolution of other oxides such as Fe2O3 and SiO_2 [41], this is also explained by several other studies [42-45,46]. In addition, according to A. Suharyanto, the use of sodium hydroxides (NaOH) can dissolve amphoteric compounds in major elements such as SnO_2, SiO_2, Al_2O_3 and Zr [47], this is related with some research [48-51].

This study aims to determine the effect of heat treatment and alkaline or acid leaching on Bangka tin slag (BTS) to increasing the contents of tantalum and niobium oxide through roasting followed by quenching with NaOH, leaching with HCl as base solvents in several temperature and concentration variables.

2. Experimental method

2.1. Materials & Apparatus

In this investigation, tin slag was taken from smelters in Bangka and Belitung Islands, Indonesia. The characterization using XRF Spectro Xepos Ametek represented above 20% SnO_2. This indicated that the tin slag was 1TS. This process employed (1) a Pyrex beaker glass 1000 ml-IWAKI TE-32 Asahi Techno Glass, (2) a measuring cup 100 ml ± 0.1ml-IWAKI CTE33 Asahi Glass, (3) a Pyrex measuring pumpkin 1000 ml ± 0.15 ml-IWAKI TE-32 Asahi Techno Glass, (4) a stirring bar, (5) aquades, (6) a magnetic stirrer IKA® C-MAG HS 7, and (7) a magnetic stirrer Yellow Line MSH Basic IKA.

2.2. Methods

First, 1TS was characterized through XRF to find its chemical compounds. These samples were, then, roasted at 900°C for 2 hours, quenched by NaOH 0.5 M, dewatered, and sieved with +100, -100+120, -120+180 mesh. The +100 mesh products of quenching were divided into two groups. Some would undergo HCl leaching with temperature variables and others with concentration variables. The first process exploited 6M HCl at 25°, 45°, and 65°C while the second one used 4M, 6M, and 8M HCl. The detail procedures can be seen in Figure 1.

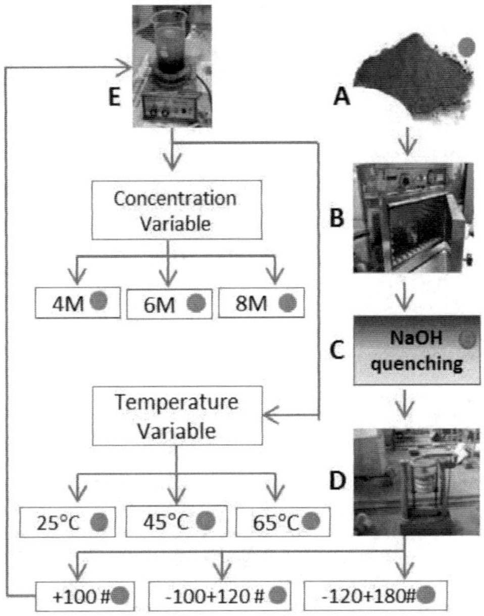

Figure 1. **A**. 1st Tin slag, **B**. Roasting, **C**. NaOH quenching, **D**. Sieving, **E**. Leaching (●) XRF Characterization (●) Dewatering

2.3. Abbreviations and Terminology

To easily follow the discussion, several abbreviations are provided. 1TS is an abbreviation for 1st tin slag while 1TS-RQS means first tin slag which has been roasted at 900°C and NaOH quenched.

Other abbreviations are TNP, MOO, and EMO. Tantalum and niobium pentoxide is abbreviated as TNP while MOO is a short-term for major other oxides. These oxides contain high concentration of SnO_2, Fe_2O_3, SiO_2, K_2O, Al_2O_3, Na_2O, TiO_2, and ZrO_2. Lastly, other elements or oxides that do not belong to both TNP and MOO are summed up as elements or minor other oxides (EMO).

2.4. Characterization and Chemical Software

X-ray fluorescent (XRF) was performed to do the characterization of sample 1TS, 1TS-RQS, ITS-RQS leached. X-ray diffraction (XRD) was performed to do the characterization of sample 1TS. Additionally, thermodynamic software was included to enrich discussion materials [52].

3. Results and Discussion

The research results are comprised of a variety of data which have been characterized by XRF. They are 1TS, 1TS-RQS, ITS-RQS leached with temperature variables, and ITS-RQS leached with concentration variables.

3.1 Composition of 1st tin slag

Table 1 illustrates the contents of Ta_2O_5 and Nb_2O_5 which are 0.29(7) % and 0.58(2) % respectively, and there is CaO that is less than 3 digits. Compared to BTS [23], this 1TS holds several oxides such as SiO_2, Al_2O_3, TiO_2, ZrO_2, and CaO which are less than BTS contents, and Fe_2O_3 and SnO_2 which are more than BTS contents.

Table 1. XRF-Characterized 1st Tin Slag Chemical Composition

Ta_2O_5 (%)	Nb_2O_5 (%)	Fe_2O_3 (%)	SiO_2 (%)	Al_2O_3 (%)	TiO_2 (%)	ZrO_2 (%)	CaO (%)	SnO_2 (%)
0,29	0.58	14.72	11.34	4.48	4.92	2.41	<0.00	22.95

1TS characterization with X-ray diffraction (XRD) is shown in Figure 2. XRD analysis result shows that tantalum was compounding with silica and titanium; $Si_{0.72}Ta_{3.28}$ peaks are seen at 2θ: 34.2°, 60.9° and 72.6°, while $Ta_{0.15}Ti_{0.85}$ peaks are at 2θ: 34.7°, 60.9° and 79.9°. But niobium, it was compounding with silica, and this is seen in Nb_5Si_3 with its peaks at 2θ: 31.3°, 50.0° and 90.3°. XRD analysis results show MOO elements in the form of AlF_3 where the peaks are seen at 2θ: 30.8°, 52.4° and 54.8°.

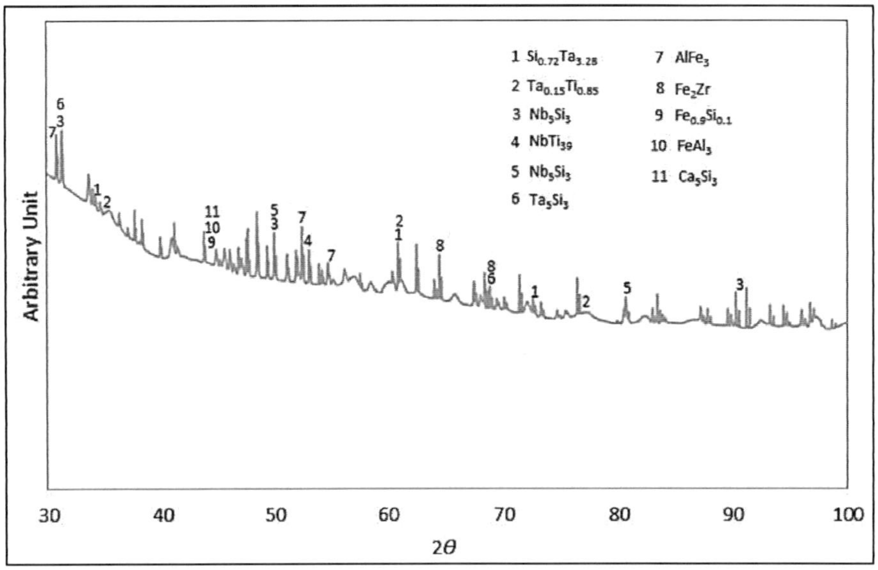

Figure 2. X-Ray Diffraction analysis of the 1st Bangka tin slag (1TS).

3.2 Roasting and sodium hydroxide quenching

In comparison with 1TS contents, those of 1TS-RQS decreased by about two-thirds. Ta_2O_5 and Nb_2O_5 the size of 100 mesh became 0.10(4) % and 0.20(0) %, the size of -100+120 mesh 0.12(6) % and 0.18(6) %, and the size of -120+180 mesh 0.14(7) % and 0.20(6) % respectively. Figure 3 (a) shows a decline in the contents of Ta_2O_5 and Nb_2O_5 after being roasted at 900°C and NaOH quenched.

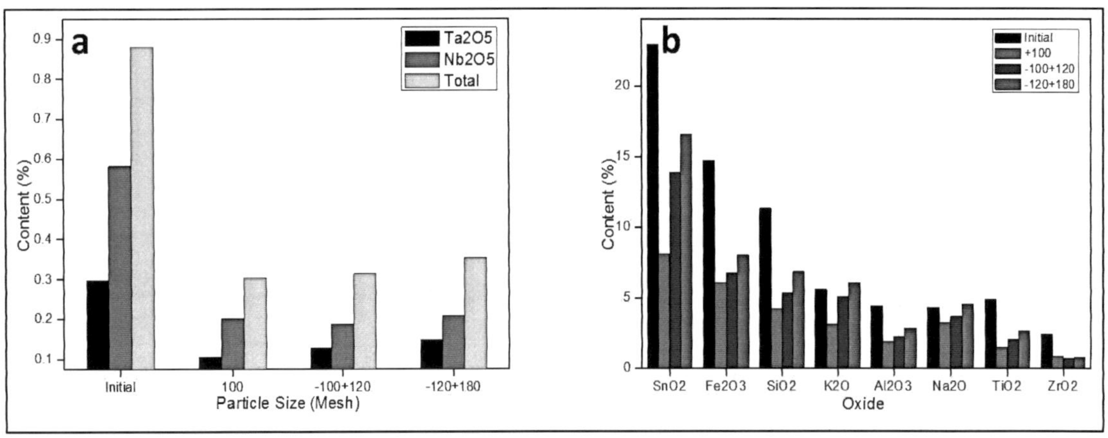

Figure 3. Roasted and NaOH quenched products after being sieved:
(a) the contents of TNP (b) the contents of MOO.

In Figure 3 (b), the contents of all MOO and SnO_2 sizes declined. In this case, those measuring 100 mesh became dominant. Among all, the 100 mesh TiO_2 underwent the highest solution, from 4.91(6) %

in 1TS to 1.48(2) % in 1TS-RQS. In other words, it only left 30 %. Other highly soluble oxides whose concentration declined were SnO_2, SiO_2, and ZrO_2. The three substances left almost merely 40%.

Ferri and silica oxides become the majority of 1TS elements. This indicates that this substance is categorized as a ceramic compound. The structure of ceramic is so brittle that it can be broken before deformation occurs. This is driven by the structure and development of a crack over the materials [53]. Tin slag composition in Table 1 shows 22.95 % SnO_2 which indicates that the tin slag is 1TS. Roasting of tin slag brings about reactions which go from the surface to the core of particles. These reactions are controlled by oxygen diffusion that comes in and sulfur dioxide that comes out through oxide layers. Roasting would develop porosity, a thermal expansion which creates a micro crack [54], and extend the surface of solids. This would encourage the optimization of leaching [55].

Roasting and sodium hydroxide quenching dissolved near two-thirds of TNP and MOO contents. HSC Chemistry simulation, as in Table 2, represents a reaction of ΔG negative to Ta_2O_5, SiO_2, TiO_2, and ZrO_2 with NaOH.

Table 2. Values of ΔG negative at 900°C and NaOH solution.

No	Reaction Equation	ΔG_{900} (kcal)
1	$Ta_2O_5 + NaOH = NaTaO_3 + H_2O$	-0.322
2	$SiO_2 + NaOH = Na_2SiO_3 + H_2O$	-16.881
3	$TiO_2 + NaOH = Na_2Ti_6O_{13} + H_2O$	-23.040
4	$ZrO_2 + NaOH = Na_2ZrO_3 + H_2O$	-2.936

Correlation between Table 2 and the result of HSC Chemistry simulation represents that ΔG negative values produce compounds in a Pourbaix diagram on pH 13.69 (8), it's means that the reaction it will be process when contact in that condition. Figure 4 shows the Pourbaix diagram of Ta_2O_5, SiO_2, TiO_2, and ZrO_2 with NaOH.

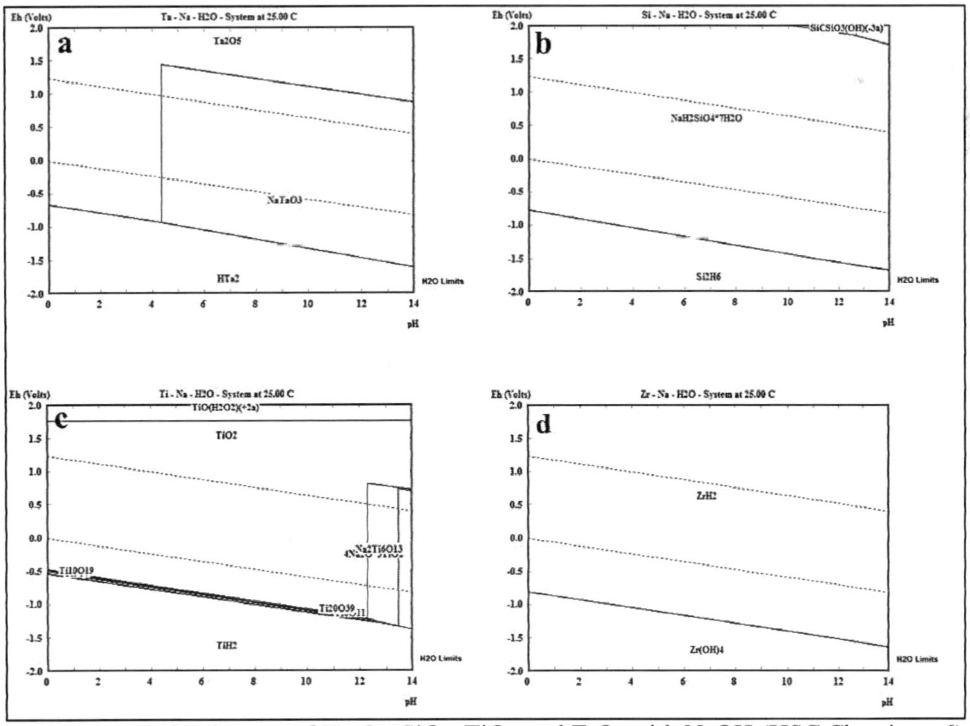

Figure 4. Pourbaix diagram of Ta_2O_5, SiO_2, TiO_2, and ZrO_2 with NaOH (HSC Chemistry 6).

IOP Conf. Series: Materials Science and Engineering **924** (2020) 012025 doi:10.1088/1757-899X/924/1/012025

3.3 Hydroxide chloride leaching

3.3.1 Leaching with temperature variables. Leaching with 45°C and 65°C temperature variables represented a significant upgrade in both TNP and MOO contents. The contents of TNP correlated with the rise of temperature. Meanwhile, leaching at 65°C resulted in the maximum contents. Figure 5 (a) shows the upgrade in TNP contents referred to (1) TNP from the 100 mesh 1TS-RQS and (2) TNP from the 25°C, 45°C, and 65°C leaching.

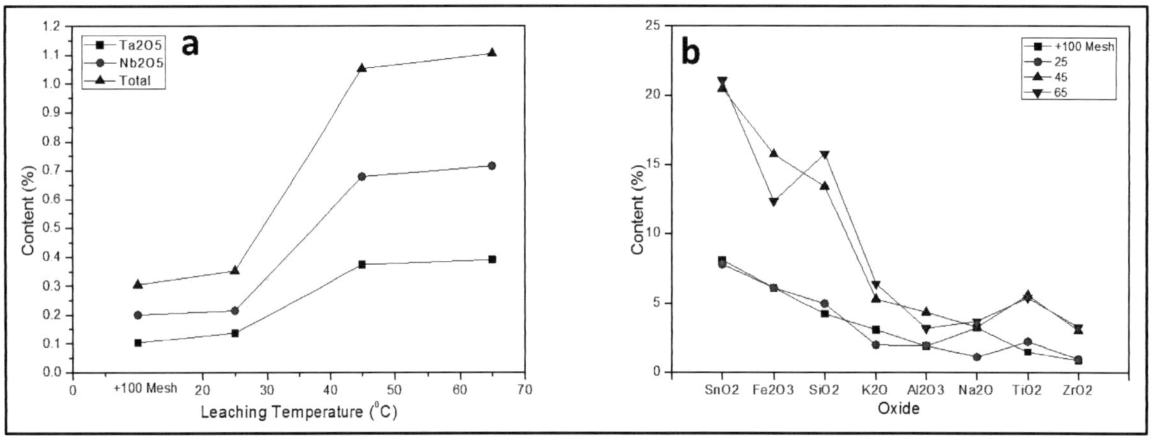

Figure 5. Leached products with temperature variables: (a) the contents of TNP (b) the contents of MOO.

Figure 5 (b) shows an increase in MOO contents leached at 25°C, 45°C, and 65°C. Among the three temperatures, 45°C and 65°C dominated the rise. The related substances were SnO_2, SiO_2, and ZrO_2.

Table 3. Values of Δ G negative of ITS-RQS solution of HCl.

Reaction Equation	ΔG_{25} (kcal)	ΔG_{45} (kcal)	ΔG_{65} (kcal)
$NaTaO_3 + HCl(g) \rightarrow Ta_2O_5+NaCl+H_2O$	-24.683	-23.604	-22.603
$Na_2SiO_3 + HCl(g) \rightarrow$ SiO2+NaCl+H2O	-48.690	-47.589	-46.503

In Table 3 and Figure 6, a thermodynamic reaction occurs. Ta_2O_5 was found in pH -0.77(8) in which 6M HCl was employed. In other words, NaOH quenched tantalum pentoxide, if HCl leached, will produce the same substance, Ta_2O_5. Dissolution at all temperature variables does not influence the tantalum compounds. However, Table 4 shows that the yield ratio shows 3.75 as the maximum value of Ta_2O_5 at 65°C.

Table 4. Yield ratio of leaching with temperature variables.

	Initial +100 #	Content 25°C	Yield Ratio 25°C	Content 45°C	Yield Ratio 45°C	Content 65°C	Yield Ratio 65°C
Ta_2O_5	0.10(4)	0.13(8)	1.32	0.37(4)	3.59	0.39(0)	3.75
Nb_2O_5	0.20(1)	0.21(5)	1.07	0.67(9)	3.39	0.71(5)	3.57
Total	0.30(5)	0.35(3)	1.16	1.05(3)	3.46	1.10(6)	3.63

Yield Ratio = content before process/content after process.

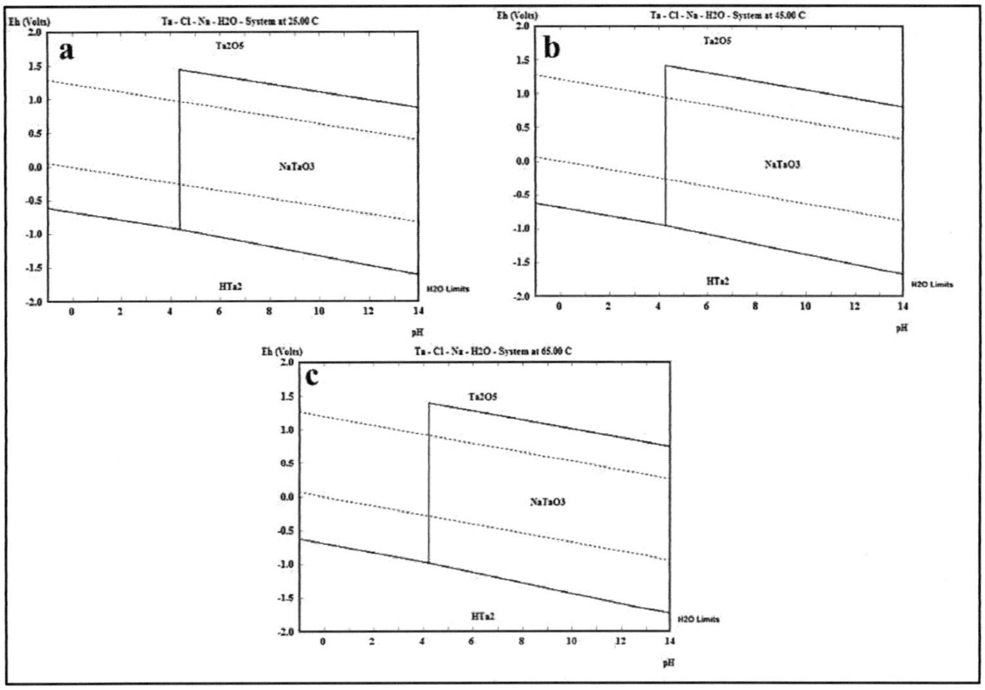

Figure 6. Pourbaix diagram of tantalum oxides in NaOH and HCl solution at (a) 25°C, (b) 45°C, and (c) 65°C.

3.3.2 Leaching with concentration variables. HCl leaching with concentration variables did not show a significant upgrade in the contents of TNP. However, MOO contents including SiO_2, Al_2O_3, and TiO_2 dominantly rose after being HCl leached with concentration of 4M. The increase in concentration did not correlate with the contents of MOO. Figure 7 (a) shows an upgrade in TNP contents referred to (1) TNP contents from 1TS-RQS measuring 100 mesh and (2) TNP contents leached with concentration of 4M, 6M, and 8M. Figure 7 (b) shows an increase in MOO contents leached at 4, 6, and 8M.

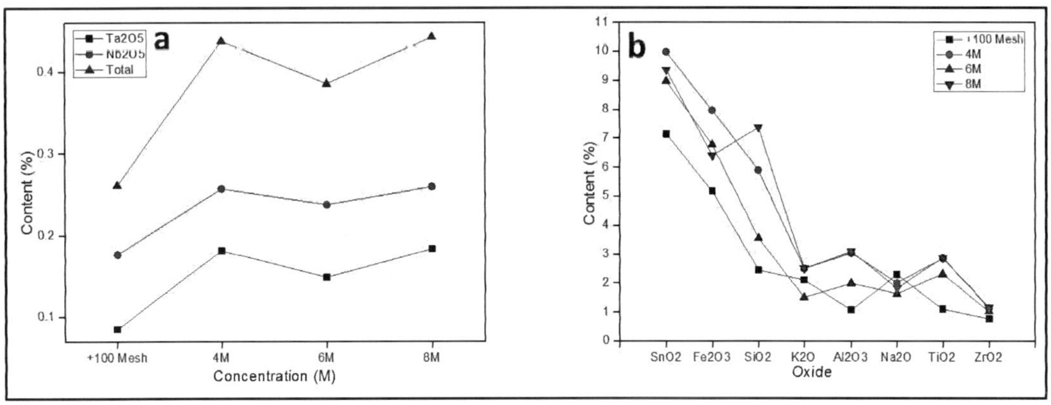

Figure 7. Leached products with HCl concentration variables: (a) the contents of TNP (b) the contents of MOO.

HCl leaching thermodynamic reactions with concentration of 4, 6 and 8 M (pH -0.602, -0.778 and -0.9030) at 25°C on tantalum oxides and silica can be seen in Figure 8. The extension of pH from -

0.602 to -0.9030 does not indicate any compound reactions. In Table 5, no concentration variables show a significant rise in the yield ratio.

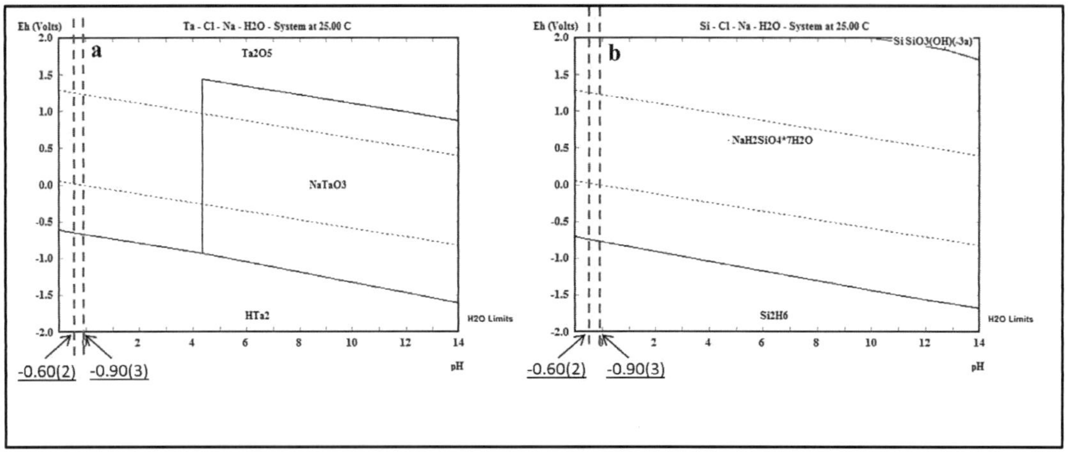

Figure 8. Pourbaix diagram of (a) tantalum and (b) silica in NaOH and HCl solution at 25°C.

Table 5. Yield ratio of leaching with concentration variables.

	Initial +100 #	**Content 4M**	**Yield Ratio 4M**	**Content 6M**	**Yield Ratio 6M**	**Content 8M**	**Yield Ratio 8M**
Ta_2O_5	0.10(4)	0.18(1)	1.74	0.14(8)	1.42	0.18(3)	1.76
Nb_2O_5	0.20(1)	0.25(7)	1.28	0.23(7)	1.18	0.25(9)	1.29
Total	0.30(5)	0.43(8)	1.44	0.38(6)	1.27	0.44(3)	1.45

Yield Ratio = content before process/content after process.

4. Conclusion

The main results of thermal treatment (roasting) followed by sodium hydroxide quenching and acid leaching by HCl obtains in this study can be summarized that roasting of tin slag in 900°C and NaOH quenching make TNP and MOO contents within residues dissolve in filtrates. While, leaching process of 1TS-RQS with HCl are increase in each temperature variables, but the maximum contents of TNP at 65°C with yield ratio of 3.75 Ta_2O_5 and 3.57 Nb_2O_5. It's also see in the thermodynamic analysis with Gibbs Energy and Pourbaix Diagram are negative. But leaching process of 1TS-RQS with HCl in some concentration variables does not indicate any dominance in yield ratio.

Acknowledgment

Special thanks to Research Centre for Metallurgy, Indonesian Institute of Sciences-LIPI, Indonesia and Mechanical Engineering Department, Politeknik Negeri Jakarta, Depok 16425, Indonesia.

References

[1] European Commisioning, " Report on Critical Raw Materials for the EU- critical raw materials profiles," 2015.
[2] M. S. P. REFRAM, "Niobium and tantalum," the European Union's Horizon 2020 research and innovation programme under grant agreement No 688993
[3] Roskill Information Services Ltd., *Tantalum: Global Industry, Markets and Outlook to 2020*, 12th ed. London: Roskill Information Services, 2016.
[4] Sidney M. Heins, "US Patent 3.129.896-Separation of Heavy Metal Contituents of Tin Slag," 3.129.896, 1964.

[5] J. A. Cenerazzo, C. E. Mosheim, Zionsville, and Corrodo E. Marvisa, "US Patent 3.585.024-Up Grading The Tantalum and Columbium Contents of Tin Slag," 1971.

[6] I. Gaballah, E. Allain, M.-C. Meyer-Joly, and K. Malau, "A Possible Method for the Characterization of Amorphous Slags Recovery of Refractory Metal Oxides from Tin Slags," *Metall. Trans.*, vol. 23, no. June, pp. 249–259, 1992.

[7] A. K. Suri, C. Subramanian, and C. K. Gupta, "Investigations on flowsheet development for the treatment of niobium-tantalum bearing cassiterite," *Miner. Process. Extr. Metall. Rev.*, vol. 9, no. 1–4, pp. 293–303, 1992.

[8] M. Y. Sulaiman, "Simultaneous determination of thorium and uranium in tin slag," *Sci. Total Environ.*, vol. 131, pp. 187–195, 1993.

[9] I. Gaballah and E. Allain, "Recycling of Strategic Metals from Industrial Slag by a Hydro-and Pyrometallurgical Process," *Resour. Conserv. Recycl.*, vol. 10, no. 1–2, pp. 75–85, Apr. 1994.

[10] E. Allain, M. Djona, and I. Gaballah, "Kinetics of Chlorination and Carbochlorination of Pure Tantalum and Niobium Pentoxides," *Metall. Mater. Trans. B*, vol. 28, no. April, pp. 223–233, 1997.

[11] I. Gaballah, E. Allain, and M. Djona, "Extraction of Tantalum and Niobium from Tin Slags by Chlorination and Carbochlorination," *Metall. Mater. Trans.*, vol. 28, no. June, 1997.

[12] C. Subramanian, A. K. Suri, and B. Atomic, "Recovery of Niobium and Tantalum from Low Grade Tin Slag - A Hydrometallurgical Approach," *Environ. Waste Manag.*, pp. 100–107, 1998.

[13] P. K. Jena and E. A. Brocchi, "Studies on the Kinetics of Carbon Tetrachloride Chlorination of Refractory Metal Oxides," *Min. Pro. Ext. Met. Rev*, vol. 22, pp. 597–613, 2001.

[14] N. T. Mudzanapabwe, O. S. Chinyamakobvu, and D. J. Simbi, "In Situ Carbothermic Reduction of a Ferro-Columbite Concentrate in The Recovery of Nb and Ta as Metal Matrix Composite from Tin Smelting Slag Waste Dump," *Mater. Des.*, vol. 25, pp. 297–302, 2004.

[15] Byung-Su Kim, And, and Y.-Y. Choi, "Kinetics of the Chlorination Reaction of Tantalum Pentoxide with Carbon Tetrachloride Gas," *Mater. Trans.*, vol. 46, no. 9, pp. 2102–2106, 2005.

[16] MS, M. Yusoff, and M. Muslimin, "Uranium and Thorium Analysis In Tin Slag by EDXRF," *J. Nucl. Relat. Technol.*, vol. 4, pp. 177–180, 2007.

[17] E. A. Brocchi and F. J. Moura, "Chlorination Methods Applied to Recover Refractory Metals from Tin Slags," *Miner. Eng.*, vol. 21, pp. 150–156, 2008.

[18] M. Meor Sulaiman. M. Y, Muslimin, "Quantitative Analysis of Uranium and Thorium in Local Zircon and Tin Slag by The EDXRF Technique," *J. Nucl. Relat. Technol.*, vol. 7, no. 1, pp. 21–28, 2010.

[19] C. Francois, "Patent_WO_2013_040694_A1_3.pdf," 2013.

[20] J. U. Odo, W. C. Okafor, S. O. Ekpe, And, and C. C. Nwogbu, "Extraction of Niobium from Tin Slag," *Int. J. Sci. Res. Publ.*, vol. 4, no. 11, pp. 1–7, 2014.

[21] S. I. Angadi, T. Sreenivas, H.-S. Jeon, S.-H. Baek, and B. K. Mishra, "A Review of Cassiterite Beneficiation Fundamentals and Plant Practices," *Miner. Eng.*, vol. 70, pp. 178–200, Jan. 2015.

[22] M. Y. Meor Sulaiman, N. S. Kamaruzaman, and A. K. Zahari, "Standardless EDXRF Analysis Methods of U and Th in Malaysian Tin Slag Waste," *Mater. Sci. Forum*, vol. 840, pp. 410–415, 2016.

[23] S. Permana, J. W. Soedarsono, A. Rustandi, and et.al, "Other Oxides Pre-removed from Bangka Tin Slag to Produce a High Grade Tantalum and Niobium Oxides Concentrate," in *IOP Conference Series : Materials Science and Engineering*, 2016, vol. 012006, p. 10.

[24] R. Smith, "An Analysis of the Processes for Smelting Tin," *Bull. Peak Dist. Mines Hist. Soc.*, vol. 13, no. 2, p. 9, 1996.

[25] R. U. A. Laura Talens Peiro, Gara Villalba Mendez, "Rare and Critical Metals as By-products and The Implications for Future Supply," 2011.

[26] A. Maksum, A. Rustandi, S. Permana, and J. W. Soedarsono, "Influence of Roasting-Quenching Pretreatment on the Rice Husk Silica Prepared by Calcination Method," vol. 020009, 2017.

[27] J. Soedarsono, M. Burgard, Z. Asfari, and J. Vicens, "Liquid-liquid extraction properties of

25,27-dicarboxy-26,28-dimethoxy-5,11,17,23-tetra-tert- butylcalix [4]arene towards the alkali ions," *Solvent Extr. Ions Exch.*, vol. 13, no. 4, pp. 755–769, 1995.
[28] J. Soedarsono, M. Burgard, Z. Asfari, and J. Vicens, "Liquid-liquid Extraction of Rare Earth (III) Ions by 25,27-dicarboxy-26,28-dimethoxy-5,11,17,23- tetra- tert-butylcalix[4] arene," in *XVII International Symposium on Macrocyclic Chemistry*, 1993.
[29] J. Soedarsono, A. Hagege, M. Burgard, Z. Asfari, and J. Vicens, "Liquid-liquid Extraction of Rare Earth Metals Using 25,27-dicarboxy-26,28-dimethoxy- 5,11,17,23-tetra- tert-butylcalix[4]arene," *J. Phys. Chem.*, vol. 100, no. 4, pp. 477–481, 1996.
[30] Y. Sohama, H. Isaka, and H. Watanabe, "Patent WO2002022895 Method for producing metal or metal compound comprising process of treating with fluorine and adjusted raw material used therein," WO2002022895, 2003.
[31] W. Wang, Y. Pranolo, and C. Y. Cheng, "Recovery of Scandium from Synthetic Red Mud Leach Solutions by Solvent Extraction with D2EHPA," *Sep. Purif. Technol.*, vol. 108, pp. 96–102, 2013.
[32] S. Permana, A. Rustandi, and R. A. Majid, "Thermodynamic Analysis With Software : A Case Study of Ugrading Rare Earth Elements Content in Bangka Tin Slag," *Far East J. Electron. Commun.*, vol. 17, no. 5, pp. 1211–1220, 2017.
[33] V. T. Luong, D. J. Kang, J. W. An, M. J. Kim, and T. Tran, "Factors affecting the extraction of lithium from lepidolite," *Hydrometallurgy*, vol. 134–135, pp. 54–61, 2013.
[34] V. T. Luong, D. J. Kang, J. W. An, D. A. Dao, M. J. Kim, and T. Tran, "Iron sulphate roasting for extraction of lithium from lepidolite," *Hydrometallurgy*, vol. 141, pp. 8–16, 2014.
[35] T. T. Hien-Dinh, V. T. Luong, R. Gieré, and T. Tran, "Extraction of lithium from lepidolite via iron sulphide roasting and water leaching," *Hydrometallurgy*, vol. 153, pp. 154–159, 2015.
[36] C. A. Pickles, J. Forster, and R. Elliott, "Thermodynamic analysis of the carbothermic reduction roasting of a nickeliferous limonitic laterite ore," *Miner. Eng.*, vol. 65, pp. 33–40, 2014.
[37] D. R. Swinbourne, "Understanding ferronickel smelting from laterites through computational thermodynamics modelling," *Miner. Process. Extr. Metall.*, vol. 123, no. 3, pp. 127–140, 2014.
[38] M. K. Jeon, J. M. Shin, J. J. Park, and G. Il Park, "Simulation of Cs behavior during the high temperature voloxidation process using the HSC chemistry code," *J. Nucl. Mater.*, vol. 430, no. 1–3, pp. 37–43, 2012.
[39] T. Roine, J. Kaartinen, and P. Lamberg, "Training simulator for flotation process operators," in *IFAC Proceedings Volumes (IFAC-PapersOnline)*, 2011, vol. 18, no. PART 1, pp. 12138–12143.
[40] E. Turpeinen, R. Raudaskoski, E. Pongrácz, and R. L. Keiski, "Thermodynamic analysis of conversion of alternative hydrocarbon-based feedstocks to hydrogen," *Int. J. Hydrogen Energy*, vol. 33, no. 22, pp. 6635–6643, 2008.
[41] H. F. Putra and N. S. Aryanti, "Upgrading tantalum and niobium oxides content in Bangka tin slag with double leaching Upgrading tantalum and niobium oxides content in Bangka tin slag with double leaching," 2018.
[42] R. A. Majid, A. Rustandi, and S. Permana, "Simulation of Tantalum and Niobium Pentoxides Extraction from Bangka Tin Slag Waste," vol. 24, no. 1, pp. 767–772, 2018.
[43] E. Allain, N. Kanari, F. Diot, and J. Yvon, "Development of a process for the concentration of the strategic tantalum and niobium oxides from tin slags," *Miner. Eng.*, vol. 134, no. January, pp. 97–103, 2019.
[44] J. W. Soedarsono and S. Permana, "Other Oxides Pre-removed from Bangka Tin Slag to Produce a High Grade Tantalum and Niobium Oxides Concentrate Other" IOP Conf. Ser.: Mater. Sci. Eng. 131 012006, 2016.
[45] W. Kartika, R. A. Majid, and D. Navanti, "Studi Pemanfataan Limbah Terak Timah 2 Bangka Sebagai Sumber Sekunder Unsur Skandium," vol. 19, no. 1, pp. 8–17, 2019.
[46] S. Permana, A. Maksum, and J. Wahyuadi, "Enrichment On Bangka Tin Slag ' S Tantalum And Niobium Oxide Contents Through Non-Fluoride Process," January 2018.

[47] A. Suharyanto, E. Sulistiyono, and F. Firdiyono, 'Pelarutan Terak Timah Bangka Menggunakan Larutan NaOH,' vol. 15314, 2014 .," vol. 15314, p. 15314, 2014.

[48] R A Majid, S. Permana and J W Soedarsono, "Peningkatan Kadar Tantalum & Niobium Oksida dari Terak Timah Bangka Menggunakan Pelarut NaOH dilanjutkan dengan HNO3 dan H3PO4," Journal of Chemical Process Engineering Vol. 4, No.01, 2019.

[49] S. F. Vincia, S. Permana, A. Maksum, and J. W. Soedarsono, "Study to Enhance Tantalum and Niobium Contents in Bangka Tin Slag by NaOH and HClO 4 Leaching" IOP Conf. Ser.: Mater. Sci. Eng. 547 012049, 2019.

[50] S. Permana, A. Maksum, and J. W. Soedarsono, "Enrichment of Tantalum and Niobium Contents in Bangka Tin Slag by NaOH and H 3 PO 4 Leaching" IOP Conf. Ser.: Mater. Sci. Eng. 547 012050, 2019.

[51] P. Harcourt, E. S. Science, V. Schools, and M. Board, "Extraction of Niobium from Tin Slag," vol. 4, no. 11, pp. 1–7, 2014.

[52] A. Roine, T. Kotiranta, and et.al, "HSC Chemistry, Version 6.0," vol. 03, no. 5641. Outokumpu Research Oy, p. 5641, 2006.

[53] P. W. R. Beaumont, "On the problems of cracking and the question of structural integrity of engineering composite materials," *Appl. Compos. Mater.*, vol. 21, no. 1, pp. 5–43, 2014.

[54] J. Chang, J. Peng, L. Zhang, and J. Cheng, "Extraction of Indium from Zinc Oxide Flue Dust by Microwave Sulfation Roasting and Water Leaching," *Drying, Roasting, Calcining Miner.*, pp. 35–42, 2015.

[55] E. O. Olanipekun, "Kinetics of leaching laterite," *Int. J. Miner. Process.*, vol. 60, no. 1, pp. 9–14, 2000.

Iodogen Method on Iodine-131 (^{131}I) Radiolabelling of Silver Nanoparticle (AgNPs) as a New Agent of Molecular Imaging

A Aries*, E Sarmini, A Nurmanjaya, M Subechi, UN Sholikhah, E Lestari, H Setiawan, S Febriana, S Sriyono, A Abidin
Center for Radioisotope and Radiopharmaceutical Technology
National Nuclear Energy Agency of Indonesia
PTRR-BATAN, Kawasan Puspitek, Gedung 11, Tangerang Selatan 15314, Indonesia

*arni-aries@batan.go.id

Abstract. The application of nanomaterials in the treatment of various types of diseases continues to increase, including the use of silver nanoparticles (AgNPs). However, there still limitation in terms of the research on labelling AgNPs using radioactive compound such as ^{131}I. The aim of this study is to carry out a method on ^{131}I radiolabelling of AgNPs by using Iodogen as an iodination reagent. The radiolabelled ^{131}I-AgNPs were then purified by using Sephadex-25 column chromatography with 0,05 M phosphate buffer solution as mobile phase for the first purification and HEPES solution for the second purification. The radiochemical purity of radiolabelled ^{131}I-AgNPs was then determined by using autoradiography scanner. ^{131}I-AgNPs with a purity 94,5±0,2121% were obtained after the purification. Stability test of the 131I-AgNPs was carried out by determining the radiochemical purity of the ^{131}I-AgNPs on the first day until the fifth day of storage in the room temperature and refrigerator. The best stability of the ^{131}I-AgNPs after purification resulted in radiochemical purity >90% until the fourth day and <90% on the fifth and subsequent days in both storages. This result shows that storage in the refrigerator can be a better choice rather than in the room temperature.

Keywords: Silver Nanopartikel, ^{131}I, ^{131}I-AgNPs, Iodogen, Labelled Compound, Autoradiography Scanner

1. Introduction

The application of nanomaterials in both organic or inorganic for the treatment of various types of diseases continues to increase. One of the most popular inorganic nanomaterials is silver nanoparticle. Silver nanoparticles are famous for their ability as broad-spectrum antimicrobials and antiviral [1,2]. The study also found the application of silver nanoparticle as photosensitizers and/or radiosensitizers, anticancer therapeutic agents, wound dressing, textiles, and biomedical devices [2,3].

The study of radioactive Iodine has been done for a long time ago. ^{131}I is one of iodine isotope that has 8 days half-life [4]. ^{131}I can produce by using neutron activation of Tellurium-130 dioxide($^{130}TeO_2$) [5-6]. ^{131}I can be applied in both diagnostic and therapy since it can emit gamma and beta [7]. ^{131}I has been widely used in medical application such as thyroid cancer staging, treatment of hyperthyroidism and differentiated thyroid cancer, treatment of metastatic tumours, neuroblastoma diagnosis dan therapy, etc [8-11]. Thus, ^{131}I still become an important isotope in nuclear medicine.

IOP Conf. Series: Materials Science and Engineering **924** (2020) 012026 doi:10.1088/1757-899X/924/1/012026

Nowadays, the application of nanomaterials in radiopharmaceuticals has attracted huge attention. In 2010, a method on rapid Iodine-125 (^{125}I) radiolabelling of silver nanoparticle for in vivo SPECT has been published by Chrastina and Schnitzer [12]. Even though they have different energy and half life, ^{125}I and ^{131}I have similar chemical properties. So that, there is a possibility to label AgNPs with ^{131}I. In terms of diagnosis and therapy, both of ^{125}I and ^{131}I can be used. The production of ^{125}I isotope is using natural or enriched Xenon-124 gas as target material [13,14]. Thus, this process need an adequate facility. Due to this requirement, there are a limitations in production process of ^{125}I specially in Indonesia. In different with ^{125}I, ^{131}I has a high production capasity in Indonesia. PTRR-BATAN can produce up to 9 Ci of ^{131}I for each batch production. So that, it is important to carry out a method on AgNPs labelled ^{131}I as new agent in molecular imaging through radioiodination to support nuclear medicine aplication in Indonesia. Radioiodination is a molecule chemically modifying process to contain radioactive iodine. Radioiodination can be done by using directly or indirectly labelling. There many popular reagents indirect labelling such as chloramine-T, iodo-beads, iodogen, lactoperoxidase-catalyzed iodination, etc [15,16]. Iodogen is 1,3,4,6,-tetrachloro-3α,6α-diphenylglycouril, an N-haloamine derivate that similar to Iodo-beads in term of oxidizing properties. Radioiodination can be stopped easily by removing the aqueous phase in iodogen reaction. Iodogen is being used frequently to prepared any kind of radiolabelled compound such as peptides and probes [15].

In this research, a method on ^{131}I radiolabelling of silver nanoparticle by using iodogen as the oxidative agent will be done. The stability test of synthesis of ^{131}I-AgNPs will be carried out in the room temperature and refrigerator storage.

2. Experimental Method

2.1 Materials

This research was used several materials such as Silver nitrate, sodium chloride, universal pH indicator that were purchased from Merck, while Sodium borohydride, phosphate buffer, polyvinylpyrrolidone (PVP) and methanol were bought from Sigma Aldrich, and Iodogen were purchased from Pierce. Silver nitrate was used as the main materil for the synthesis of the silver nanoparticle, sodium chloride and PVP as stabilizer for the silver naoparticles [15], phosphate buffer as pH adjusting agent, sodium borohydride as reducing agent, methanol as mobile phase for radiochemical purity test of ^{131}I-AgNPs, and iodogen as the iodination reagent. The equipment was used include glass equipment, magnetic stirrer, micropipette, disposable cuvette, SPHADEX-25 column and vial. The instruments were used consist of an analytical balance (Acculab), dose calibrator (Atomlabplus), Uv-Vis Spectrophotometer (Jasco32), Particle Size Analyzer (PSA), Transmission Electron Microscopy (TEM) in LIPI Physics Research Center, and Autoradiography Scanner (Cyclone-Plus).

2.2 Synthesis and Characterization of AgNPs

AgNPs were synthesis by reduction method by using sodium borohydride as the reducing agent as explained by Umi et. al. [17]. The characterization of AgNPs was determined by using transmission electron microscopy (TEM), UV-Vis spectrophotometer, and particle size analyzer (PSA).

2.3 Preparation of 10 mM HEPES solution

0,598 g of powdered HEPES was dissolved by using 250 mL of demineralized water and were stirred until homogeneous. The HEPES solution pH was then adjusted until 7 by using HNO3 or NaOH.

2.4 Preparation of Iodogen for Iodination

1 mg of iodogen were dissolved by using 1 mL of cloroform.100 μL of this solution were prepared into 2 glass vessels respectively. The first vessel was used for labelling of AgNPs by using ^{131}I and the other for the control. The solvent was evaporated slowly by using a stream of nitrogen gas. The vessels were then sealed until used.

2.5 Labelling AgNPs by Using ^{131}I
1 mL of AgNPs were added into the first vessels that had been prepared before. 20 µL of 0.5 M phosphate buffer pH 7 solution was added and followed by adding ~1mCi of $Na^{131}I$. The solution was vortex and left for 1 minute. 200 µL of 0.05 M phosphate buffer pH 7.4 solution were then added. The radiochemical purity test was then determined by using thin paper chromatography which Whatman paper 1 as stationary phase and methanol 75% as mobile phase.

2.6 Preparation of Control
20 µL of 0.5 M phosphate buffer pH 7 solution was added into the second vessels that had been prepared before and followed by adding ~1mCi of $Na^{131}I$. The solution was vortex and left for 1 minute. 200 µL of 0.05 M phosphate buffer pH 7.4 solution were then added. The radiochemical purity test was then determined by using thin paper chromatography which Whatman paper 1 as stationary phase and methanol 75% as mobile phase.

2.7 Purification 1
The Sephadex G-25 column was prepared and installed on a stative, then the tip was cut out until all the water inside the column discharged. The column was then washed out by using a 0.05 M phosphate buffer pH 7.4 solution until the filtrate that coming out of the column has pH 7.4. The ^{131}I-AgNPs solution was poured into the column and was eluted by using a 0.05 M phosphate buffer pH 7,4 solution. The filtrate was then collected 500 µL for each fraction and the radioactivity was measured respectively by using dose calibrator. The radiochemical purity of the fraction with the highest radioactivity was then determined by using paper chromatography method with Whatman paper 1 as stationary phase and methanol 75% as mobile phase.

2.8 Purification 2
The Sephadex G-25 column was put on a stative, and all of the solutions inside was discharged. The column was then conditioned by using 10 mM HEPES pH 7 solution until the filtrate that removed out of the column has pH 7. The ^{131}I-AgNPs solution was set into the column and was eluted by using 10 mM HEPES pH 7 solution. The filtrate was then collected 500 µL for each fraction and the radioactivity was measured respectively by using dose calibrator. The radiochemical purity of the fraction with the highest radioactivity was then determined by using paper chromatography method with Whatman paper 1 as stationary phase and methanol 75% as mobile phase.

2.9 Stability Test of ^{131}I-AgNPs
The stability test of^{131}I-AgNPs was done by the determination of the radiochemical purity of 3 fractions of ^{131}I-AgNPs with the highest activity and radiochemical purity which were stored in the room temperature and refrigerator. Each fraction were spotted in the Whatman paper as stationary phase and methanol 75% as mobile phase and the radioactivity was measured by using autoradiography scanner.

3. Results and Discussion
Physical examination of synthesis of AgNPs is a deep yellow-solution, clear, without precipitation. Figure 1 And 2 show the characterization of AgNPs by using Particle Size Analyzer (PSA) and Transmission Electron Microscopy (TEM), respectively. PSA result shows that the average distribution result of AgNPs is 14.5 nm.

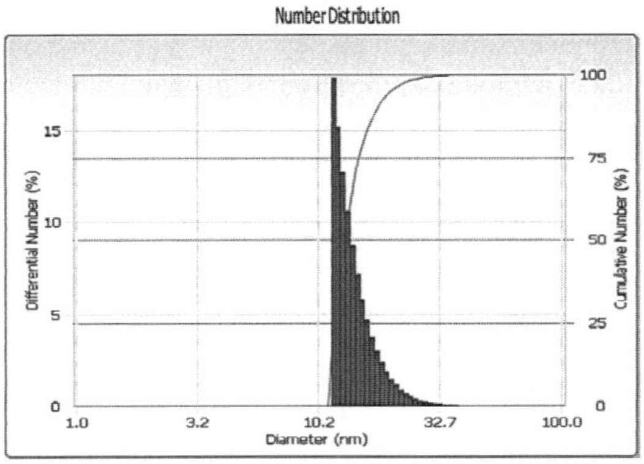

Figure 1. Characterization of AgNPs by using PSA.

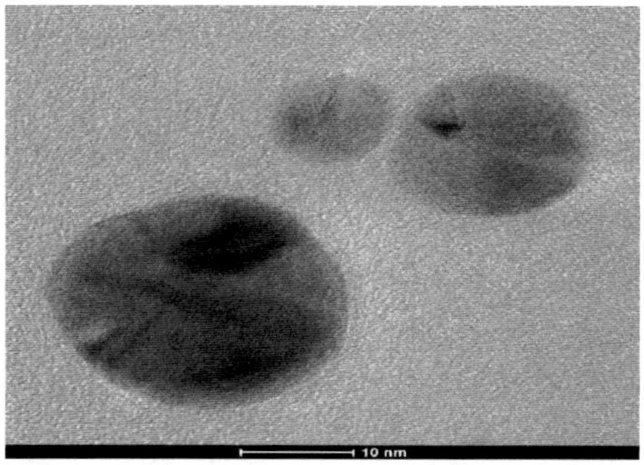

Figure 2. Characterization of AgNPs by using TEM.

The labelling process of AgNPs by using ^{131}I was proceeded by the iodination process. In this research, iodogen was used as an oxidizing agent. Radioiodine may interact with AgNPs by absorption mechanism. This labelling method has resulted in ^{131}I-AgNPs with pH 7 and radiochemical purity (RCP) as shown in Figure 3-5 and Table 1.

Table 1. Radiochemical Purity of ^{131}I-AgNPs.

Sample	Activity (µCi)	RCP (%)	Retention Factor (cm)	Appearance
Bulk ^{131}I	1,551	99.2	0.75	Clear solution
^{131}I-Iodogen	1,489	99.2 ± 0.2828	0.69	Clear solution
^{131}I-AgNPs	1,131	80.4 ± 0.2121	0.29	Yellow-clear solution

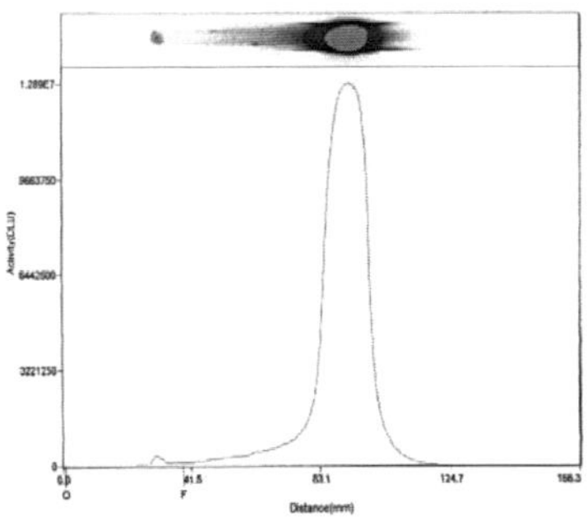

Figure 3. Autoradiogram of Radiochemical Purity of ^{131}I Bulk .

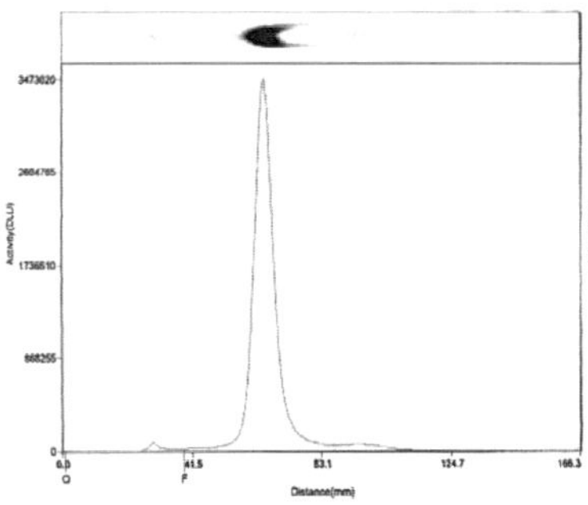

Figure 4. Autoradiogram of Radiochemical Purity of ^{131}I-Iodogen.

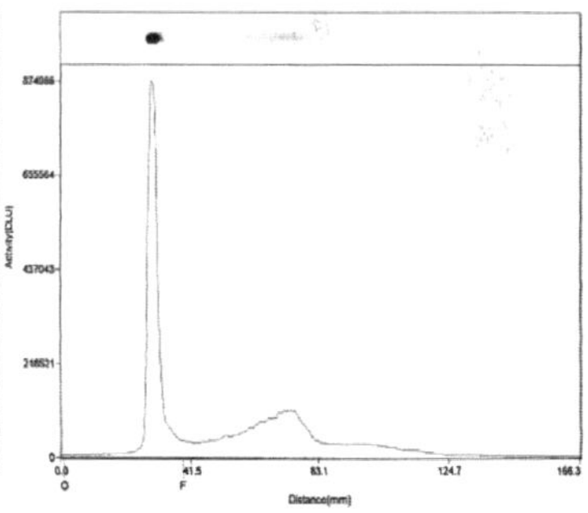

Figure 5. Autoradiogram of Radiochemical Purity of ^{131}I-AgNPs.

Radiochemical purity resulted in the labelling of AgNPs by using ^{131}I with iodogen as the oxidizing agent was 80.4%. Since the radiochemical purity of ^{131}I-AgNPs was still low, ^{131}I-AgNPs were then purified by using Sephadex-25 column with phosphate buffer 0.05 M as mobile phase. Each 500 µL of the filtrate were collected and the radioactivity were measured by using *Dose Calibrator* as shown in Figure 6.

IOP Conf. Series: Materials Science and Engineering **924** (2020) 012026 doi:10.1088/1757-899X/924/1/012026

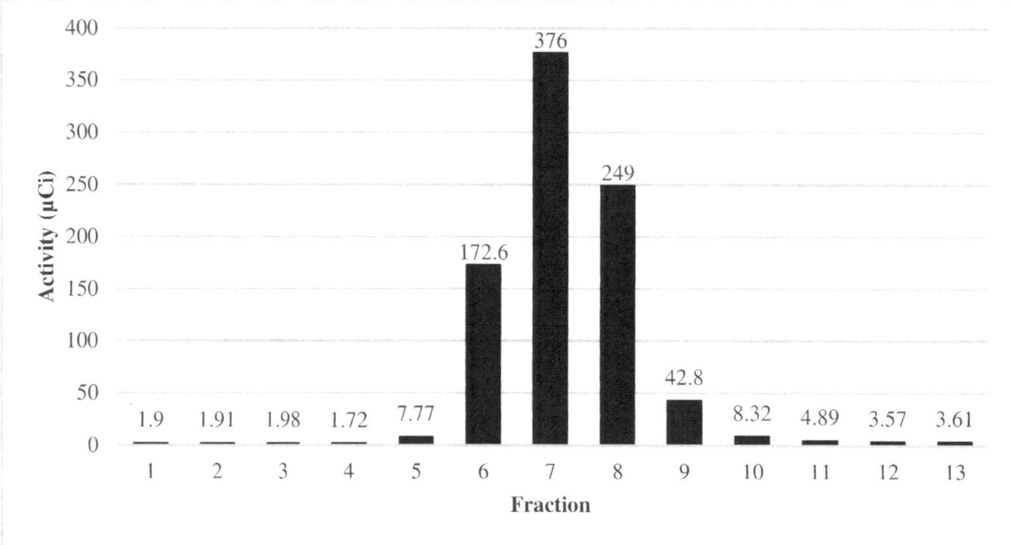

Figure 6. The Elution Profile of Purification 1 of the ^{131}I-AgNPs with phosphate buffer 0.05 M as mobile phase.

The purification resulted in three fractions with the highest activity were fraction 6, 7 and 8. The radiochemical purity test of these fractions was then determined and the result as shown in Table 2 and Figure 7-9.

Table 2. Radiochemical Purity of ^{131}I-AgNPs after purification 1.

Fraction	RCP (%)	Retention Factor (cm)	Appearance
Fraction 6	59.5 ± 3.1113	0.29	Yellow-clear solution
Fraction 7	48.8 ± 0.2828	0.29	Yellow-clear solution
Fraction 8	60.7 ± 0.0707	0.29	Yellow-clear solution

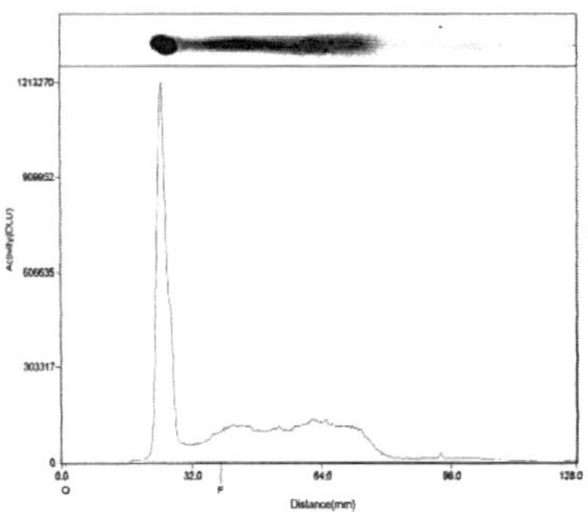

Figure 7. The Autoradiogram of Radiochemical Purity of ^{131}I-AgNPs fraction 6.

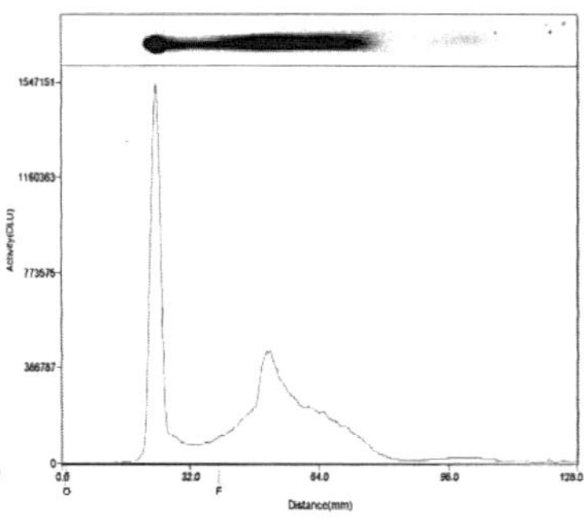

Figure 8. The Autoradiogram of Radiochemical Purity of ^{131}I-AgNPs fraction 7.

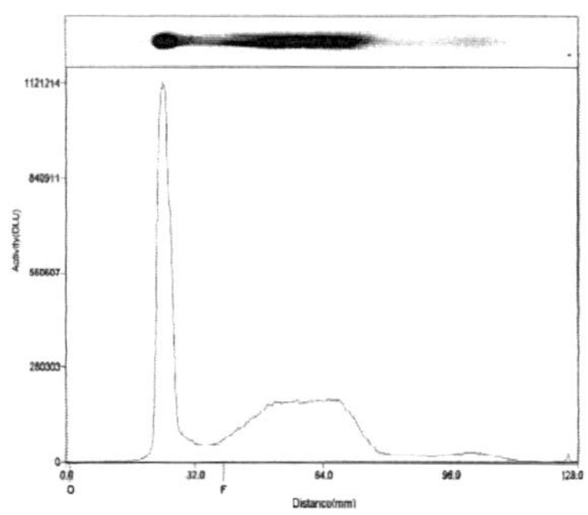

Figure 9. The Autoradiogram of Radiochemical Purity of ^{131}I-AgNPs fraction 8.

As shown in Table 2, the radiochemical purity of ^{131}I-AgNPs was decreased to less than 65%. As a result, the three fractions with the highest activity in purification 1 were mixed and purified with HEPES 10 mM pH 7 as mobile phase. Each 1 mL of filtrate was collected and the radioactivity was measured by using *Dose Calicbator* as shown in Figure 10.

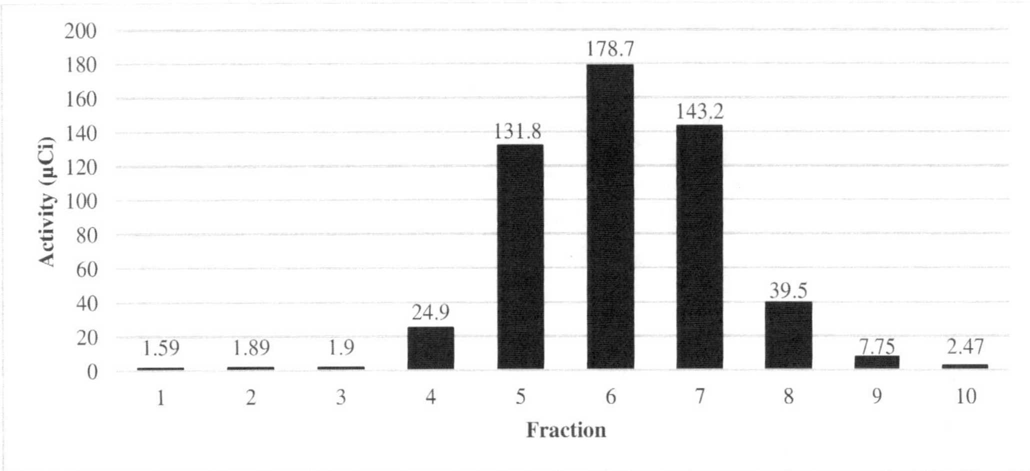

Figure 10. The Elution Profile of Purification 2 of the ^{131}I-AgNPs with HEPES 10 mM ph 7 as mobile phase.

The radiochemical purity of the fractions 5, 6, and 7 as the fraction with the highest radioactivity of more than 90% was then determined by using Autoradiography scanner. The result shows that the fraction 6 has the highest radiochemical purity of 94.6% compared to the fraction 5 and 7 as shown in Table 3.

Table 3. Radiochemical Purity of ^{131}I-AgNPs after purification 2

Fraction	RCP (%)	Retention Factor (cm)	Appearance
Fraction 5	90.4 ± 0.3536	0.18	Yellow-clear solution
Fraction 6	94.5 ± 0.2121	0.18	Yellow-clear solution
Fraction 7	91.2 ± 0.8485	0.19	Yellow-clear solution

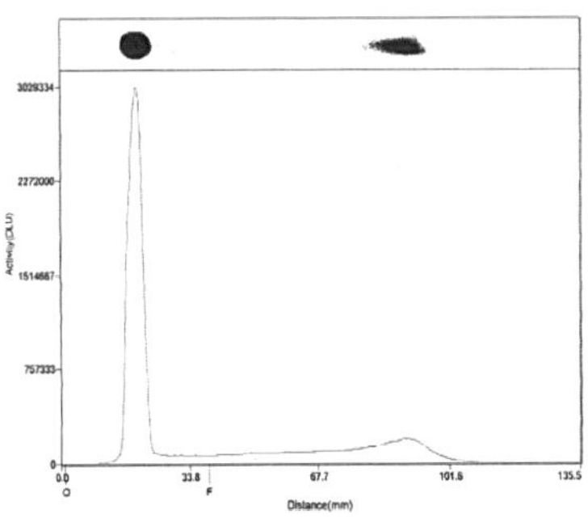

Figure 11. The Autoradiogram of Radiochemical Purity of ^{131}I-AgNPs fraction 5.

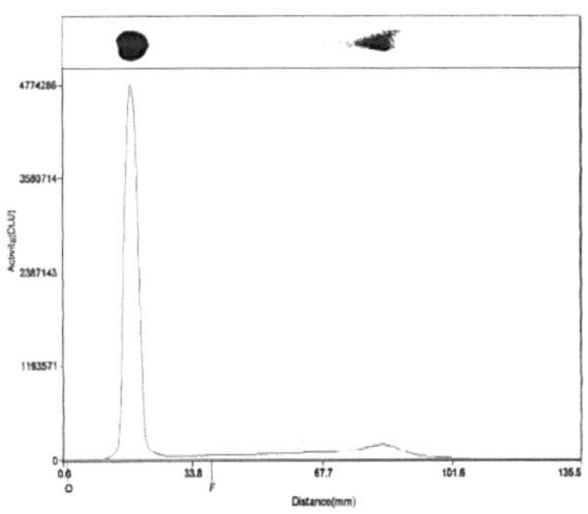

Figure 12. The Autoradiogram of Radiochemical Purity of ^{131}I-AgNPs fraction 6.

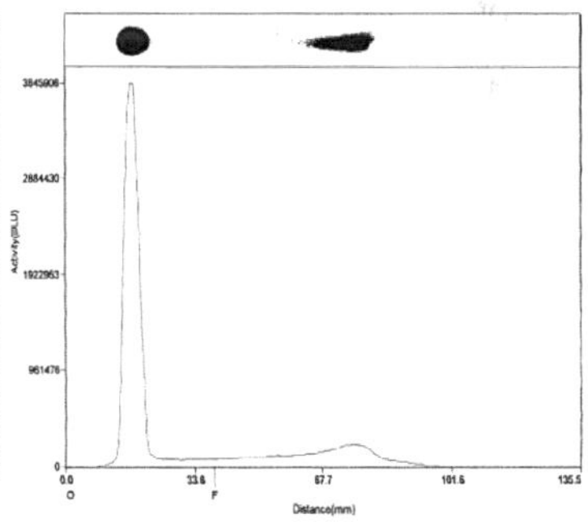

Figure 13. The Autoradiogram of Radiochemical Purity of ^{131}I-AgNPs fraction 7.

Purification 2 shows the result that fraction 6 has the highest activity of 178.7 µCi and highest radiochemical purity of 94.6% compared to the fraction 5 and 7. These fractions were then stored in the room temperature and refrigerator to determine the stability of the ^{131}I-AgNPs after labelling and purification process.

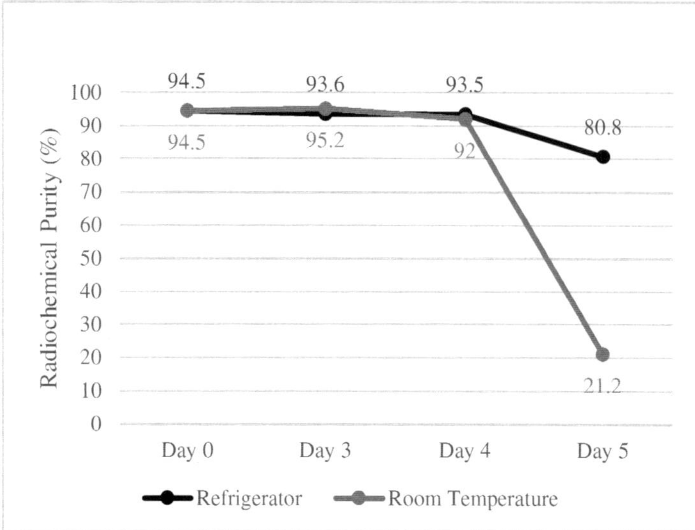

Figure 14. Comparison of Radiochemical Purity of Fraction 6 of ^{131}I-AgNPs.

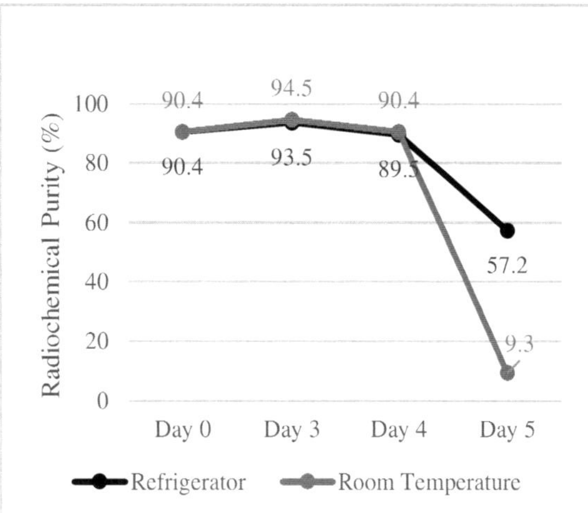

Figure 15. Comparison of Radiochemical Purity of Fraction 5 of ^{131}I-AgNPs.

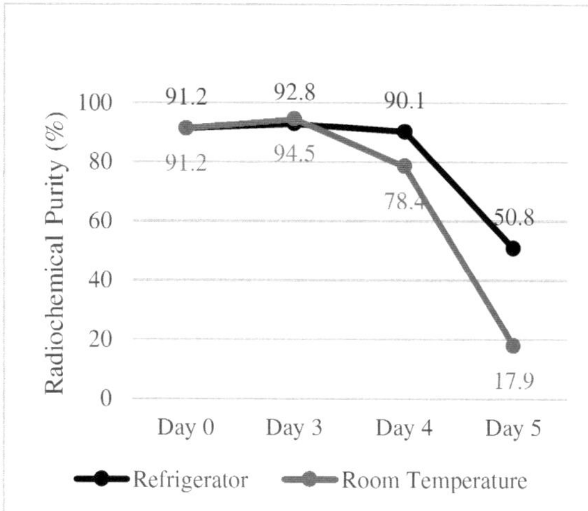

Figure 16. Comparison of Radiochemical Purity of Fraction 7 of ^{131}I-AgNPs .

Figure 14-16 shown the comparison of radiochemical purity of fraction 5-7 of ^{131}I-AgNPs after purification 2. All of the three fractions have radiochemical purity more than 90% until the 3rd day after purification in both storages. However, in the 4th day after purification, only fraction 6 that shown ^{131}I-AgNPs with radiochemical purity more than 90% in both storages. In the 5th day after purification, all of the three fractions have radiochemical purity less than 90%. All of the three fractions stored in room temperature show radiochemical purity less than 50%, while the fractions that stored in the refrigerator show radiochemical purity more than 50% with fraction 6 still have radiochemical purity of 80%. Only fraction 6 that stored in a refrigerator that shows radiochemical purity up to 80% while the others show less. This result shows that storage in the refrigerator can be a better choice rather than in the room temperature.

4. Conclusion

Iodogen as the oxidizing agent on ^{131}I radiolabelling of AgNPs resulted in ^{131}I-AgNPs with radiochemical purity 80.4 ± 0.2121 % and 94.5 ± 0.2121% after purification. Stability test shows that the best ^{131}I-AgNPs has radiochemical purity more than 90% in both storages (room temperature and refrigerator) until the 4th day after purification. Only ^{131}I-AgNPs with the highest activity that stored in a refrigerator that shows radiochemical purity up to 80% while the others show less. ^{131}I-AgNPs can be used until the 4th day after purification as long as the radiochemical purity still more than 90%. Storage in the refrigerator can be a better choice rather than in the room temperature.

References

[1] G Franci, et al. 2015. *Review* Silver Nanoparticles as Potential Antibacterial Agent *J. Mol.* **20** pp 8856-8874
[2] XF Zhang, ZG Liu ,W Shen and S Gurunathan. 2016. *Review* Silver Nanoparticles: Synthesis, Characterization, Properties, Applications, and Therapeutic Approaches. *Int. J. Mol. Sci.* **17** p 1534
[3] L Wei, J Lu, H Xu, A Patel, ZS Chen, G Chen. 2015. Review Silver Nanoparticles: Synthesis, Properties, and Therapeutic Application. *J. D. Dis. Tod.* **20** pp 595-601
[4] SJ Goldsmith. Iodine Therapy of Differentiated Thyroid Carcinoma: Redesigning the Paradigm. 2017. *J. Mol. Ima. Rad. The.* **26** pp 74-79
[5] A Didi, A Dadouch, HE Bekkouri. Feasibility Study For Production Of Iodine-131 Using Dioxide Of Tellurium-130. 2016. *Int J Pharm Pharm Sci* **8** pp 327-331
[6] R Haffner, WH Miller, S Morris. Verification of I-131 Yields From The Neutron Irradiation Of Tellurium. Applied Radiation and Isotopes. 2019. *J. App. Rad. Iso.* **151** pp 52-61
[7] A Yordanova, et al. Theranostics In Nuclear Medicine Practice. 2017. *J. Onc. Tar. The.* **10** pp 4821–4828
[8] A Wyszomirska. Iodine-131 For Therapy Of Thyroid Diseases. Physical And Biological Basis: Review. 2012. *J. Nuc. Med. Rev.* **15** pp 120–123
[9] X Yi, et al. Imaging-Guided Combined Photothermal and Radiotherapy to Treat Subcutaneous and Metastatic Tumors Using Iodine-131-Doped Copper Sulfide Nanoparticles. 2015. *J. Adv. Fun. Mat.* **25** pp 4689-4699
[10] AM Avram. An Important Diagnostic Tool for Thyroid Cancer Staging and Risk Stratification. 2012. *J. Nuc. Med.* **53** pp 754-764
[11] SE Sharp , AT Trout, BD Weiss, and MJ Gelfand. MIBG in Neuroblastoma Diagnostic Imaging and Therapy. 2016. *J. Rad Grap.* **36** pp 258–278.
[12] A Chrastina and JE Schnitzer. Iodine-125 radiolabeling of silver nanoparticles for *in vivo* SPECT imaging. 2010. *Int. J. Nan.* **5** pp. 653–659
[13] HQ Ho, et all. Feasibility Study of Large-scale Production of Iodine-125 at the High Temperature Engineering Test Reactor. 2018. *J. App. Rad. Iso.* **140** pp 209-214.
[14] R Awaludin et all. 2009. Iodine-125 Radioactivity in Production Test Using Xenon-124 Enriched Targets. *J. Sai. Tek. Nuk. Ind.* **X(1)** pp 1-9 (in Indonesia)
[15] GT Hermanson. Bioconjugate Techniques Third Edition. 2013. Academic Press.
[16] GB Saha. Fundamentals of Nuclear Pharmacy Sixth Edition. 2010. Springer Science and Business Media, LLC.
[17] UN Sholikhah, et al. Stability of Silver Nanoparticles as Imaging Materials. 2016. *J. Pure App. Chem. Res.* **5** pp 173-177

Estimation of Muon Stopping Site in $CoCr_2O_4$ Using Density Functional Theory

Fiqhri Heda Murdaka[1], Kohji Nakamura[2], and Agustinus Agung Nugroho[3*]
[1,3] Faculty of Mathematics and Natural Sciences, Institut Teknologi Bandung, Jl. Ganesha No 10, Bandung 40132, Jawa Barat, Indonesia
[2] Department of Physics Engineering, Mie University, Tsu, Mie 514-8507, Japan

*nugroho@fi.itb.ac.id

Abstract. Spinel material Cobalt Chromite ($CoCr_2O_4$) is a ferrimagnetic material with T_C of about 93 K. This compound has attracted much attention due to the presence of electrical polarization which related to spiral magnetic transition at T_S = 26 K. [Physical Review B, 70, 214434, 2004]. This spiral magnetic transition phenomena could be explored by experiment using Muon Spin Rotation, Relaxation, and Resonance (μSR). Positive muon in μSR plays a role as a local magnetic probethus, we need to determine the muon stopping site information. In this study, we investigate the muon stopping site of $CoCr_2O_4$ with ferrimagnetic ground state using spin polarized Density Functional Theory implemented in Full-potential Linearized Augmented Plane Wave (FLAPW) method. The exchange-correlation effect was considered in the scheme of GGA+U_{eff} approximation. Based on the minimum energy of electrostatic potential, we obtain the muon stopping site around the three Oxygen ions in the tetrahedral site of Cobalt.

Keywords **:** Muon stopping site, electrostatic potential, $CoCr_2O_4$, DFT, FLAPW

1. Introduction

Cobalt Chromite, $CoCr_2O_4$, is a ferrimagnetic material with T_C of about 93 K. There has been a high interest in this compound due to the presence of electrical polarization which related to spiral magnetic transition T_S, at around 26 K [1]. The presence of the clamping between the magnetization and electrical polarization makes $CoCr_2O_4$ belong to the class of multiferroic material [2]. The coexistence of ferromagnetism and ferroelectricity in multiferroic material could be applied to some potential devices such as a magnetic field sensor [3], modulator [4], memory devices [5], gyrator [6], and energy harvesting device [7].

$CoCr_2O_4$ crystallizes in the cubic spinel structure with the lattice constant of 8.3346 Å [8] as shown in Fig.1. The A site of the spinel occupied by Co^{2+} ion while the B site occupied by Cr^{3+} ion [1]. Due to the magnetism of Co^{2+} cation in A site, the geometric frustration inside $CoCr_2O_4$ cannot be released by lowering the crystal symmetry. The alternative way to get rid the geometrical frustration is through the temperature dependent magnetic structure mechanism [9]. The magnetic structure of $CoCr_2O_4$ is changing from collinear ferrimagnetic to be a complex noncolinear spiral magnetic configuration at 26 K [10]. Due to the complexity of noncolinear spiral configuration, this paper only considers the collinear ferrimagnetic structure (the temperature region from 93 K down to 26 K).

The magnetic transition phenomena in $CoCr_2O_4$ can be measured by experiment using Muon Spin Rotation, Relaxation, and Resonance (μSR) technique. Since the μSR is a local probe to measure the internal magnetic fields, therefore it is necessary to find the positions of the muon stop in the material. This paper reports the muon stopping site in $CoCr_2O_4$ crystal based on the local minima of the electrostatic potential obtained from density functional theory (DFT) implemented in Full-potential Linearized Augmented Plane Wave (FLAPW) method.

2. Computational method

The positive muon which is commonly used in the experiment would prefer to stop in the minimum of the electrostatic potential near the negative ions [11]. In term to know the muon stopping site, firstly we need to calculate the electrostatic potential felt by the muon inside of $CoCr_2O_4$ using:

$$V_\mu(\vec{r}) = -\frac{e^2}{2}\int \frac{n(\vec{r}')}{\vec{r}-\vec{r}'}dr' + \sum_i \frac{Z_i e^2}{\vec{r}-\vec{R}_i}, \quad (1)$$

where the first term in this expression is a Hartree term with $n(\vec{r}')$ is the density of electron, and the second term is a nuclei term. $V_\mu(\vec{r})$ value inside of materials can be calculated using ab-initio strategy [11-13].

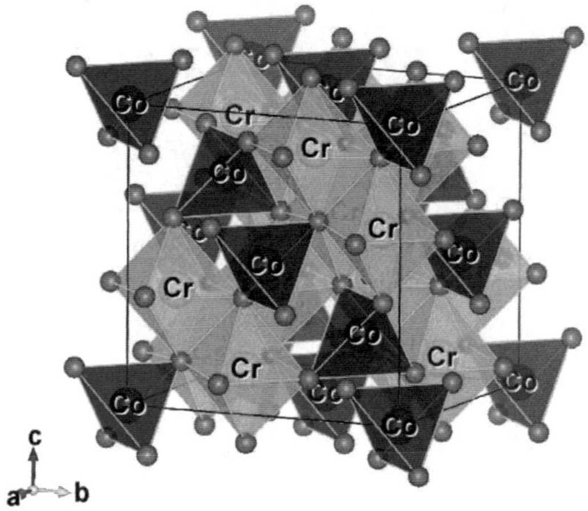

Figure 1. Crystal structure of $CoCr_2O_4$ conventional cell.

The electrostatic potential of the muon inside $CoCr_2O_4$ was obtained from equation (1) utilizing the electron density provided by DFT calculation implemented in FLAPW method [14-16]. The Linearized Augmented Plane Wave (LAPW) function used a cutoff of $|\boldsymbol{k} + \boldsymbol{G}| \leq 3.9$ a.u and muffin-tin sphere radii of 2.0, 2.1, and 1.6 a.u for Co, Cr, and O ions, respectively. These radii were considered to avoid overlap among the muffin-tin spheres and to ensure the loss of electron from the core is less than 0.05 with the respect of the total electron. We used generalized gradient approximation (GGA) [17] for exchange-correlation and *Ueff* value for strong correlation effect [10]. The *Ueff* term is consist of a subtraction between U value denoting as the Coulomb interaction and J as Hund's coupling [18]. The U value for Co and Cr ion was set to be 5 eV and 3 eV, for the J value both of Co and Cr was set to be 1 eV [10]. The minimum electrostatic potential of the muon was calculated based on Eq. 1 for the $CoCr_2O_4$ conventional cell as show in Fig. 1 with the mesh of 101×101×101.

3. Results and discussion

In order to determine the local minima of the electrostatic potential sites of the muon inside the $CoCr_2O_4$, we draw an isosurface potential of 0.15 eV greater than the global minimum potential in all positions of the grid. We obtain the isosurface forms a ring around three Oxygen ions of tetrahedral side of Cobalt as shown in Fig. 2. It is noted that the distance between Oxygen ions at the tetrahedra side of Cobalt (3.20604 Å) is bigger than at the octahedral side of Chromium (2.95243 Å). Figure 3 shows the tetrahedra symmetry of the Cobalt ions and the tetrahedra symmetry of the ring at the Cobalt sides.

The possibility to find another local minimum around the Cobalt site was carried out by varying the isosurface potential value from 0.1 to 1 eV. Figure 4 shows that the result of increasing the

IOP Conf. Series: Materials Science and Engineering **924** (2020) 012027 doi:10.1088/1757-899X/924/1/012027

isosurface potential value. It is interesting to note that there are no other minima outside the cluster. In order to find the local minimum, we will focus only the position inside the ring cluster and yields tominimum positions with a distance of 2.01 Å from the Cobalt site. All of these six positions are also located near three Oxygen positions around the Cobalt ion.

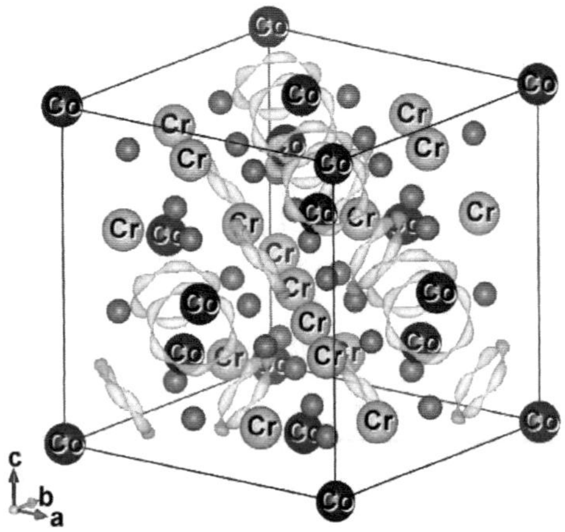

Figure 2. Electrostatic potential isosurfaces with 0.15 eV greater than the global minimum value.

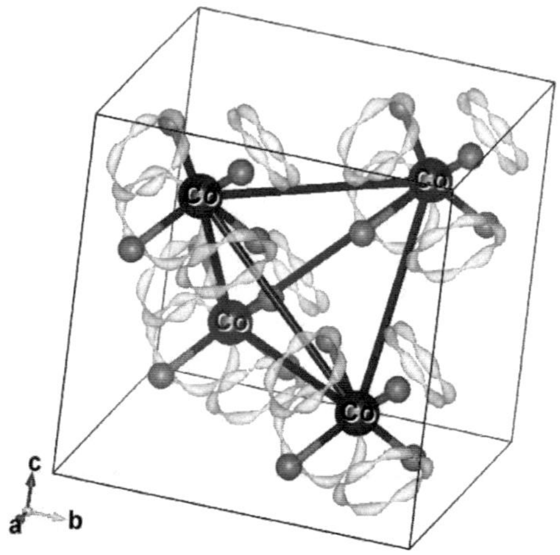

Figure 3. A tetrahedra symmetry applies for both the ring of electrostatic potential isosurfaces and Cobalt ion.

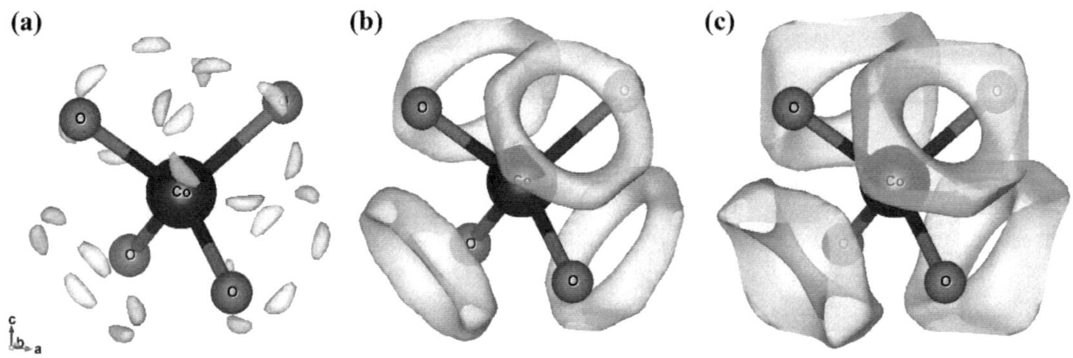

Figure 4. Ring-shaped clusters of electrostatic potential isosurfaces around A site (Co ion) with isosurfaces value of (a) 0.1, (b) 0.5, and (c) 1 eV greater than the global minimum value of electrostatic potential energy.

All of those six positions described above cannot be considered as final muon stopping sites because there is a possibility that muon can still move away from the minimum positions due to the arrangement of the crystal structure in the presence of muon. This will be resolved by considering the relaxation of the local structure caused by muon and the muon zeropoint motion (ZPM) [11,13,19].

4. Conclusion

The position of the minimum electrostatic potential position in $CoCr_2O_4$ has been determined using density functional analysis based on FLAPW method with correlation effects. Electrostatic potential isosurface plot from the resulted calculation shows that there are six minimum positions with 2.01 Å radius from the Cobalt tetrahedral site.

Acknowledgment
Authors want to acknowledge the World Class University sandwich programs from ITB and riset ITB program for financial support. FHM also acknowledge Magnetic Research Group in Mie University of Japan for the hospitality and computing resources during the research process.

References

[1] K. Tomiyasu, J. Fukunaga and H. Suzuki, "Magnetic short-range order and reentrant-spin-glass-like behavior in CoCr2O4 and MnCr2O4 by means of neutron scattering and magnetization measurements," *Physical Review B,* vol. 70, p. 214434, 2004.

[2] Y. Yamasaki, S. Miyasaka, Y. Kaneko, J. P. He, T. Arima and Y. Tokura, "Magnetic reversal of the ferroelectric polarization in a multiferroic spinel oxide," *Physical Review Letters,* vol. 96, p. 207204, 20016.

[3] C. W. Nan, M. I. Bichurin, S. Dong, D. Viehland and G. Shrinivasan, "Multiferroic magnetoelectric composites: Historical perspective, status, and future directions," *Journal of Applied Physics,* vol. 103, p. 031101, 2008.

[4] H. L. Mette, "Solid-state magnetoelectric modulator and switch". U.S Patent 3,435,379, 25 March 1969.

[5] W. Eerenstein, N. D. Mathur and J. F. Scott, "Multiferroic and magnetoelectric materials," *nature,* vol. 442, no. 7104, p. 759, 2006.

[6] J. Zhai, G. C. Jassduke, J. De Vreugd, J. Li, D. Viehland, A. V. Filippov, M. I. Bichurin, D. V. Drozdov, G. A. Semenov and S. X. Dong, "Magnetoelectric gyrator," *he European Physical Journal B,* vol. 71, no. 3, p. 383, 2009.

[7] R. Gupta, M. Tomar, A. Kumar and V. Gupta, "Performance of magnetoelectric PZT/Ni multiferroic system for energy harvesting application," *Smart Mater. Struct.,* vol. 26, p. 035002, 2017.

[8] P. G. Casado and I. Rasines, "Preparation and crystal data of the spinel series Co1+ 2sCr2−3sSbsO4 (O⩽ s⩽ 23)," *Polyhedron,* vol. 5, no. 3, pp. 787-789, 1986.

[9] N. Menyuk, K. Dwight and A. Wold, "Ferrimagnetic spiral configurations in cobalt chromite," *Journal de physique,* vol. 25, no. 5, pp. 528-536, 1964.

[10] D. Das and S. Ghosh, "Density functional theory based comparative study of electronic structures and magnetic properties of spinel ACr2O4 (A=Mn, Fe, Co, Ni) compounds," *Journal of Physics D,* vol. 48, p. 425001, 2015.

[11] F. Bernardini, P. Bonfa, S. Massidda and R. De Renzi, "Ab initio strategy for muon site assignment in wide band gap fluorides," *Physical Review B,* vol. 87, p. 115148, 2013.

[12] E. Suprayoga, A. A. Nugroho, A. O. Polyakov, T. T. M. Palstra and I. Watanabe, "Search for potential minimum positions in metal-organic hybrids,(C2H5NH3) 2CuCl4 and (C6H5CH2CH2NH3) 2CuCl4, by using density functional theory," *Journal of Physics: Conference Series,* vol. 551, p. 012054, 2014.

[13] I. Watanabe, E. Suprayoga, N. Adam, S. S. Mohm-Tajudin, A. F. Rozlan, D. Puspita, R. Asih, F. Astuti, M. D. Umar, J. Angel, S. Sulaiman and M. I. Mohamed-Ibrahim, "Muon Site Estimations by DFT Calculations in Metal and Insulating Systems," *Materials Science Forum,* vol. 827, pp. 347-354, 2015.

[14] E. Wimmer, H. Krakauer, M. Weinert and A. J. Freeman, "Full-potential self-consistent linearized-augmented-plane-wave method for calculating the electronic structure of molecules and surfaces: O 2 molecule," *Physical Review B,* vol. 24, no. 2, p. 864, 1981.

[15] K. Nakamura, T. Ito, A. J. Freeman, L. Zhong and J. Fernandez-de-Castro, "Intra-atomic noncollinear magnetism and the magnetic structures of antiferromagnetic FeMn," *Physical Review B,* vol. 67, no. 1, p. 014405, 2003.

[16] K. Nakamura, T. Ito, A. J. Freeman, L. Zhong and J. Fernandez-de-Castro, "Enhancement of magnetocrystalline anisotropy in ferromagnetic Fe films by intra-atomic noncollinear magnetism," *Physical Review B,* vol. 67, p. 014420, 2003.

[17] J. P. Perdew and W. Yue, "Accurate and simple density functional for the electronic exchange energy: Generalized gradient approximation," *Physical review B,* vol. 33, no. 12, p. 8800, 1986.

[18] S. L. Dudarev, G. A. Botton, Y. Savrasov, C. J. Humphreys and A. P. Sutton, "Electron-energy-loss spectra and the structural stability of nickel oxide: An LSDA+U study," *Physical Review B,* vol. 57, p. 1505, 1998.

[19] E. Suprayoga, A. A. Nugroho, D. Onggo, A. O. Polyakov, T. T. M. Palstra and I. Watanabe, "3D long-range magnetic ordering in (C2H5NH3) 2CuCl4 compound revealed by internal magnetic field from muon spin rotation and first principal calculation," *Physica B: Condensed Matter,* vol. 545, pp. 76-79, 2018.

Study on Manufacturing Company Sector Industrial Consumption Goods in Indonesia

Meiryani*, Fransiska Aninda Yuanita, and Mohamad Heykal
University of Bina Nusantara, Jl. K H. Syahdan No. 9 Kemanggisan–Palmerah Jakarta Barat 11480

*meiryani@binus.edu

Abstract. This research is using quantitative method. The data withdrawal techniques in this research are purposive sampling. The data are secondary data that are gained from the companies documentary that are already exist and making progress at this time. Research data are analyzed with multiple linear regression which are processed with Microsoft excel and SPSS 22. The result of this research show us that account receivables turnover doesn't effect the company liquidity data that are taken from 2016 to 2018 which are filtered in such way for the purpose of this research. It is caused by cash turnover in the company are no yet compatible with the recorded data of account receivable turnover. Meanwhile the cash turnover give positive effects towards the liquidity of consumption industry sector which are the Y variable. The bigger the cash turnover, will give the company more power to accomplish payment for short term debt which give us meaning that the company are liquid.

Keywords **:** Account Receivables Turnover, Cash Turnover, Liquidity

1. Introduction

The success of a company can be measured by the financial level that can be achieved. Company is required to be able to manage the company's cash flow so that the company can fulfill its obligations and to be able to overcome any problems that might occur or that always arise, especially in the accounts receivable turnover and cash turnover at the company. One of the most important things is the company's cash flow because it covers cash receipts and disbursements. In this competitive situation, companies often sell on credit, which is an alternative way that can be used to increase sales volume, which in turn can increase company liquidity. Credit sales is one of the most important policies, because credit sales is one of the strategic policies in conducting a sale, this policy is used to increase sales volume even though this policy can create receivables on the company's balance sheet. Generally receivables arise because of credit sales. Credit sales policy is held because in this intense competition the company is required to be able to reach market positions. Companies are required to have a credit sales strategy in order to increase sales volumes where this increase can result in increased profitability. However, the increasing sales on credit will lead to a growing number of receivables and expenses. Costs borne by the company will also increase. Receivables turnover in a company shows how the company is in managing receivables, if the management of receivables is good, the receivables turnover and cash flow smoothly and the company is said to be able to manage company receivables. However, if the management of receivables is not good, it will have an impact on the company's profit and liquid levels. Basically it can be concluded that receivables in a company are very important influence on the survival of the company. If the company does not have good receivables management, it will have an impact on the company's operations so that it can cause losses to the company. That way the company must be able to implement good receivables management to avoid bad debts so that the company can survive and compete with other companies [1].

Cash is the most current asset element or it can be said that cash is the most liquid working capital, so that with the availability of sufficient cash the company will have no difficulty in fulfilling due obligations. Information about a company's cash flow is useful for users of financial statements to be able to assess the company's ability to generate cash and assess the company's needs to use the cash flow. Fundamental cash flow for the existence of the company and shows whether or not the company is able to pay all its obligations. Cash inflows and cash outflows are cash inflows which are the sources of cash obtained while cash outflows are cash requirements for payments [2]. The cash flow statement is a report that presents an overview of all cash inflows and outflows during a certain period. Cash inflows and outflows are recorded in the cash flow statement. High accounts receivable turnover will result in higher capital and the company can be said to be liquid. If the receivables turnover is low, the existing capital condition will also be said to be low. The company must be really careful in investing the company's funds with the aim of maintaining the company's liquidity. Working capital is capital that is used to carry out company operations. Working capital is defined as investments that are invested in current assets or short-term assets, such as cash, banks, securities, receivables, inventories and current assets. Basically working capital is always in a state of rotation as long as the company is in a business state.

Bank Indonesia stated that Century Bank was experiencing liquidity problems. The global financial crisis has made Bank Century worse. Century Bank also experienced a clearing failure that was previously disclosed due to technical problems. Based on the analysis, Bank Indonesia decided that Century Bank was taken by the government through the Indonesia Deposit Insurance Corporation (LPS). This is triggered by a number of factors, the factor in question is the potential for the Fed to increase interest rates twice again until the end of 2018 and also the potential weakening of the Rupiah against the US Dollar because the US Dollar is expected to continue to strengthen until the end of 2018, inevitable from fears of the impact of the escalation of trade wars between the United States and other trading partner countries. This causes a problem that occurs in the company against rising debt which indicates a low level of company liquidity. The main cause of the lack and incapability of the company to pay its obligations is actually due to negligence of the company's management in carrying out its business. For further explore clearly the accounts receivable turnover and cash flow turnover of the company's liquidity level, the authors are interested in conducting research and raising the title "Analysis of the Effects of Accounts Receivable Turnover and Cash Turnover on Company Liquidity". The aim of this study is to know how receivable turnover and cas flow turnover influence company liquidity.

2. Theoretical Framework

Conducted a study with the title Effect of Accounts Receivable Turnover and Cash Turnover on Likuiidtas in CV. Sinar Karya Pekan Baru, the results of the study indicate that Accounts Receivable Turnover has no significant effect on liquidity, while cash turnover has a significant effect on liquidity[3]. A study with the title Effect of Accounts Receivable Turnover and Cash Turnover on Liquidity, the results showed that Partially, receivable turnover has no effect on liquidity and cash turnover also has no effect on liquidity[4]. However, simultaneously, receivables and cash affect liquidity. Conducted a study entitled The Effect of Cash Flow and Receivables Turnover on Liquidity, the results showed that partial cash flow and accounts receivable turnover had a significant effect on liquidity with each of the large, low, and medium influences[5]. Research conducted by [6,7] states that accounts receivable turnover has a significant effect on liquidity (current ratio). Based on previous research above, the first hypothesis of this study is: H1: the higher the accounts receivable turnover, the higher the company's liquidity. Research conducted by Suharti and Yuniati (2018) states that cash turnover has a significant effect on liquidity (current ratio). Based on previous research above, the second hypothesis of this research is: H2: the higher the cash turnover, the higher the company's liquidity.

According to previous research conducted by [8,9] examining the Effects of Cash Turnover and Accounts Receivable Turnover on Liquidity Levels in Automotive Companies listed on the Indonesia Stock Exchange in the period 2008-2011 there were cases at PT Astra International Tbk. where there was an increase in current debts by an average of Rp 34,778,250 each year and the highest increase in the average number of companies in 2010-2011 with the highest dominant increase in 2011 amounting to Rp 5,274,006 with a percentage level of 67% 12 automotive companies listed on the Indonesia Stock Exchange. This causes a problem that occurs in the company against rising debt which indicates a low level of company liquidity. [10,11] it examines the Effects of Cash Flow, Accounts Receivable Turnover, and Working Capital Turnover on Company Liquidity in Telecommunications Companies listed on the Indonesia Stock Exchange Period 2010-2015. From this research, it can be concluded that cash flow in telecommunications companies significantly influences company liquidity, while accounts receivable turnover does not affect liquidity, but accounts receivable turnover can affect cash flow which is known that cash flow affects company liquidity [12,13].

3. Research Methodology

This study was conducted to determine the effect of accounts receivable turnover and cash turnover on the liquidity of companies engaged in the Manufacturing of the Consumer Goods Industry sector which are listed on the Indonesia Stock Exchange. in the implementation of this research will use a quantitative research type approach which is carried out through data collection from the Indonesia Stock Exchange and the research method used is descriptive research. In this study the data used are quantitative data. Quantitative data is data in the form of numbers or numbers that can be processed or analyzed using mathematical or statistical calculation techniques. The type of data used in this study is secondary data. Secondary data is data obtained or collected from sources that already exist. The data is usually obtained from libraries or previous research reports/documents. In this study the data used are the financial statements of the Manufacturing Companies in the Consumer Goods Industry sector which are listed on the Indonesia Stock Exchange for the period of 2016 - 2018. In this study, the sample is 31 manufacturing companies of Consumer Goods Industry sectors that are listed on the Indonesia Stock Exchange for the period of 2016-2018. The population in this study were all manufacturing companies in the Consumer Goods Industry sector which were listed on the Indonesia Stock Exchange during the 2016-2018 period, amount to 31 companies. In this study the sampling technique was carried out using the purposive sampling method, which is a sampling method with certain criteria.

Table 1 Results of sample selection [3].

Information	total
Number of Manufacturing companies in the Consumer Goods Industry sector are listed on the IDX	42 Companies
Companies listed on the Indonesia Stock Exchange for the period of 2016-2018	42 Companies
Companies that publish 2016-2018 financial statements	31 Companies
Company Samples	31
Total Observation for 3 years	93

Source : www. idx.com

From the results of the selection above it can be seen that the sample in this study amounted to 31 Manufacturing Companies in the Consumer Goods Industry sector listed on the IDX. The

IOP Conf. Series: Materials Science and Engineering **924** (2020) 012028 doi:10.1088/1757-899X/924/1/012028

following are the Manufacturing Companies of the Consumer Goods Industry sector which were sampled in this study.

Table 2. List of Samples.

NO	CODE	COMPANY NAME
1	ICBP	PT. Indofood CBP Sukses Makmur Tbk
2	INDF	PT. Indofood Sukses Makmur Tbk
3	MYOR	PT. Mayora Indah Tbk
4	ROTI	PT. Nippon Indosari Corporindo Tbk
5	SKBM	PT. Sekar Bumi Tbk
6	SKLT	PT. Sekar Laut Tbk
7	RMBA	PT. Bentoel International Investama Tbk.
8	ULTJ	PT. Ultrajaya Milk Industri and Trading Company Tbk
9	ADES	PT. Akasha Wira International Tbk
10	FAST	PT. Fast Food Indonesia Tbk
11	SMAR	PT. Smart Tbk
12	SIPD	PT. Sierad Produce Tbk
13	UNVR	PT. Unilever Indonesia Tbk.
14	MAIN	PT. Malindo Feedmill Tbk.
15	IIKP	PT. Inti Agri Resources
16	GGRM	PT. Gudang Garam Tbk.
17	ALTO	PT. Tri Banyan Tirta Tbk
18	CEKA	PT. Wilmar Cahaya Indonesia Tbk.
19	DLTA	PT. Delta Djakarta Tbk.
20	MLBI	PT. Multi Bintang Indonesia Tbk.
21	HMSP	PT. Handjaya Mandala Sampoerna Tbk.
22	TBLA	PT. Tunas Baru Lampung Tbk.
23	BTEK	PT. Bumi Teknokultura Unggul Tbk.
24	BUDI	PT. Budi Starch & Sweetener Tbk.
25	TSPC	PT. Tempo Scan Pacific Tbk.
26	KINO	PT. Kino Indonesia Tbk.
27	MBTO	PT. Martina Berto Tbk.
28	TCID	PT. Mandom Indonesia Tbk.
29	KAEF	PT. Kimia Farma (Persero) Tbk.
30	KLBF	PT. Kalbe Farma Tbk.
31	PYFA	PT. Pyridam Farma Tbk.

Source : www. idx.com

Data analysis method: Classical Assumption Test, the classic assumption test in this study uses the normality test, the multicollinearity test, the Heteroscedasticity Test. Data Presentation Method is quantitative research, so using data presentation methods in the form of numbers, tables or diagrams as well as the appointment of hypotheses. Statistic test: (1) Hypothesis Testing also tested the hypothesis test. Data obtained from the results of the above data collection can be processed according to the type of data then presented in the form of tables and numerical statistical methods; (2) T Test (Partial): according to Ghozali (2012: 98) t-test different tests are used to test how far the influence of the independent variables used in this research individually in explaining partially dependent variables. The basis for decision making used in the t test is as follows: (1) If the significance value is> 0.05, the hypothesis is rejected. The hypothesis rejected has the meaning that the independent variable does not significantly influence the dependent variable; (2) If the significance probability value <0.05, the hypothesis is accepted. The hypothesis cannot be rejected which means that the independent variable influences the dependent variable significantly.

4. Result and Discussion

4.1 Description of Research Object

The object used in this study is manufacturing companies in the consumer goods industry sector which are listed on the Indonesia Stock Exchange during the period of 2016 to 2018. The sample selection method used is purposive sampling, where the sample must meet established criteria. The procedure for selecting samples is presented in Table 3 as follows:

Table 3. Sample Selection Procedure.

Information	Company	Data
Number of Manufacturing companies in the Consumer Goods Industry sector are listed on the IDX	42	126
Companies that publish 2016-2018 financial statements	31	93
Number of samples before outlier	31	93
Data Outlier		29
Number of samples after outlier		64

Source : www. idx.com

Table 3 shows that there are 31 companies that can be sampled in this study. The amount of data that can be used in this study previously there were 93 research data from the financial statements of each company during the period 2016 to 2018. When a normality test was conducted on the data, it was found that some data had to be transformed or commonly referred to as outlier data. The cause of transform data is needed for research because the data that has been calculated and analyzed produces negative data where negative data can affect the normality of the data. After transforming the data, 29 outliers data were found out of the total 93 data obtained. After finding 29 outlier data, the next decision was to eliminate 29 data. Outlier of a total of 93 data so that the total data becomes 64 data.

Table 4. Distribution of Sample Companies by Industry

NO	CODE	COMPANY NAME
1	ICBP	PT. Indofood CBP Sukses Makmur Tbk
2	INDF	PT. Indofood Sukses Makmur Tbk
3	MYOR	PT. Mayora Indah Tbk
4	ROTI	PT. Nippon Indosari Corporindo Tbk
5	SKBM	PT. Sekar Bumi Tbk
6	SKLT	PT. Sekar Laut Tbk
7	STTP	PT. Siantar Top Tbk
8	ULTJ	PT. Ultrajaya Milk Industri and Trading Company Tbk
9	ADES	PT. Akasha Wira International Tbk
10	FAST	PT. Fast Food Indonesia Tbk
11	SMAR	PT. Smart Tbk
12	SIPD	PT. Sierad Produce Tbk
13	UNVR	PT. Unilever Indonesia Tbk.
14	MAIN	PT. Malindo Feedmill Tbk.
15	IIKP	PT. Inti Agri Resources
16	AISA	PT. Tiga Pilar Sejahtera Food Tbk
17	ALTO	PT. Tri Banyan Tirta Tbk
18	CEKA	PT. Wilmar Cahaya Indonesia Tbk.
19	DLTA	PT. Delta Djakarta Tbk.
20	MLBI	PT. Multi Bintang Indonesia Tbk.

21	SIPD	PT. Sierad Produce Tbk.
22	TBLA	PT. Tunas Baru Lampung Tbk.
23	BTEK	PT. Bumi Teknokultura Unggul Tbk.
24	BUDI	PT. Budi Starch & Sweetener Tbk.
25	TSPC	PT. Tempo Scan Pacific Tbk.
26	KINO	PT. Kino Indonesia Tbk.
27	MBTO	PT. Martina Berto Tbk.
28	TCID	PT. Mandom Indonesia Tbk.
29	KAEF	PT. Kimia Farma (Persero) Tbk.
30	KLBF	PT. Kalbe Farma Tbk.
31	PYFA	PT. Pyridam Farma Tbk.

Source: spss processed data

4.2 Descriptive Statistics

Descriptive Statistics Method is Statistics used to analyze data by describing or drawing data that has been collected as it is without intending to make conclusions that apply to the public or generalizations [9]. Descriptive statistics of the variables show a description of the research data in the form of minimum values, maximum values, mean values, and standard deviations of each variable presented in Table 5 below:

Table 5. Descriptive Statistics

	Descriptive Statistics				
	N	**Minimum**	**Maximum**	**Mean**	**Std. Deviation**
Receivables	64	1,36	13,21	7,3977	2,98684
Cash	64	1,73	85,19	26,2642	25,99960
Liquidity	64	,61	4,51	1,7589	,99367
Valid N (listwise)	64				

Source: SPSS data processing results 22

The descriptive statistical test results contained in Table 5 show that: (1) The first independent variable shows that the data from the Accounts Receivable Turnover has the lowest value of 1.36, the highest value of 13.21, an average value of 7.3977 and a standard deviation that shows the range of data in the sample of 2.98684; (2) The second independent variable shows that the data from Cash Turnover has the lowest value of 1.73, the highest value of 85.19, an average value of 26.2642 and a standard deviation that shows the range of data in the sample of 25.99960; (3) The dependent variable taken from Liquidity data has the lowest value of 0.61, the highest value of 4.51, an average value of 1.7589 and a standard deviation that shows the range of data in the sample of 0.99367.

4.3 Classic assumption test

The classic assumption test is carried out to obtain an accountable regression. The classic assumption test in this study uses the normality test, the multicollinearity test, the heteroscedasticity test, and the autocorrelation test.

4.3.1 Normality test. To test the normality of the data in this study Kolmogrov-Smirnov was used. To accept or reject the hypothesis, namely by: If the probability> 0.05, then the distribution of the regression model is normal, but if the probability <0.05, then the distribution of the regression model is not the norm.

Table 6. Normality Test Results before the outlier.

One-Sample Kolmogorov-Smirnov Test		
		Unstandardized Residual
N		93
Normal Parameters[a,b]	Mean	0E-7
	Std. Deviation	1,65223196
Most Extreme Differences	Absolute	,190
	Positive	,190
	Negative	-,151
Kolmogorov-Smirnov Z		1,835
Asymp. Sig. (2-tailed)		,002

a. Test distribution is Normal.
b. Calculated from data.
Source: SPSS data processing results 22

Based on Table 7 of the normality test results with Kolmogrov-Smirnov, it can be seen that the p-value (Asymtotic Significance) of unstandardized residuals is 0.002 where the p-value (Asymtotic Significance) <p-value significance level (0.002 <0.05), so that it can it is said that the data is not normally distributed.

Table 7. Normality Test Results after outliers

One-Sample Kolmogorov-Smirnov Test		
		Unstandardized Residual
N		64
Normal Parameters[a,b]	Mean	0E-7
	Std. Deviation	,96446495
Most Extreme Differences	Absolute	,167
	Positive	,167
	Negative	-,098
Kolmogorov-Smirnov Z		1,338
Asymp. Sig. (2-tailed)		,056

a. Test distribution is Normal.
b. Calculated from data.
Source: SPSS data processing results 22

Abnormal data must be tested again using an outlier test where this test is used to dispose of data that has extreme high. From the results of the outlier test found 29 data that must be discarded so that the data can be normally distributed. After the outlier test and removing 29 data that have extreme high, the total data is 64 data. After the outlier test, the total p-value (Asymtotic Significance) of the Unstandardized Residual is 0.056. It can be said that is greater than p-value (probability/significance value) significance level of 0.05 (0.056> 0.05), so that overall the data is stated to have a normal distribution or a normal distribution of data.

4.3.2 Multicollinearity Test. Multicollinearity test aims to test whether a regression model has a correlation between independent variables. A good regression model should not occur correlation between independent variables. Multicollinearity testing is seen from the amount of VIF (Variance Inflation Factor) and tolerance. Tolerance measures selected independent variables that are not explained by other independent variables. So, a low tolerance value is equal to a high VIF (because VIF = 1/tolerance). The cutoff value commonly used to indicate the presence or absence of multicollinearity

Based on Tolerance value: (1) Tolerance> 0.10 then there is no multicollinearity; (2) Tolerance <0.10, multicollinearity occurs. Based on VIF value: (1) VIF <10.00 then multicollinearity does not occur; (2) VIF> 10.00 then multicollinearity occurs. From the output results it can be seen that the VIF value is smaller than 10, and the tolerance value is close to 1, this means that among the independent variables in this study there is no relationship or no relationship with each other. It can be concluded that this regression model does not have Multicollinearity.

Table 8. Multicollinearity Test Results.

Model		**Unstandardized Coefficients**		**Standardized Coefficients**	**t**	**Sig.**	**Collinearity Statistics**	
		B	Std. Error	Beta			Tolerance	VIF
1	(Constant)	1,959	,372		5,261	,000		
	Account Receivable	,005	,042	,016	,124	,902	,977	1,024
	Cash	-,009	,005	-,238	-1,891	,063	,977	1,024

a. Dependent Variable: Likuiditas
Source: SPSS data processing results 22

4.3.3 Heteroscedasticity Test. Heteroscedasticity test aims to test whether in the regression model there is an inequality of variance from residuals of one observation to another. If the variance from one observation residuals to another observation remains, then it is called homoscedasticity and if different is called heteroscedasticity. test heteroscedasticity can use the Spearman Rank correlation coefficient test that is correlating between absolute residual regression results with all independent variables.

Table 9. Heteroscedasticity Test Results.

Model		**Unstandardized Coefficients**		**Standardized Coefficients**	**t**	**Sig.**
		B	Std. Error	Beta		
1	(Constant)	,867	,206		4,198	,000
	Account Receivable	,002	,023	,010	,078	,938
	Cash	,004	,003	,194	1,525	,132

a. Dependent Variable: RES2
Source: SPSS data processing results 22

The table above shows that there is an independent variable, namely the accounts receivable turnover and the cash flow from the significance results of 0.938 and 0.132 where the value is above 0.05. Thus, it was concluded that the independent variable of this study did not occur heteroscedasticity where in the data found no similarities in variance.

4.3.4 Autocorrelation Test. Autocorrelation testing is done by using the Durbin Watson test, namely the upper limit (du) and the lower limit (dL). The testing criteria are as follows: (1) If 0 <d <dL, then a positive autocorrelation occurs; (2) If dL <d <du, then there is no certainty whether or not autocorrelation occurs; (3) If d-dL <d <4 then a negative autocorrelation occurs; (4) If 4-du <d <4-dL, then there is no certainty whether or not autocorrelation occurs; (5) If du <d <4-du, then there is no positive or negative autocorrelation.

Table 10. Autocorrelation Test Results.

Model	R	R Square	Adjusted R Square	Std. Error of the Estimate	Durbin-Watson
1	,160[a]	,026	-,007	,81321	1,888

Source: SPSS Data Processing Results 22

From the above output it can be seen that the value of Durbin Watson is 1,888. By using a sample of 64 and 2 independent variables as many as 2 variables, the value of du in the Durbin Watson table is 1.6601. Because the value of Durbin Watson is greater than the value of du and greater than 4-du, it is free from autocorrelation.

The formula = Du <d <4-du = 1.6601 <1,888 <2.3399. There is no positive or negative autocorrelation.

4.3.5 Multiple Linear Regression Test. Multiple regression analysis is used to determine the effect of two or more independent variables with the dependent variable.

Table 11. Results of Multiple Linear Regression Tests.

Model		Unstandardized Coefficients		Standardized Coefficients
		B	Std. Error	Beta
1	(Constant)	1,959	,372	
	Account Receivable	-,005	-,042	-,016
	Cash	,009	,005	,238

The results of multiple regression testing are done by looking at the value of the variable regression coefficients presented in the following table: (1) Constants (a) equal to 1,959, meaning that if Accounts Receivable Turnover, Cash Turnover is missing or the value is 0, then the liquidity value is 1,959; (2) The variable regression coefficient for the use of Receivables (X1) is -0.005, meaning that if the accounts receivable turnover is less than 1 unit, the liquidity decreases by -0.005 units. The coefficient is negative, meaning that there is no equal relationship between Receivables Turnover and liquidity; (3) Cash variable regression coefficient (X2) of 0.009 means that if cash flow is increased by 1 unit, then liquidity will increase by 0.009 reference. The coefficient is positive, meaning that there is a direct relationship between cash flow and liquidity.

4.3.6 Coefficient of Determination. R Square Adjusted (determination coefficient) is the adjusted R Square value, this value is always smaller and this number can have a negative number. Regression with more than two independent variables is used R Square as the coefficient of determination. Then from this the researcher will use R Square to assess the coefficient of determination test.

Table 12. Determination Coefficient Test Results.

	Summary Model			
Model	R	R Square	Adjusted R Square	Std. Error of the Estimate
1	,241[a]	,058	,027	,98015

a. Predictors: (Constant), Cash, Receivables
Source: SPSS Data Processing Results 22

Based on the test results obtained an R Square value of 0.058 (5.8%) can be interpreted that the independent variable, namely accounts receivable turnover and cash turnover together has an influence on the dependent variable namely liquidity by 5.8%, while the remaining 94.2% is influenced by other factors outside of the variables in this study. This means that accounts receivable turnover and cash turnover only have an effect of 5.8% on liquidity in each of the manufacturing companies in the consumer goods industry registered on the Indonesia Stock Exchange in the period 2016-2018. Quite small in influencing company liquidity, while for other factors outside the variables studied had an influence of 94.2% on company liquidity. The company is seen whether or not liquid is not only seen from the receivables turnover and cash turnover, but can also be seen from other factors that are more influential.

4.3.7 Hypothesis testing. This study also tested the hypothesis test. Data obtained from the results of data collection above can be processed according to the type of data and then presented in the form of tables and numerical statistical methods.

In this study the hypothesis of the researcher that partially accounts receivable turnover and cash flow has a significant influence on company liquidity. Then we will see in the following table using the t-test.

Table 13. T Test Results (Partial) Receivables.

	Model	Unstandardized Coefficients		Standardized Coefficients	t	Sig.
		B	Std. Error	Beta		
1	(Constant)	7,123	,769		9,266	,000
	Account Receivable	,156	,381	,052	,410	,683

a. Dependent Variable: Likuiditas
Source: SPSS Data Processing Results 22

From the results of the above output, it is seen that the significance of the accounts receivable turnover is 0.683, which means that it is greater than 0.05, which indicates that the accounts receivable turnover has no effect on liquidity.

The value of sig <0.05, which means the first hypothesis regarding receivables affects the company's liquidity is not significant.

Table 14. T Test Results (Partial) Cash.

	Model	Unstandardized Coefficients		Standardized Coefficients	t	Sig.
		B	Std. Error	Beta		
1	(Constant)	37,318	6,504		5,738	,000
	Cash	6,284	3,226	,240	1,948	,047

a. Dependent Variable: Likuiditas
Source: SPSS Output 22

From the results of the above output, it can be seen that the significance of cash flow is 0.047, which means that cash flow has an influence on liquidity because the significance is less than 0.05. The value of sig <0.05, which means the second hypothesis regarding cash significantly influences company liquidity.

5. Conclusion

The results of this study provide evidence that receivables turnover does not significantly affect company liquidity. Receivables are actually working capital that is always spinning. This receivable turnover period starts at the time the cash is issued to obtain inventory and then the inventory is sold on credit so that it will cause receivables where the receivables will change back to cash when the payment of the receivables occurs. That is because cash from sales on credit has not been in the hands of the company, so that the accounts receivable turnover does not affect the company's liquidity. From the results of the study in table 4.3, accounts receivable turnover has a minimum value and the lowest average value which concludes that the accounts receivable turnover has little effect on company liquidity. The lower receivables turnover will not affect the company can be said to be liquid or not because the circulation of receivables only affects the company's capital turnover in which the company's capital turnover will continue to go hand in hand with the sale and purchase of goods made by the company. Where receivables turnover has decreased, the liquidity rate will not decrease and vice versa. Where accounts receivable turnover has a positive effect but does not significantly influence liquidity. The increase that occurred on cash had a direct impact on liquidity. Based on the results of the analysis of this study using a sample in the Manufacturing Industry Company Consumer Goods Industry Listed on the Stock Exchange in 2016-2018, then the following are conclusions that can be given by researchers based on the results of testing all hypotheses. Receivables turnover has no effect on the liquidity of manufacturing companies in the consumer goods industry sector listed on the Indonesia Stock Exchange in the 2016-2018 period, where the results of significance figures greater than 0.05 and based on these values can be concluded H1 is rejected because partially the receivables turnover has no effect. significant effect on liquidity. Cash flow turnover has a significant positive effect on the liquidity of manufacturing companies in the consumer goods industry listed on the Indonesia Stock Exchange in the 2016-2018 period, where the significance level is less than 0.05 and based on this value, it can be concluded that H2 is accepted because partially cash turnover has significant effect on liquidity.

Acknowledgment
This work is supported by the Research Technological Transfer Office (RTTO) Binus University as a part of Hibah Binus Penelitian Internasional Binus (PIB) Research Grant to Binus University entitled "Quality of Accounting Information Systems and Their Emerging Contributions of Public Listed Companies"

References

[1] Horne, James C. Van and John M Wachowicz, Jr. "Principles Financial Management".Jakarta: *Salemba Empat*. 2012
[2] Agus Harjito and Martono. "Financial management". 2nd edition. Econosia,Yoyakarta. 2012.
[3] J. Fred Weston, Eugene. "Fundamentals of Financial Management". Jakarta:Erlangga. 2012.
[4] Carl S. Warren. "Introduction to Indonesian Adaptation Accounting". Jakarta: Salemba Empat. 2014.
[5] Ahmadi and Hermawan. "E-Business & E-Commerce". Yogyakarta: Andi. 2013.
[6] Harahap, Sofyan Syafri. "Crisis Analysis of Financial Statements". Raja Grafindo Persada. Jakarta. 2008.
[7] Subramanyam. K. R and John J. Wild. "Financial Statement Analysis". Translator Dewi Y. Jakarta: Salemba Empat. 2014.
[8] Kartikahadi, Hans, et al. "Financial Accounting Based on IFRS-Based IFRSs". Jakarta: IAI. 2016
[9] Sugiyono. Qualitative, Quantitative, and Educational Education Research Methods. Bandung R&D: Alfabeta. (2010).
[10] J. Fred Weston, Eugene. "Fundamentals of Financial Management". Jakarta: Erlangga. 2012
[11] Meiryani, Azhar Susanto and Dezie Leonarda Warganegara. "The Issues influencing of environmental accounting information systems : An Empirical investigation of SMEs in Indonesia". *International Journal of Energy Economics and Policy*. 2019. 9 (1) :282-290.
[12] Meiryani & Lusianah. "The Influence of Business Process on Accounting Information System Quality". *Pertanika Journal of Social Sciences and Humanities*. 2018. 26, pp. 209-218.

[13] Meiryani. "The Factors That Affect the Quality of Accounting Information System Empirical Testing in the State-Owned Enterprises". *Journal of Theoretical and Applied Information Technology*. 2018. Vol. 96. No. 7

IOP Conf. Series: Materials Science and Engineering **924** (2020) 012029 doi:10.1088/1757-899X/924/1/012029

The Effects of Rice Husk Ashes Filled Recycled Polyethylene Composites towards Physical, Mechanical Properties and Its Degradation Characteristics

Chasri Nurhayati[*], **Tri Susanto**
Palembang Institute for Industrial Research and Standardization, Ministry of Industry, Indonesia Jl. Perindustrian II No. 12 KM 9 Palembang, 30152, Indonesia

*chasrinurhayati@yahoo.com

Abstract. Plastic waste is dominated by 46% Low Density Polyethylene (LDPE) and High Density Polyethylene (HDPE). The plastics waste are persistent pollutants that is not biodegradable waste. Recycled polyethylene (RP) from the plastic waste would potentially be used as polymer composites for pavement or road construction application. In order to improve physical, mechanical and water absorption properties, some fillers need to be added. The agro-industrial waste of rice husk ashes (RHA) containing highly silica SiO_2 may be utilized as filler in this composites. The objective of this research is to observe the effect of RHA addition in to the recycled plastic composites towards its physical, mechanical properties and degradation characteristics. The RHA loading was designed weight/weight of RP as follows 0%; 5%; 10%; 15%; 20%, while degradation duration was 0, 30 and 60 days. The recycled ethylene was smelted and mixed with the RHA at temperature 120 °C; and then pressed using 200 $kg.cm^2$ hydraulic press as pavement block (20x10x6 cm dimension). The results showed that the addition of RHA leads to increase the water absorption capacity, whereas it reduces the compressive strength. The RHA filler addition into composites has less significantly effects on thickness and surface-physical appearance (cracking profile). The 60 days degradation at ambient temperature with the sun lighting in a burial soil has a slight effect on the decreasing of compressive strength, thickness, and physical appearance (small crack on the surface) due to the thermal oxidative degradation. In addition, the burial soil exacerbate the bio-degradation mechanism due to the presence of high water absorbed within the RP-RHA composites. To sum up, the RHA may be potentially used as filler to improve water absorption capacity. Anti-oxidant and plasticizer types need to be researched to improve its physical mechanical properties.

Keywords: compressive strength, water absorption, recycled polyethylene, rice husk ashes filler

1. Introduction

In a modern life, plastic waste from household activities, offices, markets, social activities and other sources becomes a serious problem. Generally, plastic waste in only burned and put into the ground as landfill without any treatments. Therefore, reuse and reutilization plastic waste in order to minimize soil and air pollution would become as an alternative [1]. The plastic waste was mostly dominated by HDPE and LDPE Polyethylene about 40-48%, polypropylene and polystyrene approximately 30-35%, polyvinyl chloride (PVC), polyethylene terephthalate (PET), Acrylonitrile-butadiene styrene (ABS) about 7%, 5%, 5% respectively [2].

The HDPE and LDPE is the biggest quantity of plastic waste that becomes main sources of environmental pollution. Polyethylene waste grows at rate 12% and around 140 million tons of synthetic polymers are produced worldwide annually. The burning of those polyethylene cause a serious air pollution due to dioxin and some gas pollutants, while land fill would affect the soil quality due to the micro plastics in burial soil [1, 3].

Indonesia has a large number of paddy field and rice production about 56.54 million tons. Rice husk ash in paddy field waste is about 20% of total rice production. Currently, rice husk ash waste has been utilized for refractory and ceramic industries, fuel, and others [4, 5]. Rice husk ash containing 87-97% amorphous SiO_2 and other minerals that is potentially used as filler in cement, concrete or adsorbent in water treatment [6].

The performance of structural concrete with recycled plastic waste as a partial replacement for sand has been studied, the results show that replacing 10% of sand from the volume of recycled plastic is a viable proposition that has the potential to save 820 million tons of sand every year [7].The addition of about 10% volume sand and cement into polyethylene pavement concrete blends leads to increase compressive strength and hardness [8]. Furthermore, temperature process on blending sand and polyethylene into pavement give significant results, it is stated that the process at 300 o C has higher harness and compressive strength compared to blending at 200 o C. Meanwhile, the bulk density experience decreasing as high sand replacement in polyethylene blending [3, 7].

Research on rice husk ash for cement replacement in concrete have been widely developed. It is found that 10% RHA or more may cause the increasing of optimum moisture content while decreasing maximum dry density [9]. In addition, 20% RHA loading in high performance fine-grained concrete could improve compressive strength, tensile strength, and chloride penetration resistance [10].

In terms of recycling plastic waste for pavement or road construction, physic mechanical properties of recycled polyethylene needs to be improved by adding some fillers. Some fillers like calcium carbonates, kaolin and recycled silicon have been researched [6, 9, 11, 12]. However, the research of using RHA in HDPE and LDPE blending is still limited particularly for pavement or road construction.

Bearing in mind, the necessity of technology for recycling process of HDPE and LDPE using agro-industrial waste such as rice husk ashes filler, this research would contribute technology of blending that composites for pavement and road construction application. The key important parameter is the water absorption of the composites. It indicates the capability of the composites to allow water steep into the ground so the surface run-off could be reduced more effectively. Therefore, this recycled plastic and RHA composites could be applied in road construction as a replacement of concrete or sand pavement. Therefore, this work would investigate the effects of rice husk ashes (RHA) as filler on recycled polyethylene towards physical, mechanical properties and Its degradation characteristics. The research result would contribute knowledge and practical information on developing the recycled plastic products for pavements.

2. Experimental method

The materials used are rice husk (paddy field), plastic waste and water. The equipment hydraulic press (capacity 200 $kg.cm^2$), temperature sensors, stopwatch, digital calliper, smelted furnace, reactor, magnifying glass, and water bath.

Rice husk was burnt at 150 °C until it becomes ashes and the ground it into 150 mesh. The plastic waste was separated only for LDPE & HDPE plastic (Recycled Plastic) and then washed and cleaned by soaking the LDPE-waste using water and then dry it. The blending process of RP and RHA is in accordance with previous research that used injection moulding machine with compressive strength about 200-220 $kg.cm^2$ [8, 12]. The composite of Recycled Plastic (RP) and rice husk ashes (RHA) as filler was prepared by melting and blending the RP, RHA and Cu (catalyst) in a furnace reactor at temperature 120 °C for about 15 minutes. After that the composites of RP and RHA was pressed by hydraulic press at 200 $kg.cm^2$ for about 3-5 minutes using metal moulding 10 x 20 cm thickness 6 cm. Finally, the composites was soaked in water for about 8 minutes before testing. In this research, the composition of RP and RHA was designed by varying the percentage weight/weight of (PR:RHA) as follows (100:0); (95:5); (90:10); (85:15); and (80:20) %.

The degradation characteristics was observed by burying the composites in a burial soils (media of garbage dump soil), while 1 cm is on the surface that could contact with the air, sunny lighting and ambient temperature. The physical appearance, compressive strength, water absorption and thickness were measured in duration of burial of 0, 30, and 60 days. Physical appearance was observed using magnifying glass on the surface of the composites, while the thickness was measured using digital calliper. Compressive strength was tested in according to ASTM C 936 and the water absorption testing was carried out based on ASTM D570-95. The testing of those properties were conducted in three repetitions in order to obtain high accuracy data.

3. Results and discussion

3.1. Physical Appearance

Table 1 shows the physical appearance of composites RP/RHA in 0, 30 and 60 days degradation. After 30 days degradation, the surface of each composites using RHA as filler was not deteriorated significantly. The surfaces become brittle and rough with a few voids and cracks. The surface of recycled plastic composites without RHA remain stable and does not experience serious attack rather than composites with higher RHA. It could be inferred that the higher RHA loading, the surface attacks become more serious rather than without RHA. This degradation may be explained due to the existence of bacteria/ microorganism and air/ozone attack during degradation process [3, 9]. The oxidation on surface cause cracks on surface and brittle due to the lack of antioxidant agent in the composites [11, 13]. In addition, the process-ability and dispersion of the filler within the low density and high density polyethylene matrix may not very fine, so that the surface attack could not be hindered.

Table 1. Physical-Surface Appearance of Composites of RP/RHA in 0, 30, and 60 days degradation.

RP/RHA (w/w) %	0 Day Degradation	30 Days Degradation	60 Days Degradation
100/0	No cracks, smooth surface	brittle and rough with a few voids and cracks	brittle and small rough with a few voids
95/5	No cracks, smooth surface	brittle and rough with a few voids and cracks	more brittle and possess a rough with voids and more cracks in surface
90/10	No cracks, smooth surface	brittle and rough with a few voids and cracks	more brittle and possess a rough with voids and more cracks in surface
85/15	No cracks, smooth surface	brittle and rough with a few voids and cracks	more brittle and possess a rough with voids and more cracks in surface

3.2. Thickness

Table 2 shows that the thickness of the composites were not significantly changes due to the longer degradation days. The composites of RP/RHA (85/15) % experience the highest shrinkage after 60 days. The RHA as filler give positive physical properties contribution in the composites and improve its strength. It is predicted that the longer degradation periods would influence the decreasing of thickness of the composites, particularly the higher RHA contained in the PR/RHA composites.

Table 2. Thickness of Composites of RP/RHA in 0, 30 and 60 days degradation

RP/RHA (w/w) %	0 Day Degradation	30 Days Degradation	60 Days Degradation
100/0	6 cm	6 cm	6 cm
95/5	6 cm	6 cm	6 cm
90/10	6 cm	6 cm	5,99 cm
85/15	6 cm	5,99 cm	5,99 cm

3.3. Compressive Strength

Figure 1 shows the compressive strength of the RP/RHA composites with RHA variation loadings in 0, 30 and 60 days degradation. Composites without RHA has the highest compressive strength (94.08 MPa), while the higher loading of RHA cause the decreasing of compressive strength slightly. The rice husk ashes containing 87-97 % amorphous SiO_2 that would acts as reinforcing filler in polymer matrix such as natural rubber and styrene rubber [4, 7, 10]. For cement or concrete application, RHA filler contributes positively to not only compressive strength but also for its hardness. Conversely, in this recycled polyethylene composites, the RHA may not finely disperse within the polymer and its process ability is very limited due to the difference polarity, density and size between filler and polyethylene [2, 9, 11, 12]. Those physical and chemical barrier may probably limit the chemical and physical bonding between hydroxyl groups of silica with the ester functional of polyethylene [4, 7, 10]. Therefore, the physical surface appearance and its compressive strength decrease due to the RHA addition.

As shown in Figure 1, the longer degradation under sun lighting in a burial soil do not significantly affect the compressive strength of PR/RHA composites. After 60 days degradation, the compressive strength decrease slightly for about 2-5 MPa. This phenomena may be explained as the effect of thermo-oxidative ageing that exacerbate the degradation both in surface and within the composites. Figure 1 also shows that the longer degradation duration with the higher RHA loading cause the decreasing of compressive strength due to the low filler dispersion and process-ability.

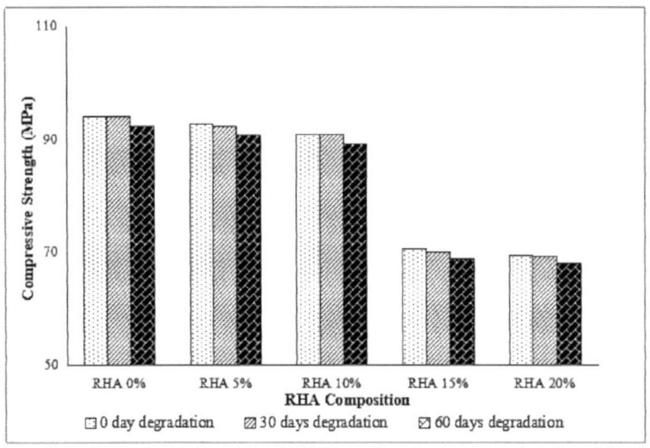

Figure 1. Compressive Strength of PR/RHA Composites in 0, 30, 60 days degradation.

3.4. Water absorption

For pavement ground and road construction application, the water absorption capacity is the key important properties to determine its utilization and function. The water absorption test is carried out by immersing the whole test sample in water for about 24 hours, weighing the weight in wet state, drying approximately in ambient temperature °C until weighing at two weighing disputes not more than 0.2% of the previous weighing.

Figure 2 illustrates the water absorption capacity of the RP/RHA composites and degradation profile within 0, 30, and 60 days. It can be seen that the highest water absorption (3.14%) was obtained in 20% RHA loading substituted the recycled polyethylene, while the composites without RHA filler has very low water absorption only 0.52%. This result indicates that RHA loading contributes positively in order to improve water absorption capacity of the composites. The nature of polyethylene is impermeable to water and does not dissolve in water at ambient temperature, so that polyethylene without filler is not inevitable for road or pavement construction. However, polyethylene has high crystallinity and strong intermolecular attraction so that its mechanical strength is excellent [7, 9, 11, 14].

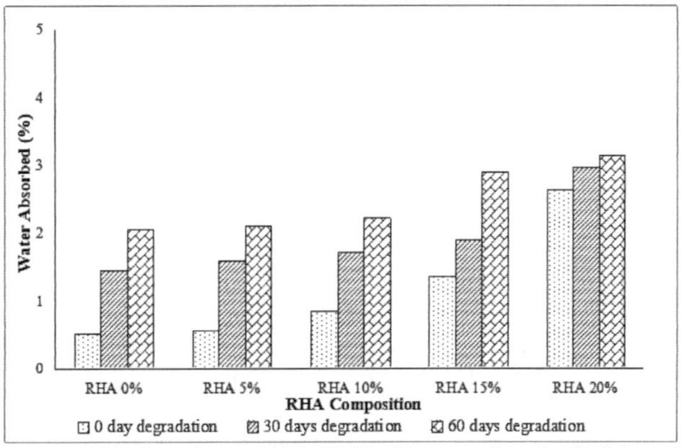

Figure 2. Water Absorption of RP/RHA Composites in 0, 30 and 60 days.

The presence of RHA improves the hydrophilicity of the RP composite and allows more water absorption due to ease of hydrogen bonding between silica and hydroxyl of rice husk ashes and water molecules. Figure 2 shows that the longer degradation increase the water absorption of the composites. It could be seen that the RHA addition also increase the water absorption capacity during 30 and 60 days degradation. It could be explained that high RHA contain would absorb more water into the composites and enable microbial attack during degradation in a burial soil. Biodegradation on RP/RHA may happen due to the existence of microbial from burial soil and water activity within the composites [3, 10, 13, 15]. Thermo-oxidative ageing and biodegradation along with the surface attack by sun lighting allows more water absorbed and leads to some cracks, weight loss and possibly breakdown the polyethylene within the matrix [3, 15].

4. Conclusion

Composite of recycled plastic that consist of LDPE and HDPE using rice husk ash as filler has been successfully synthesized. The RHA loading causes the surface of the composites become more brittle and rough with a few voids. It is indicated that the higher RHA concentration would affect the physical appearance of the composites. The compressive strength of the composites tend to decrease due to the higher loading of RHA. This phenomena could be explained that the process-ability and dispersion of RHA filler within the polyethylene matrix may need to be improved. Conversely, the water absorption capacity of the RP/RHA composites increase slightly due to the RHA addition. It is suspected that silica and hydroxyl groups from RHA may allow water absorption through hydrogen bonding. Further research would be study anti-oxidant and plasticizer types to enhance the dispersion of RHA filler and recycled polyethylene in order to improve its mechanical, thermal ageing and bio degradation properties.

Acknowledgment

The authors would like to express their gratitude to the Palembang Institute for Industrial Research and Standardization for financial support, laboratory instruments and other facilities. The authors also gratefully acknowledge to all team members for the hard work during finish the research.

References

[1] Fortelný, I., D. Michálková, and Z. Kruliš, *An efficient method of material recycling of municipal plastic waste.* Polymer degradation and stability, 2004. **85**(3): p. 975-979.

[2] Usha, R., T. Sangeetha, and M. Palaniswamy, *Screening of polyethylene degrading microorganisms from garbage soil.* Libyan Agric Res Cent J Int, 2011. **2**(4): p. 200-4.

[3] Hanfi, I., A. Rohani, and M. Razaina, *Effect of soil burial on properties of linear low density polyethylene/thermoplastic sago starch blends.* Pertanika Journal of Science & Technology, 2011. **19**(1): p. 189-197.

[4] Hung, T.D., *Affects of mass burning scales on properties of rice husk ash as raw material to produce geopolymer.* Int. J. Emerg. Technol. Adv. Eng, 2013. **3**: p. 422-428.

[5] Nair, D.G., et al., *A structural investigation relating to the pozzolanic activity of rice husk ashes.* Cement and Concrete Research, 2008. **38**(6): p. 861-869.

[6] Roy, A., *Soil stabilization using rice husk ash and cement.* International journal of civil engineering research, 2014. **5**(1): p. 49-54.

[7] Lasco, J.D., M. Madlangbayan, and M.B. Sundo, *Compressive Strength and Bulk Density of Concrete Hollow Blocks (CHB) with Polypropylene (PP) Pellets as Partial Replacement for Sand.* Civil Engineering Journal, 2017. **3**(10): p. 821-830.

[8] Oyetunji, O., O. Olaoye, and I. Dada, *PRODUCTION OF PAVING BLOCK FROM RECYCLED POLYETHYLENE.* LAUTECH Journal of Engineering and Technology, 2014. **8**(2): p. 182-187.

[9] Suharty, N.S., et al., *Tensile Properties and Biodegradability of Rice Husk Powder-filled Recycled Polypropylene Composites: Effect of Crude Palm Oil and Trimethylolpropane Triacrylate.* Journal of Physical Science, 2014. **25**(2).

[10] Le, H.T., S.T. Nguyen, and H.-M. Ludwig, *A study on high performance fine-grained concrete containing rice husk ash.* International Journal of Concrete Structures and Materials, 2014. **8**(4): p. 301-307.

[11] Thorneycroft, J., et al., *Performance of structural concrete with recycled plastic waste as a partial replacement for sand.* Construction and Building Materials, 2018. **161**: p. 63-69.

[12] Udawattha, C., H. Galabada, and R. Halwatura, *Mud concrete paving block for pedestrian pavements.* Case studies in construction materials, 2017. **7**: p. 249-262.

[13] Zheng, Y., et al., *Improving mechanical properties of recycled polypropylene-based composites using Taguchi and ANOVA techniques.* Procedia CIRP, 2017. **61**: p. 287-292.

[14] Nguyen, D.H., et al. *Effects of rice husk biochar and sugar-mill by-products on methane consumption from two different soils.* in *Proc Aust Soc Sugar Cane Technol*. 2013.

[15] Luyt, A., J. Molefi, and H. Krump, *Thermal, mechanical and electrical properties of copper powder filled low-density and linear low-density polyethylene composites.* Polymer Degradation and Stability, 2006. **91**(7): p. 1629-1636.

Magnetization of Biodiesel (Cooking Oil Waste) to Temperature and Pressure Combustion in Diesel Engine

T H Nufus[1], S Lestari[2], A Ulfiana[1*], and M Manawan[1]

[1]Mechanical Engineering, Politeknik Negeri Jakarta, Prof. Dr. G.A Siwabessy Street, Kampus UI, Depok, Jawa Barat 16424, Indonesia.

[2]Electrical Engineering Politeknik Negeri Jakarta, Prof. Dr. G.A Siwabessy Street, Kampus UI, Depok, Jawa Barat 16424, Indonesia.

*andi.ulfiana@mesin.pnj.ac.id

Abstract. Research related to the provision of magnetic field strength in the flow of fuel that can cause more complete combustion has been reported by many researchers and most of these observations were carried out by measuring machine performance using a dynamometer. Given the dynamometer prices are very expensive then another way to prove the occurrence of complete combustion is through the combustion chamber indicator. The hypothesis is the greater magnetic field passed by fuel, the more perfect the combustion is. In addition, the combustion chamber temperature also increases and thermal efficiency increases. The fuel used is biodiesel derived from cooking oil waste, biodiesel 100% (denoted as B0), a blend of biodiesel (20%)-diesel (80%) (denoted as B20), a blend of biodiesel (40%)-diesel (60%) (denoted as B40), a blend of biodiesel (30%)-diesel (30%) (denoted as B70), and diesel 100% (denoted as B100). The magnetic field used is 500 Gauss, 900 Gauss and 1500 Gauss. Measurement of combustion temperature is using R type temperature sensor, which is connected to the NI USB interface to computer and LabVIEW for data acquisition. 13 HP capacity of diesel engine was used. The results obtained is 13 % temperature increase in the combustion chamber of the engine and 7 - 12% thermal efficiency increase equipped with a magnet compared to a non-magnetic engine.

Keywords: cooking oil waste, temperature, magnetic field, engine performance.

1. Introduction

Fuel oil is a non-renewable energy source that will eventually run out in the future. Efforts to overcome this situation are by saving fuel use or finding alternative energy sources. Several studies have been carried out in order to save energy through increasing combustion efficiency, including mixing additives in fuels which increase octane and cetane values, effective combustion processes and increased engine power [1]. Fuel magnetization, by installing a permanent magnet on the fuel line to the combustion chamber, decrease 9% to 30% of fuel consumption, reduced 5% to 32% of HC exhaust emissions and reduce 5 to 34.3% of CO exhaust emissions [2-9]. However, both emissions produce chemical additives that contain metal elements that are harmful for human health. Other studies tried using installed electromagnetic fields on gasoline or diesel engines to minimize fuel consumption by 12.8% to 30% and reduce levels of HC exhaust emissions by 44-58% and CO decrease by 35-80% [10-12]. This electromagnetic field does not contain harmful element and safe to used in diesel engine vehicles [7,8,13,14].

The hypothesis is based on the theory of electric magnets by Bio-Savart, where when the magnetic field is given to a substance, its atoms will be oriented parallel to the direction of the magnetic field, and the vibration produced by this electromagnetic field causes wide distance between molecules hence fuel and oxygen will react easily, therefore a more perfect combustion occurs. The purpose of this study is to analyze the effect of temperature and pressure on a diesel engine performance.

2. Experimental Methods

2.1. Materials

The data was obtained by direct observation and measurement in the field. The experiment was conducted in the Energy Conversion and Heavy Equipment Laboratory, Jakarta State Polytechnic, on June 2018-August 2018. Instruments were calibrated and installed as shown in Figure 1. A 3-cylinder generator (Perkin 403D-15) with a power of 13.5 HP. The independent variables in this test are fuel composition (B0, B10, B30, B50, and B100 type) and variations in electromagnetic field (500 Gauss, 900 Gauss and 1500 Gauss). The dependent variable for this performance test is fuel consumption, temperature, and combustion chamber pressure.

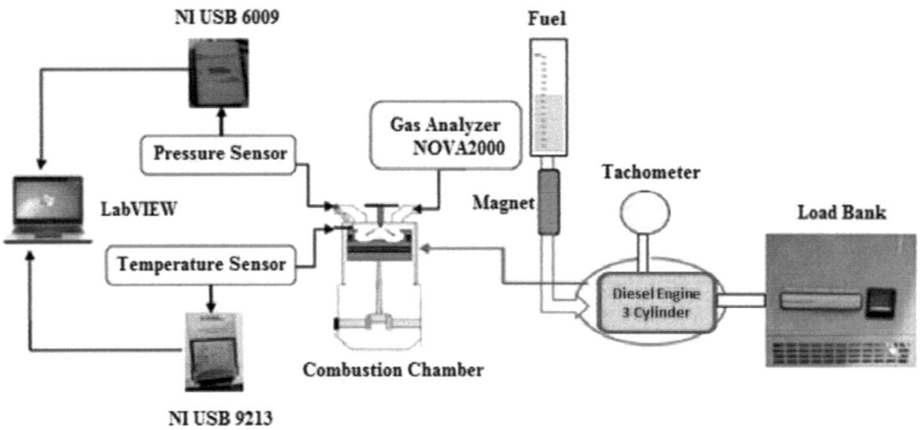

Figure 1. Flowchart of the generator set testing with load bank.

2.2. Sample preparation

The magnitude of fuel consumption, temperature, pressure and exhaust emissions was observed. Diesel engine rotation was measured with a Digital Tachometer. Temperature and pressure sensors were positioned to insert into the combustion chamber. Data result from temperature and pressure sensors were stored in data bases through NI USB 6009 and NI USB 9213 interface. The results were displayed on a laptop equipped with LabVIEW program. The test was started by turning on Perkin P135-4 diesel engine at 1550 rpm and left for ± 10 minutes to get the engine's normal working temperature.

Fuel consumption is calculated based on the difference in the fuel level on measuring cup that connected to the fuel tank, per time unit. Measurement and recording of fuel consumption was carried out five times, on each time fuel and electromagnetic field changes. The temperature and pressure of the combustion chamber are observed through a pressure sensor capable of measuring up to 750 Psi and a R type temperature sensor capable of measuring up to 1750°C. Both sensors are connected to the data logger for pressure and temperature on each NI USB 6009 and NI USB 9213 interface. The data sampling is 400 data/second. The results of temperature (T) and pressure measurement data are used to calculate the thermal efficiency of the engine using Equation (1).

$$H_{th} = 1 - \frac{1}{k}\left(\frac{T_4-T_1}{T_3-T_2}\right) \quad (1)$$

where :

η_{th} is thermal efficiency; T_1 is temperature at point 1 (K); T_2 is temperature at point 2 (K); T_3 is temperature at point 3 (K); T_4 is temperature at point 4 (K); Cp is heat capacity at constant pressure (kJ/kg.K); Cv is heat capacity at constant volume (kJ/kg.K); k = (Cp/Cv) is specific heat ratio.

Thermal efficiency is the amount of utilization of heat energy stored in fuel to be converted into effective power by an internal combustion motor. Thermal efficiency is calculated using P-V and T-S diagrams.

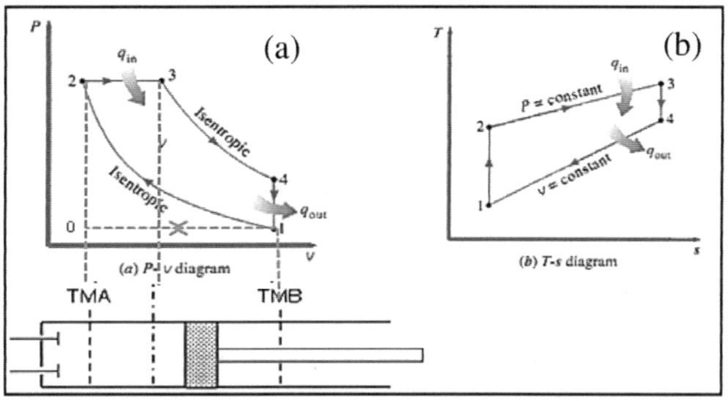

Figure 2. (a) P–V and (b) T–S Diagram.

Fuel condition data results were compared before and after magnetization. The data then processed quantitatively in the form of several graphs, namely the graph of thermal efficiency of the load, graph of the temperature of the combustion chamber against time, and graph of combustion pressure against time.

3. Results and discussion

Diesel engine performance is indicated by fuel consumption, specific fuel consumption, thermal efficiency and exhaust emissions. Therefore, research data leads to these indicators. Every data is presented and analyzed in graphical form. Pressure and Temperature in the combustion chamber will make an oxidation reaction between fuel and air (oxygen) which produce heat. The fuel will burn perfectly when there is sufficient oxygen supply (O_2). The amount of oxygen reaches 20.9% of the air, 79% is nitrogen (N_2) and the rest are other elements such as argon, helium, neon, krypton, carbon dioxide, hydrogen, xenon, and others. Complete combustion of hydrocarbon compounds will form carbon dioxide and water vapor, while incomplete combustion forms carbon monoxide.

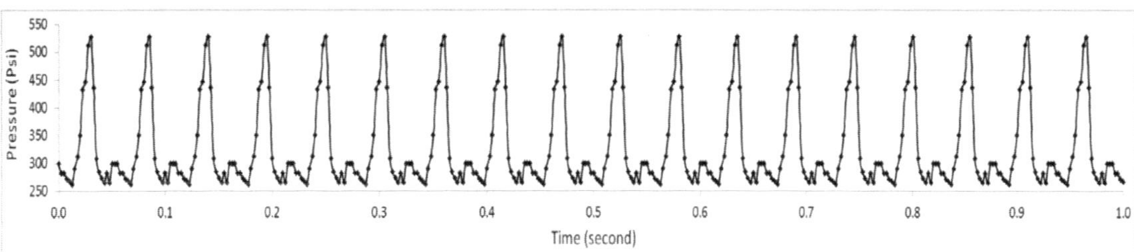

Figure 3. Relationship between pressure in the combustion chamber and time at 2200 rpm.

IOP Conf. Series: Materials Science and Engineering **924** (2020) 012030 doi:10.1088/1757-899X/924/1/012030

The measurement of combustion chamber pressure was done to calculate thermal efficiency which is one of the engine performance indicators. Figure 3 shows the relationship between pressure in a combustion chamber and time for diesel fuel without magnetization. The graph data is obtained at 2200 rpm engine speed or equal to 36.67 rps, considering that 1 combustion cycle = 2 revolutions, for the engine speed of 2200 rpm there are 19 cycles in one second, hence the time of each cycle is 0.05 seconds and pressure of 525 Psi.

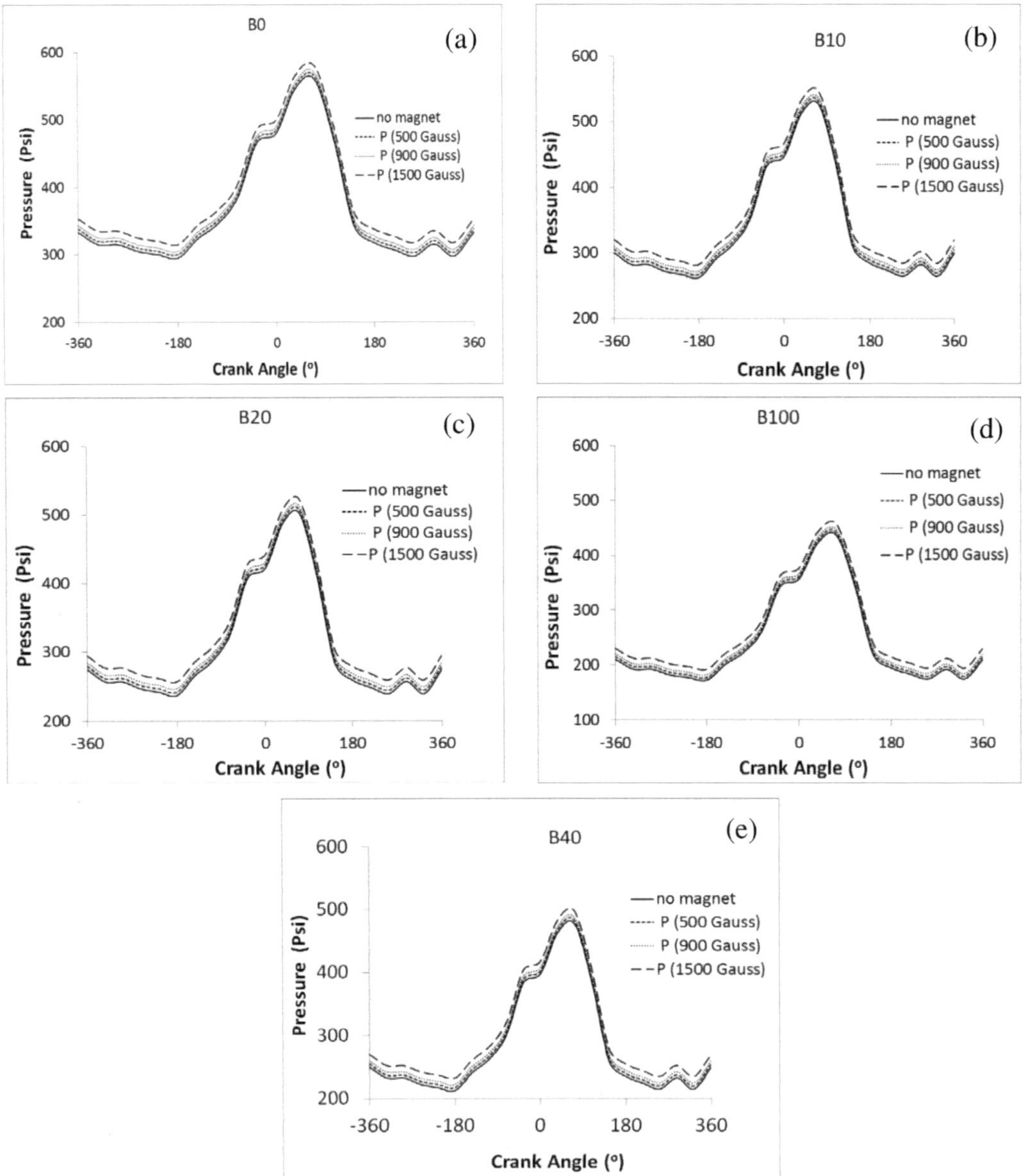

Figure 4. Relationship between pressure in the combustion chamber and the crank shaft angle on various types of fuel (a) B0 (b) B10 (c) B20 (d) B40 & (e) B100.

Figure 4 shows the relationship between pressure in the combustion chamber and the crank shaft angle on various types of fuel. This graph is obtained by converting Figure 3, which is on the x-axis from time to angle of the crank shaft. It shows that the greater the magnetic field exerted on the fuel the greater pressure that occurs in the combustion chamber. This is caused by the magnetized fuel causing the fuel molecule to vibrate and becomes unstable. This has an effect on the reduced attraction between the atoms causes the molecular affinity energy to be smaller, then the fuel molecule is more and easier to react with oxygen, and finally the combustion process becomes more homogeneous and the fuel burns more as a result the pressure becomes greater. More bioethanol mixture in gasoline causes the pressure to decrease, because although bioethanol has a high octane value, the power produced by the engine is getting smaller.

At the compression pressure there is a slight turn. This shows that at the time of the spraying of fuel has a lower pressure than the compressed air so that the pressure at that time did not rise and the time is very fast

Figure 5 presents a graph of the relationship between temperature in the combustion chamber and time, the same as in Figure 2 that the graph is obtained at engine speed of 2200 rpm or equal to 36.67 rps, given that 1 combustion cycle = 2 cycles, so for 2200 rpm engine speed there are 19 cycles in one second, and each cycle takes about 0.05 seconds and a temperature of 1050^{0}C.

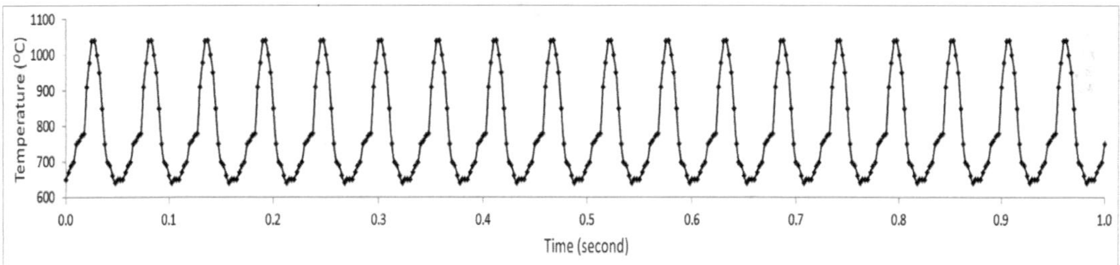

Figure 5. Relationship between temperature in the combustion chamber and time.

Figure 6 presents a graph of the relationship between the temperature in the combustion chamber and the crank shaft angle, this graph is obtained by converting Figure 5 is on the x-axis of time into the crank shaft angle. The greater the magnetic field applied to the fuel, the greater the temperature that occurs in the combustion chamber. This is caused by the magnetized fuel which causes the fuel molecule to vibrate more and becomes unstable, this has an effect on the reduced force attraction between atoms which results in smaller energy affinity of molecules, and more fuel molecules and easier to react with oxygen, finally the combustion process becomes more homogeneous and more fuel is burned and the pressure becomes greater. More bioethanol mixture in gasoline causes the temperature to decrease, because although bioethanol has a high octane value, the power produced by the engine is smaller.

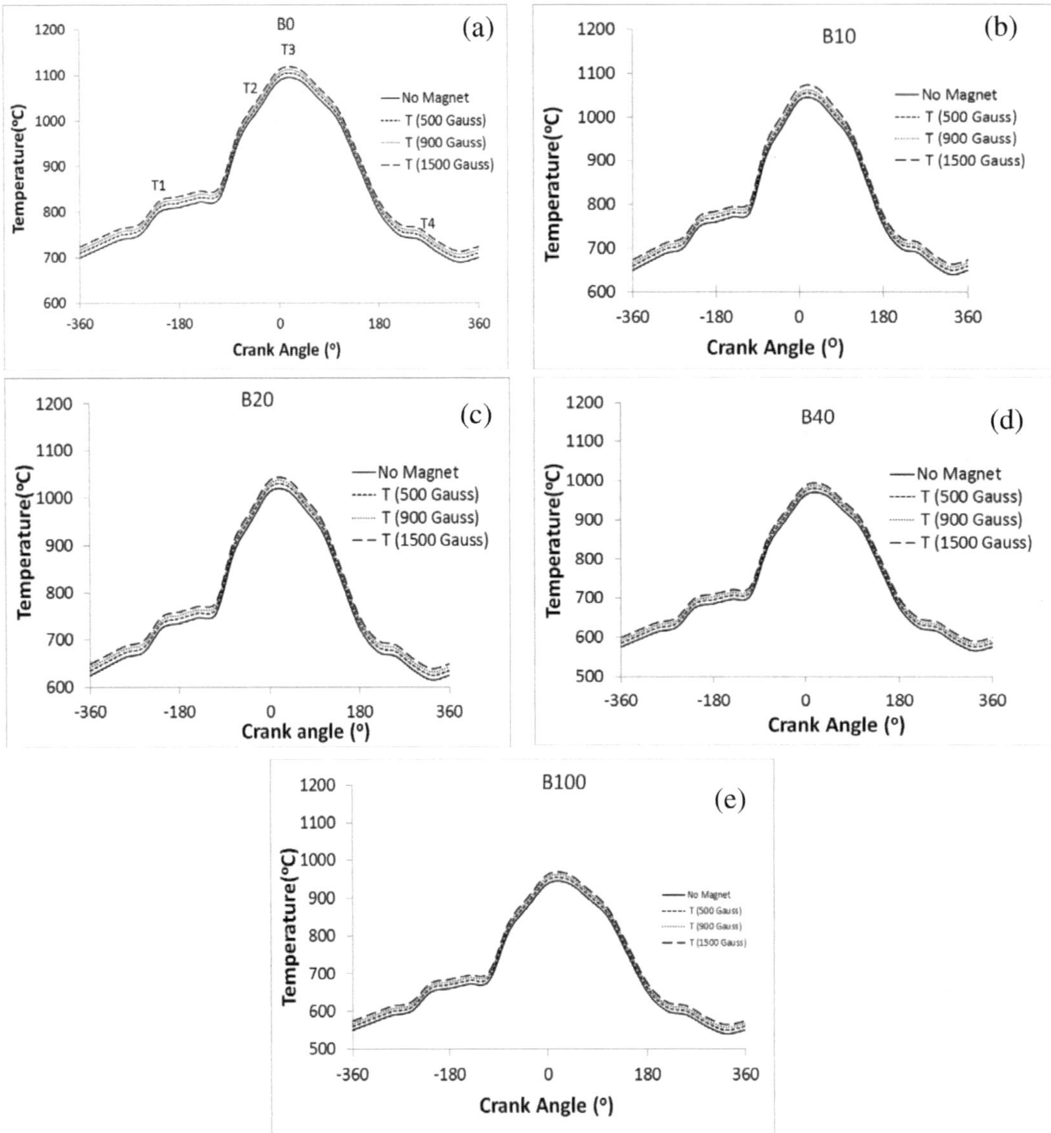

Figure 6. Relationship between temperature in the combustion chamber and crank shaft angle with fuel variation, (a) B0 (b) B10 (c) B20 (d) B40 and (e) B100.

Figure 7 present a graph of the relationship between pressure and temperature with respect to crank shaft angle. The temperature value is determined with the help of the intersection of the temperature graph with respect to time and dashed lines. One cycle has 4 steps, therefore one cycle in Figure 7 is divided by 4 equal dashes line hence the T1(°C) value is the lowest temperature because air from outside enters the combustion chamber. Under these conditions both temperature and pressure are at the lowest value. T2 (°C) is the compression temperature where all valves are closed, so T2> T1. Furthermore T3 (°C) is the combustion temperature at the pressure and the temperature is at the highest position and T4 (°C) is the exhaust temperature, where at this position the pressure will start to fall, while the temperature is higher than the temperature T1. Furthermore, the value of k is calculated using equation (1).

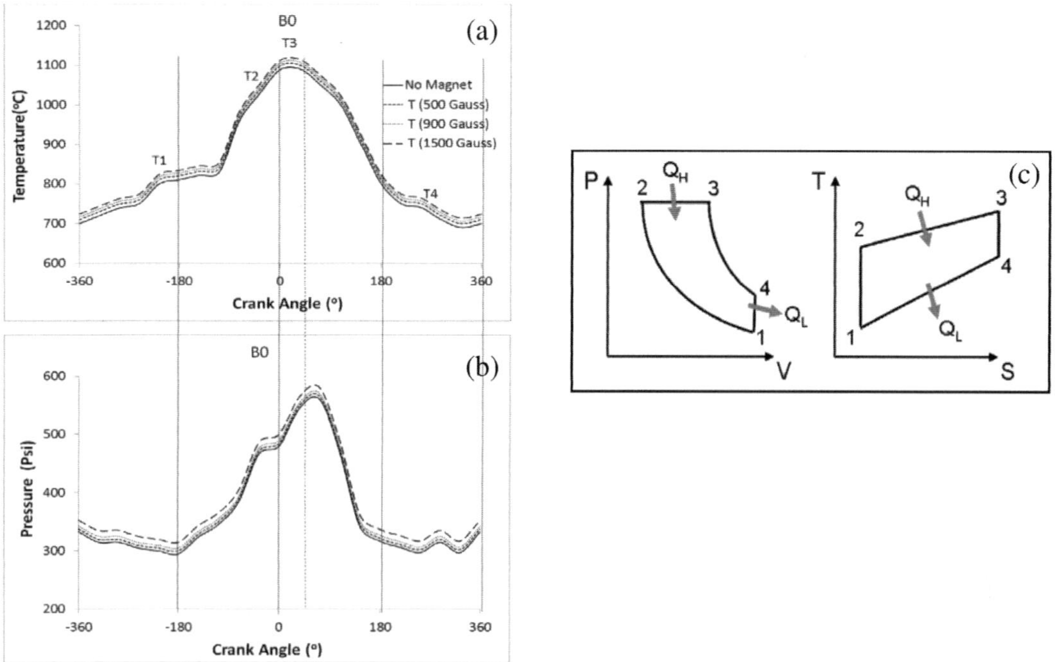

Figure 7. (a) Graph of 1 diesel cycle expressed by (a) pressure with respect to crank angle (b) temperature with respect to crank angle and (c) PV & TS graphs.

Figure 8 shows the relationship between thermal efficiency and the types of fuels that are magnetized and non-magnetized. The magnetized fuel has a higher thermal efficiency than non-magnetized fuel. Fuel magnetization functions to break down clumping fuel molecules into non-clumping and polarize fuel molecules. This has the effect of mixing the fuel and air more homogeneously therefore the fuel that burns more and combustion becomes perfect, and thermal efficiency increases.

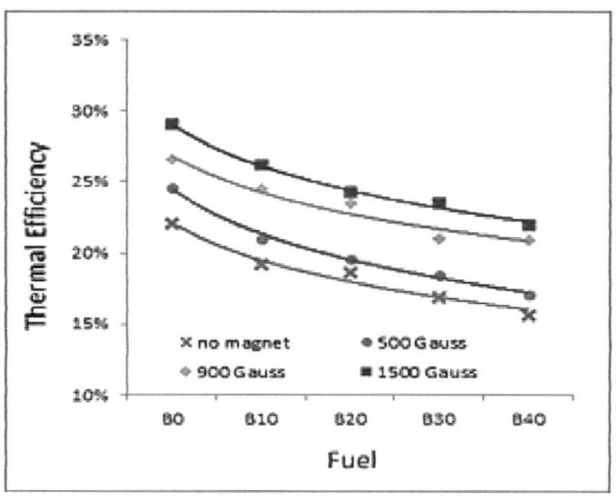

Figure 8. Relationship between thermal efficiency with fuels.

The increase of the percentage of biodiesel mixture in diesel fuel causes a decrease in the value of thermal efficiency. The decrease in the value of thermal efficiency due to the high viscosity value of biodiesel makes its ability to form a homogeneous mixtures with air to be low. This causes combustion to be less perfect and thermal efficiency becomes lower. The increased in thermal efficiency ranges from 7 to 12%

The addition of the percentage of biodiesel mixture in diesel fuel causes a decrease in the value of thermal efficiency. The decrease in the value of thermal efficiency due to the high value of biodiesel viscosity makes its ability to form a homogeneous mixture with air to be low. This causes combustion to be less than perfect and the thermal efficiency becomes lower. The increase in thermal efficiency ranges from 9.2-18.3%. The results of the thermal efficiency calculation are shown in Figure 8.

4. Conclusion

Fuel magnetization causes an increase in temperature and pressure in the combustion chamber increase by 13% and in thermal efficiency by of 7 to 12%. In conclusion, fuel magnetization increase engine performance.

Acknowledgment

This research was supported by the Ministry of Research and Technology of Indonesia and Jakarta State Polytechnic

References

[1] Fayyazbakhsh A and Pirouzfar V 2016 Investigating the influence of additives-fuel on diesel engine performance and emissions: Analytical modeling and experimental validation *FUEL* **171** 167-177

[2] Govindasamy P and Dhandapani S 2007 Experimental investigation of cyclic variation of combustion parameters in catalytically activated and magnetically energised two-stroke SI engine *Journal of Energy & Environment* **6** (4) pp 561-569

[3] Faris AS *et al* 2012 Effects of magnetic field on fuel consumption and exhaust emissions in two-stroke engine *Elsevier Energy Procedia* **18** pp 327–338

[4] Jain S and Deshmukh S 2012 Experimental Investigation of Magnetic Fuel Conditioner in I.C. Engine *IOSR Journal of Engineering* **2** (7) pp 27-31

[5] Singh AK and Solank RM 2013 Investigation of fuel saving in annealing lehr through magnetic material fuel sarver *International Journal of Science and Research* **6** (14) pp 178-180

[6] Patel P *et al* 2014 Effect of magnetic field on performance and emission of single cylinder four stroke diesel engine *Journal of Engineering (IOSRJEN)* **4** (5) pp. 28-34

[7] Kumar PV *et al* 2014 Experimental study of a novel magnetic fuel ionization method in four stroke diesel engines *International Journal of Mechanical Engineering and Robotics Research*, **3** (1)

[8] Kolhe AV *et al* 2014 Performance and combustion characteristic of DI diesel engine fueled with jatropha methyl esters and its blends *Jordan Journal of Mechanical and Industrial Engineering* **8** (1) pp 7-12

[9] Chaware K 2015 Review on effect of fuel magnetism by varying intensity on performance and emission of single cylinder four stroke diesel engine *International Journal of Engineering Research* **3** pp 174-178

[10] Okoronkwo C *et al* 2010 The effect of electromagnetic flux density on the ionization and the combustion of fuel *American Journal of Scientific and Industrial Research* **1** (3) pp 527-534

[11] Habbo AR *et al* 2011 Effect of magnetizing the fuel on the performance of an S.I. engine". *Journal Al Rafidain Engineering* **6** (19) pp 84-90

[12] Siregar H and Nainggolan R 2012 Electromagnetic fuel saver for enhanching the performance of the diesel engine *Global Journal of Research in Engineering Mechanical and Mechanics Engineering. Global Journal Inc (USA)* **12** (6) pp 1-4

[13] Gaikwad DR and Dange HM 2014 Experimental investigation of four stroke si engine using oxyrich air energizer for improving its performance *International of Technology Enhancement Enginnering Research* **2** (7) pp 2347-2354.

[14] Salih AM and Al-Rawaf MA 2015 The effect of increasing of diesel fuel temperature upon the engine performance by using two magnetic fields *International Journal of Engineering Research and General Science* **4** (3) pp 170-185

IOP Conf. Series: Materials Science and Engineering **924** (2020) 012031 doi:10.1088/1757-899X/924/1/012031

Vehicle Identification Systems using Virtual Line Sensors and Speed Up Robust Features

Samuel Mahatmaputra Tedjojuwono*, Nathalia Devina Widjaja, and Antonius Kurniawan
Bina Nusantara University, Jakarta, Indonesia 11480

*smahatmaputra@binus.edu

Abstract. This work proposes an improved and robust algorithm to a virtual line approach object movement detection using computer vision technique. The improved performance will be emphasized on how one line could detect multiple object, determines their sizes and could at most classify the object into group similarities. The algorithm later will identify the vehicles based on their license plate using speed up robust features - SURF technique. The method incorporating Bresenham's points plotter algorithm to determine the points coordinate in a line array, which later calculating movement within the given points and group neighboring points with pixel values indicating movement that are in close proximity as one object. This could then calculate the number of points laid within the group which then could determine the size of the object. As a result of the proposed work, a robust object classification technique that is implemented as a computer application that can be used in a live environment context.

***Keywords*:** vehicle classification, computer vision, virtual sensors

1. Introduction

Vehicles classification has been an important issue in road management systems. Its implementation could vary from surveillance system, traffic control, crowd monitoring system, and commercial/ marketing purposes. In surveillance system, traffic/ crowd monitoring system the needs to know how many people or vehicles exist in one place at a given time would assist in traffic control system and law enforcement. Current widely used implementation is by using weight in motion sensors that could be invisible and costly to be installed on every road [1]. One of the main aims of this classification is to identify users of the roads to further implement structure and regulation to increase the road lifespan. In the commercial context, the need has been extended to the area of knowing how many people visiting certain isle of products in a supermarket for example, which could improve the supply chain decision-making process. With these more extensive needs the use of computer vision or video analytics techniques such as virtual sensors turn out to be more promising and efficient in term of cost. The proposed work would aim to improve an original virtual line sensor previously implemented on a video analytics computer application [2]. The earlier work implemented could detect one movement at a time although there are more than one movement occur at the same time within the line array. It also does not inform the size of the object that passed through the line, thus in the context of heterogeneous objects the previous system could not classify the objects. The significance of the proposed research is to produce a new algorithm of robust object classification using Breseham's [2] integral point plotter, which could identify several objects passing through one virtual line at the same time. It is as well could classify the object size thus determine whether the object – in the context of live traffic system – are buses, cars, and motorbikes and later on identify them using license plate matching algorithm

A recent work on robust vehicle classification used sparse representation and Bayesian state inference framework to trace the object [3]. However, the approach does not set specific class to where a detected vehicle should be grouped. Another approach used histogram of oriented gradients (HOG) features [4] to classify vehicles into two classes: trucks and cars. The method comprises of two steps: multi directional vehicle detection and the classification stage. The results vary from 95.1% until 98.2%, however implemented using one hidden layer neural network. Another work was achieved through the implementation of Gabor wavelet transform combined with feature dimension reduction using PCA [5], which could reach the average of 90% correct rate. Although the work was tested only on passenger cars to identify their manufacturers the method is promising both in speed and accuracy, thus could potentially solved other classification problems. Deep learning or also known as CNN (convolutional neural network) is other promising approach that could do pixel level mapping. Related work [6] implemented several types of network such as AlexNet and VGG-16 with promising results to classify car from high resolution satellite image, however low precision on van, truck, and pick up classification. Both methods could produce above 90% precision on car however averagely produce around 60% for the rest of the classes. The work as well did not test against bike or other smaller vehicles, which could introduce certain level of learning challenges. Other work would like to implement computer vision-based object classification even to the extent of classifying human age and gender based on face recognition [7]. The work implemented adaptive features and SVM using RBF kernel with a promising result of 90.8% on both male and female age-based classification. However, this approach will be computationally expensive to be implemented in live traffic road systems.

2. Experimental method

The proposed methodology was divided into several modules some of which adopted from the previous original work [2]. The modules involved are: frame grabber to get the video's frames [2]; pre-processing (blurring / smoothing) to reduce noise [2]; generating line arrays which will detect line placement from the user, and registering all integral points within the line arrays [2]; newly proposed movement detection by generating background reference, background subtraction to calculate delta movement and to register movement within certain position; and lastly the proposed detecting valid objects through identifying beginning and end of an object movement to later classify the object based on the summary of delta movement an object has.

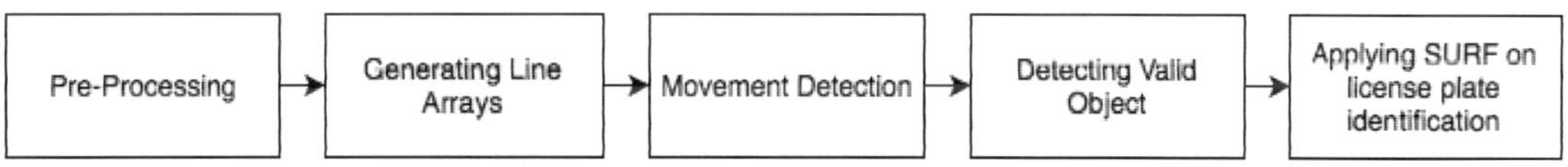

Figure 1. The proposed method that comprise of pre-processing, generating line arrays, movement detection, and later on detecting valid object through that will be classified into several classes.

The newly proposed work uses mean filtering technique where the background is derived from certain number n of frame F. Previous work [2] implemented frame F compose of the pixel value that laid on a two-dimensional matrix using x and y dimension at a given time t. The approach had resulting in a smaller granularity level of movement detection, thus to the level of picture element (pixel). This will be effective to count one object per line however could potentially introduce wrong vehicle classifications - by its size - in a context of fractal motion area for one object (figure 2-a).

Figure 2. (a) Comparison between different motion detection granularity level. It describes a fractal motion detection area for an object as a result of detail pixel level detection, (b) The contrary newly proposed method implementing grid motion detection resulting in a solid motion area.

In order to achieve a solid motion area for an object (see figure 2-b), current method proposed an improvement through grid motion detection [8,9] by dividing an Image F into several smaller two-dimension area called grid β. A grid β could comprise of several pixels such as 8x8 or 16x16 pixels (see figure 3).

Figure 3. Grid area motion detection illustrated by checkered patterns. A grid would be a 8x8 or 16x16 pixels two dimensional matrix, which would serve as a more generalized motion detection point in comparison to one pixel.

Using this approach, we could have coarser granularity in detecting movement of an object, thus resulting in a more generalized and solid (less fractal) detection area of an object. The following equation represents a frame:

$$F(x,y,t) \tag{1}$$

Whereas background G at a given time t is represented through the following equation:

$$G(x,y,t) \tag{2}$$

This background matrix values G produced from the mean value of a certain number of n previous frames, where i is the index of every frame between 0 to n-1. The background G is generated through the following equation:

$$G(x,y,t) = \frac{1}{n}\sum_{i=0}^{n-1} F(x,y,t-i) \tag{3}$$

After acquiring the background then current frame F(x,y,t) could be compared with G(x,y,t) through subtraction operation and compared with a certain threshold T. The process is explained through the following equation:

$$|F(x,y,t) - \frac{1}{n}\sum_{i=0}^{n-1} F(x,y,t-i)| > T \tag{4}$$

Movement detection on every registered point within a line array acquired by taking into account all pixel value with delta values for each point derived from equation (4) above that is higher than T. Referring to the previous work [2], from this point onwards all detection within the a line array will be accounted as an object movement – although in fact there is possibility that a virtual line sensors could be crossed by two or more vehicles/ objects. This in return will give a wrong result of only one object instead of precisely detecting how many objects crossing the line array. The previous method as well has not classified the vehicle crossing the line through several common classes such as cars and bikes. Current method proposing object classification by detecting how many valid objects going through a line *L*, measure them by their size and classify them to several common classes. The following sub topic will describe the process of detecting valid object crossing virtual line sensors.

To detect a valid object the algorithm will mark the beginning *tstart* and end *tstop* of objects crossing the virtual line. This is done through implementing variable *startobj* that will be set to true if there exist one point's delta □ value within the registered member within line *L* that is higher than a certain given threshold *T*. On the other hand, if there are no delta value of any points within the registered points member in the line *L* then the *startobj* will set to false. Within *tstart* and *tstop* of object movement detected object classification will be carried out. Object classification process is done through normalizing all neighboring points that have no movement values although it lies within a moving detected object, or so-called outliers Φ. These outliers o will be set to true by using a two-dimensional operator consist of three positions P_{i-1}, P_i, and P_{i+1}. The operator will traverse all points P within line L and comparing current Pi delta value to its neighboring points P' while normalizing all the outliers Φ. After normalizing all the outliers then the object classification process can be carried out. A class structure of Object is needed to accommodate size field, and two behaviors to access the data as a standard setter and getter methods.

Classification process was done through marking the beginning of an object *ostart h=true* and the end of an object *oend h=false*. The main difference of *ostart - oend* pair and *tstart – tstop* previously mentioned is the last pair are used to mark the start and the end of movement detected, which could involve more than one object segments. To the contrary, *ostart* and *oend* are used to mark the start and the end of every object segment detected within a lifespan of a *tstart-tstop* pair. The *oend* will be set *h=false* if there exist condition where delta value of P_i lower than the threshold however previous value of *oend* have been set to true. An object *O* will have a construct of size *O.size* data encapsulated with its

$$Osize = \sum_{\Delta=Ostart}^{Oend} \Delta \tag{5}$$

behaviors to get and set its field's value. The field *O.size* value will be obtained through summarizing all the delta D value within a pair of *ostart* and *oend*.

By using this *O.size* an object *O* could be classified according to its class. The work implementing two classes indicated by threshold value of car class and bike / bicycle class. All objects *O* that has *O.size* larger than a given threshold *Tcar* will be put into a list structure of car; everything else will be put to the list of bike class.

Figure 4. Sample values of one point from *tstart* and *tstop* of the detection comprise of 4 *moveCount* after twelve frames. It shows pixel values at the same integral point coordinate after 12 frames of images, with 4 detected movement highlited.

The number of frame will be counted for all significant delta movement detected at each point within a line L, summarized and store as *O.size* in an object class. This process will be started at the time of *tstart*, and summarizing the number of movements at *tstop*. All of these summaries will be stored in a list *moveCount*. Figure 4 above shows sample values of one point from *tstart* and *tstop* of the detection comprise of 4 *moveCount* after twelve frames. Taking into account several neighboring points that have significant *moveCount* will gives us indication of one object crossing the virtual line sensor. Figure 5 describing the neighboring movement detection of two distinct objects separated by a gap. All of the *movecount* values stored in a *movecount* list. The list maintains all of the aggregated movecount of the points member *P* of the line *L* between the beginning and end of motion detection event. The *moveCount* list was created at the same length as the virtual line in number of integral points.

All summarized count will be stored in the list based on the line dimension and coordinate of the points within line array. Thus, a continuous sequence of movement detection point is an indicator of an object and the number of movements will indicate the size of the object crossing the virtual line. Although two or more objects crossing the line at roughly the same time, the proposed algorithm could differentiate those vehicles by recognizing the gap between the continuous sequences of delta movement within the *moveCount* list. The result of this process will be object *O* that will be classified to some of common classes *listObjectCar* and *listObjectBike*. Another universal class as well was established to cross check the classification result.

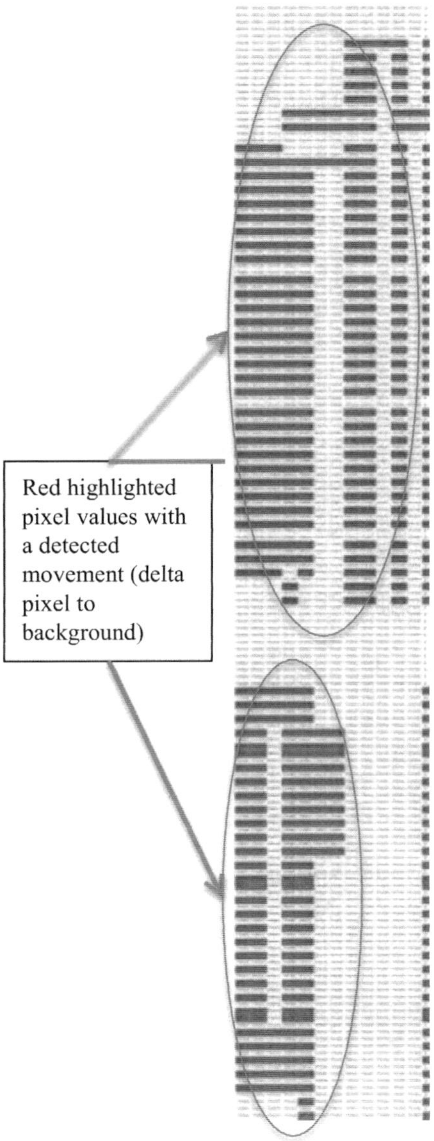

Figure 5. The neighbouring movement detection of two distinct objects separated by a gap. *Pixel* values in hexadecimal collected from 12 frames of video shows 2 object movements.

```
public void classifyObjects()
{
  int[] arrNORMALIZEDCount = new
  int[arrayMoveCount.Length];
  // Normalizing the movement counting array :
  fill all the 0 between detected values:
  for (int i = 1; i < arrayMoveCount.Length-1;
  i++) //for each +1 from begin and -1 from the
  end of the array count movement
  {
    if (arrayMoveCount[i-1] > TMOVEMENT ||
    arrayMoveCount[i] > TMOVEMENT ||
    arrayMoveCount[i+1] > TMOVEMENT) // fill all
    the 0 between detected values
    {
      arrNORMALIZEDCount[i] = 1;
    }
  }

  // calculating how many cars within the
  normalized counted movement (1 car = continuous
  1 values for several points:
  Boolean h = false;
  int beginObj_idx = 0;
  int endObj_idx = 0;
  for (int i = 0; i < arrayMoveCount.Length; i++)
  {
    int c = arrNORMALIZEDCount[i];
    if (c == 1 && h==false)
    {
      h = true; // Beginning of new object
      beginObj_idx = i;
    }

    if (c == 0 && h == true)
    {
      h = false;
      endObj_idx = i;
      MyObject o = new MyObject(endObj_idx -
      beginObj_idx); // create new objects
      //Adding universal objects, it means all
      objects
      if (o.getSize() > TSIZE)
      {
        listObject.Add(o);// end of object
      }
      if (o.getSize() > TSIZECAR)
      {
        listObjectCar.Add(o);
      }
      else
      {
        listObjectBike.Add(o);
      }}}}
```

Figure 6. The classify objects functions that will first normalize the outliers Φ, detect the beginning and the end using *h* of an object *O*, counting the delta movement of the line array *normalizedCount*, register the size of *o.size* and finish by classification process to class list. This is to achieve valid object classification.

SURF calculates the Hessian Matrix Determinant for localizing the key points to decide whether a point is an extreme value given that x = (x, y) for figure I, this is then will be used to match the vehicles license plate. The Hessian Matrix is defined as follows:

$$H(x, \sigma) = \begin{pmatrix} L_{xx}(x,\sigma) & L_{xy}(x,\sigma) \\ L_{yx}(x,\sigma) & L_{yy}(x,\sigma) \end{pmatrix} \tag{6}$$

The method for evaluating license plate matching is described through the following steps: Detects features and descriptors by detecting the location of the key points of the matrix image from the Image Object and Moving Image using the Detector Feature library which will assign key points in the form of key point matrices. Descriptors of image objects and image samples are then matched using the Matching Descriptors from the FLANN library. A good matching method will consist of the main points with the closest similarity distance to each other, using the nearest neighbor distance ratio. And determine whether the object is found; the determinant of whether the actual object can be found is when the list of good matches is greater than 100.

3. Result and discussion

The work was tested in two different conditions daylight and night for non-highway road to capture both cars and bikes. In contrast with previous study [2] the newly proposed implementation of virtual line sensor will only need one-line array to count multiple objects crossing the line at the same time and classifying them based on certain threshold size of car and bike. The modified version of grid motion detection implementation has resulting in a more solid (less fractal) motion detection area for one object or vehicle. This is caused by the generalization of several pixel points within one grid [8] (i.e. 8x8, or 16x16 grid) to inform motion for any movement detection in the point members.

Figure 7. Implementation of one virtual line sensor that could identifies more than one vehicle crossing the line at the same time, and classifies them into two classes: car and bike.

The grid motion implemented to support measuring object by its object detection area size and to accommodate the requirement of ability to count multiple objects concurrently crossing the line. The following illustration compares the previous work of virtual line sensors [2,10] that used multiple lines to cover different parts of the road. The reason to have many lines covering multiple areas is the line could only detect one moving object each time. Thus, to cover more are then many lines to the size of one vehicle's wide would be needed covering entire road width. By stretching the line too long will introduce possibility of information lost, as two vehicles could possibly cross the stretched line at the same time but only one will be counted.

To the contrary the newly proposed method (figure 8-b) needs only one virtual line sensors to detect multiple objects at once. This is where the grid motion detection [8] will be really needed. The technique will minimize fractal effect caused by a detail motion area [2] to the level of pixel point. Although later on during object validation every point in a line will be accounted towards one object-detection, however the noise gaps between points could already be minimized through the process of grid generalization. The proposed technique will group all close motion detected points (neighboring points) together as one object. A separated motion detection points by a gap to a certain number of pixels in length, will be indicated as different object. Therefore, minimizing the unnecessary gap or noise was implemented through grid motion area technique, before later on classifying and counting the number of objects based on their classes. The result was satisfactory which could reach between 82% to 97% of classification rate.

Figure 8. Image (a) shows a multiple virtual line sensor – four lines implemented in the previous work [2], (b) shows current proposed single virtual line to detect and count multiple vehicles crossing the line at the same time.

Figure 9. Tested against low light condition at night to calculate how objects correctly classified as car and bike. Image (a) shows difficult case caused by noise of headlight of cars. Implemented towards low light environment, image (b) shows a better sample taken from the back of vehicles.

The proposed method as well tested during low light condition at night. The position of the camera is purposely placed facing the back of the vehicles. This is to avoid head light reflection on the road surface that could be misunderstood as another object. The challenge was steeply increased at night condition. Although the generalization technique has been implemented, parts of the vehicles that are dark were wrongly misunderstood as the background (dark road surface), therefore not included as motion detected point. When this happen to a larger vehicle such as a car, the gap within the object

IOP Conf. Series: Materials Science and Engineering **924** (2020) 012031 doi:10.1088/1757-899X/924/1/012031

would create wrong classification and counting process of two distinct smaller vehicles. Some of which would be classified as two motorbikes. Similar result as well showed within bike classification at night. A bike with gaps detected –usually between the bike and its taillight- would suffer from wrong counting as two different bikes.

Table 1. Vehicles classification result on daylight.

Class	TRUE	FALSE	Percentage	Total
Car	28	1	97%	29
Bike	32	7	82%	39

Table 1 shows the result of the first case in daylight context, during the duration of a minute video sample where detected 28 cars and 32 motorbikes. The success rate was higher for class car which could reach 97%, whereas bikes reach slightly lower at 82% successful classification rate. This is a promising result in comparison with the original work of virtual lines [2] that could detect only one object per line although the success rate for this one object-one-line context was higher. The following graph depicting the data within table 1 above.

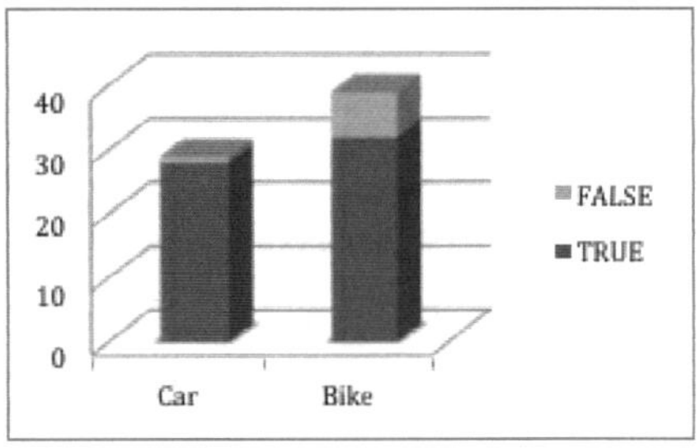

Figure 10. Result in graphs depicting classification result in daylight. Accuracy level could reach 97%.

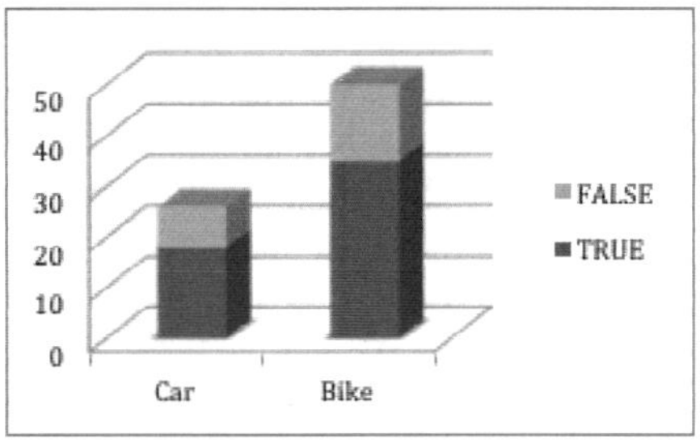

Figure 11. Result in graphs depicting classification result at minimum light. The accuracy level could reach 69% for car and 70% for bike classification.

The testing result using night video sample as well echoed at the above graph. The main cause for the low accuracy rate at night is minimum light condition that gives minimum motion detection as a result of close similarity between background value and the foreground vehicle movement.

Table 2. Vehicles classification result at night.

Class	TRUE	FALSE	Percentage	Total
Car	18	8	69%	26
Bike	35	15	70%	50

The outputs produced by the following research are prototype computer applications to identify vehicles in moving images (video) based on vehicle number plates using the Speeded Up Robust Features method; and to support the achievement of integrated and automated transportation management through the application of information technology (IT) and artificial intelligence.

Figure 12. Virtual line sensors detecting cars.

Figure 13. Detected car will be further recognized by SURF on its plate number.

The first test case is applied to vehicle license plate samples with clear images and observed video during the day (good lighting). The Speeded Up Robust Features method can detect the location of the license plate in the observed video. Figure 12 shows the process of detected cars using virtual line sensors as frame grabber function. The result of license plate identification is shown on figure 13 with accuracy of 94% during daylight.

4. Conclusion

The proposed improved method to the previous original work of virtual line sensors which is aiming to improve functionality of line sensors from only counting to classification. The proposed work has improved the efficiency of previous method of multiple lines to detect multiple area, to only one line could cover the entire road width and could detect and classify multiple objects crossing one virtual line. The accuracy of vehicle classification and counting was higher at daylight than at night. Several handicap of night scheme such as headlight reflection on the road and lack of contrast between the background and moving vehicles in a low light condition have been the main contributor to the result. The work reinforces the initial conclusions on the application of SURF in detection applications for motorized vehicle numbers. However, this method is not suitable to be applied to the context of generalization in order to localize non-specific vehicle license plates. Future development could explore more on solving the virtual line implementation during night and solving the headlight reflection problem alongside with a way to improve contrast of background detection during low light live traffic situation.

References

[1] B. Fritz, B. M. Beau, "Absolute acceleration sensor for use within moving vehicles." U.S. Patent No 10,195,989, 2019.
[2] S.M. Tedjojuwono, "Virtual Lines Sensors for Moving Object and Vehicle Counters", IAENG International Journal of Computer Science, vol. 44, no. 4, pp 421 -431, 2017.
[3] M. Xue, L. Haibin, "Robust Visual Tracking and Vehicle Classification via Sparse Representation", IEEE Transactions on Pattern Analysis and Machine Intelligence, vol. 33, no. 11, 2011.
[4] L. Kang, M. Gellert, "Fast multiclass vehicle detection on aerial images", IEEE Geoscience and Remote Sensing Letters, no. 12.9, pp. 1938-1942, 2015.
[5] T. Yong, Z. Congzhe, G. Renshu, L. Peng, and Y. Bin, ""Vehicle Detection and Recognition for Intelligent Traffic Surveillance System", Multimedia Tools and Applications, Springer, vol. 76, no. 4, pp 5817–5832, 2017.
[6] A. Nicolas, L.S. Bertrand, and L. Sébastien, "Segment-before-Detect: Vehicle Detection and Classification through Semantic Segmentation of Aerial Images", Remote Sensing, vol. 9, no. 4, pp. 368, 2017.
[7] K. Vladimir, G. Alexander, G. Maxim, and S. Lev, "Audience Analysis System on the Basis of Face Detection", Proceedings of the International Multi Conference of Engineers and Computer Scientists, vol. I, 2013.

[8] R. Vladimir, I. Haroon, S. Mubarak, "Detection and tracking of large number of targets in wide area surveillance", European conference on computer vision. Springer, Berlin, Heidelberg, pp 186-199, 2010.

[9] A. Alireza, P. Paulo, N. Urbano, "Detection and tracking of moving objects using 2.5 d motion grids", 18th International Conference on Intelligent Transportation Systems. IEEE, pp. 788-793, 2015.

[10] S. M. Tedjojuwono, "Fast Performance Indonesian Automated License Plate Recognition Algorithm Using Interconnected Image Segmentation", Intelligence in the Era of Big Data, Springer Berlin Heidelberg, Vol: 516, pp. 289-300, 2015.

Synthesis and Characterization of Anode for Lithium Ion Battery

Deswita, Wahyudianingsih, and Evi Yulianti
Centre for Advanced Materials Science and Technology
Indonesia National Nuclear Energy Agency
Bld. 71 Serpong PUSPIPTEK-15314

deswita@batan.go.id;
wahyu@batan.go.id;
yulianti@batan.go.id

Abstract. Anode sheet manufacturing for lithium batteries coated with Polyvinylidene fluoride polymer (PVDF) has been done. The anode sheet is made using a desktop bubble remover from graphite base material by varying the PVDF content of 10%, 15% and 20%, which is called as G1, G2 and G3 samples. The anode material from the mixture of graphite and PVDF is then coated again using PVDF polymer electrolyte with variations in content as much as 10%, 15% and 20%, respectively called as GL1, GL2 and GL3 samples. Characterization of the anode product was measured using XRD diffraction, scanning electron microscope (SEM) equipped with energy dispersive spectroscopy (EDS). The conductivity of the anode sheet was measured using LCR meter. The result shows that the PVDF polymer phase is only observed in GL1, GL2, and GL3 samples. The conductivity value (σ) of the anode sheet in the samples G1, G2 and G3 is obtained in the order of 10^{-3} S.cm^{-1} to 10^{-2} S.cm^{-1} slightly greater than that of the standard anode sheet (≈10^{-3} S.cm^{-1}). The conductivity (σ) of the anode sheet tends to decrease with the addition of the PVDF polymer for the GL1, GL2 and GL3 samples to 10^{-4} S.cm^{-1}. The maximum conductivity in this study was obtained in the G2 sample, with a PVDF polymer content of 15%. SEM / EDS observations show that the fluorine (F) element increases with the addition of the PVDF for GL2 sample.

Keywords: anode sheet, PVDF polymer, conductivity, lithium battery

1. Introduction

Lithium-ion batteries (Li-ion) that can be recharged (rechargeable) become one of the important components for portable electronic equipment such as mobile phones (HP) and computer devices. Li-ion batteries have advantages when compared to Ni-Cd or Ni-MH batteries, namely energy density, higher operating voltage and safer. However, research and development of Li-ion battery materials are still being developed, especially to reduce the thickness and weight of materials and study the improvement of cyclical stability [1–5].

In Li-ion battery systems, the most widely used anode material is graphite. However, graphite has a disadvantage, namely the highest energy density which is still limited to only 372 mAh.g^{-1}) [4]. The electrochemical properties of natural graphite anodes, involving initial efficiency, reversible capacity and cycle stability have been improved due to higher graphitization degree, higher purity, more suitable particle size distribution, and so on. Recently, natural graphite has come to be considered as a promising anode material for lithium ion batteries due to its high reversible capacity, appropriate charge/discharge profile and low cast. However, low first cycle efficiency and poor cycle stability has limited its practical use [3,4,6,7].

In the process of fabricating Li-ion battery, polymer binders are very important ingredients to bind the active ingredients of the electrodes. The properties of adhesives and chemical binders have a major influence on battery performance, especially on battery servicing capabilities. The polymers used to produce anodic coating on the Li-ion battery are polyvinylidene fluoride (PVDF). PVDF is easily obtained material, has high crystallinity that can be used in assembly lithium battery. PVDF interacts with electrolytes and is stable at fairly high temperatures. Furthermore, PVDF shows good chemical resistance in common organic-based carbonate [3,6-9]. The combination of binder PVDF and NMP solvents are used for binders and solvents because PVDF is soluble in NMP and allows preparation of slurry with the right viscosity. Production (solvent-based) process for anodic coatings involves such as mixing powdered active ingredients with carbon, dissolving PVDF with NMP solvent and mixing the powder with a solution to form slurry. The above production processes need to be done carefully because NMPs are heterocyclic, volatile organic compounds (VOCs) and flammable between volume percent 1.3 and 9.5. However, NMP has a high boiling point, which is around 200°C.

The stability and performance of graphite-based anodes with PVDF binders in lithium-ion battery operation has been tested very well for electrode resistance both mechanically and electrochemically [6,11,12]. So PVDF has been used for decades to produce lithium-ion batteries commercially. In this study, the anode material used in Li-ion battery technology is materials, graphite and polymer binders whose main function is to improve the mechanical stability of the anode layer and the conductivity of Li-ion batteries. The purpose of this study was to create and learn PVDF electrolyte addition anodes for Lithium battery applications and study the effect of adding coatings on graphite stability and performance, which are closely related to particle size and structure in improving graphite material performance.

2. Methods

2.1. Materials

High purity ingredients used are Polyvinylidene fluoride (PVDF), liquid $LiPF_6$ and solids, graphite, N-methyl pyrrolidone (NMP), carbon and copper sheets (Cu). The tools used in this study include compact tape casting film coater, compact precision disc cutter, large glove box, desktop hydraulic crimping machine, desktop bubble remover, hot rolling press, vacuum oven.

2.1.1. Sample Preparation. In this study, the methods carried out through two stages. First, the synthesis of the anode layer is composed of basic materials of carbon (10 weight %), Graphite (80 weight %) and PVDF polymer electrolyte (10 weight %). The PVDF polymer electrolyte was firstly dissolved with NMP until homogeneous. Carbon powder, graphite and PVDF + NPM with variations in composition 10 weight %, 15 weight% and 20 weight% are mixed together in a desktop bubble remover to form a slurry-shaped anode sheet [2]. The anode layer in the form of slurry is placed above the copper sheet (Cu) using a compact tape casting coater film and then dried in an oven. Furthermore, the resulting anode sheet was as called the G1, G2 and G3 samples, respectively.

Second, the surface of the anode sheet from the G1, G2 and G3 samples coated with PVDF + NPM polymer electrolyte solution again with variations in content of 10 weight %, 15 weight % and 20 weight %. After that the samples were dried in an oven at 90°C and compacted using a hot rolling press. The coated samples were called as GL1, GL2 and GL3 samples. The resulting anode coating product was characterized using X-ray diffractometer, scanning electron microscope and energy dispersive spectroscopy (EDS).

3. Results and Discussion

The XRD pattern of the samples G1, G2 and G3 from the anode before being coated with PVDF and the samples GL1, GL2 and GL3 after being coated with PVDF are shown in Figures 1 and 2. In Figure 1, it is clear that the samples G1, G2 and G3 have formed crystals perfectly. This can be seen from the intensity of samples. The two peaks were observed at 2θ = 26.63 and 43.49 with each Miller index of (006) and (002). From open data crystallography (COD) no. 96-901-2705 found that the material is graphite. In Figure 1, the crystalline phase of PVDF was not observed in XRD data even though the addition of PVDF polymer electrolyte content was up to 20 weight % by weight.

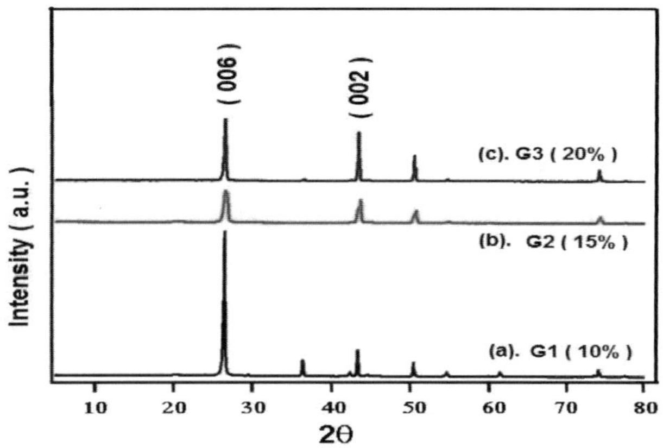

Figure 1. XRD diffraction pattern of the uncoated anode from (a) G1, (b) G2 and (c) G3 samples.

Figure 2 shows the XRD pattern of the anode sheet material after being coated with PVDF content as much as 10 weight %, 15 weight %, 20 weight % from GL1, GL2, and GL3 samples, respectively. The analysis results showed that PVDF crystals were observed at 2θ = 20.19. The high peak PVDF phase increases with increasing PVDF polymer content. Data obtained from measurements using XRD then adjusted to JCPDS data. This value is close to the Miller index value (110) in the JCCPDS data base. While the graphite phase peaks were observed at 2θ = 26.63 and 43.49.

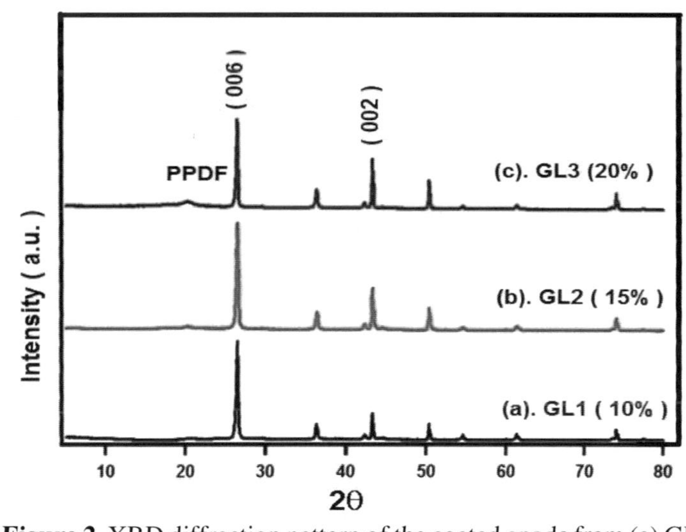

Figure 2. XRD diffraction pattern of the coated anode from (a) GL1, (b) GL2 and (c) GL3.

The anode which is coated again with PVDF electrolyte solution makes the anode sheet more elastic. PVDF which covers the top layer of the anode will experience strong inter-particle forces with graphite. In addition, the voltage between the graphite and PVDF molecules increased so that at a PVDF concentration of 15 weight % the anode layer was not strong enough to bind copper sheets. At 15% PVDF concentration is the maximum concentration where the stress between graphite molecules and PVDF reaches the strongest state.

3.1. Impedance and Conductivity of Anodes

Anode impedance measurement is carried out using instrument electrochemical impedance spectroscopy (EIS). The measurement results are frequency quantities, real impedance values (R_s), and imaginary impedance values (X) which are then plotted curves between the real impedance (Z_{real}) and the imaginary impedance ($Z_{imajiner}$) as shown in Figure 3 and Figure 4 for each sample without and with PVDF coating.

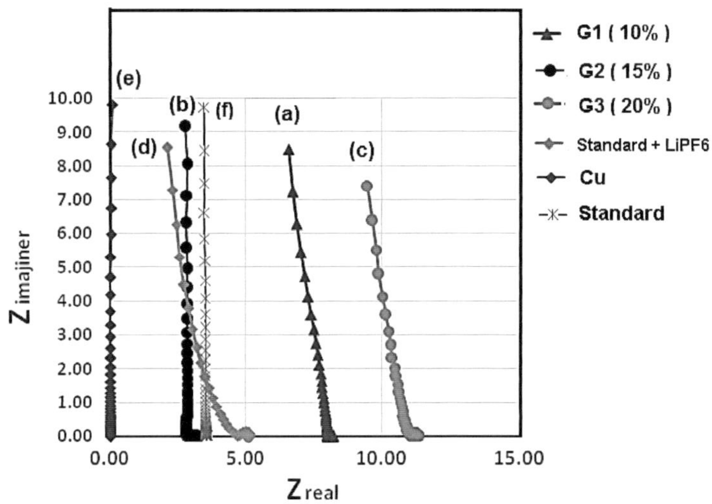

Figure 3. Impedance of anoda *non-coating* PVDF of (a) G1, (b) G2, (c) G3, (d) standard + $LiPF_6$ (e) Cu sheet and (f) Standard Cu.

Figure 3 show that the anode has a pure resistance value or pure R only. The Cu sheet and the standard anode only have a resistance value so that the shape of the curve is perpendicular to the X axis. The shape of the standard anode curve + liquid $LiPF_6$ indicates that the intrinsic properties of the anode are affected by the number of electrolytes. At the anode with variations in PVDF concentration, a shift in the curve indicates that the anode property is affected by PVDF. PVDF provides resistance to graphite but does not significantly affect the shape of the curve because the distribution is fairly even in the composite and the number is small so that the curve remains straight. This shows that the intrinsic properties of the anode remain dominated by graphite and carbon in the composite.

In Figure 4, an anode sample curve coated with PVDF 10 weight % (GL1), 15 weight % (GL1) and 20 weight % (GL3) show a different curve with the anode curve without being coated from sample G in Figure 3. This is because the PVDF concentration the excess adds barriers between molecules and barriers between PVDF molecules with graphite. PVDF will interact with graphite via a hydrogen bridge. At 10 weight % coated anode (GL3 sample), PVDF has a higher viscosity and density that is difficult to interact with graphite in the lower layer, so that the impedance curve shows the nature of graphite alone.

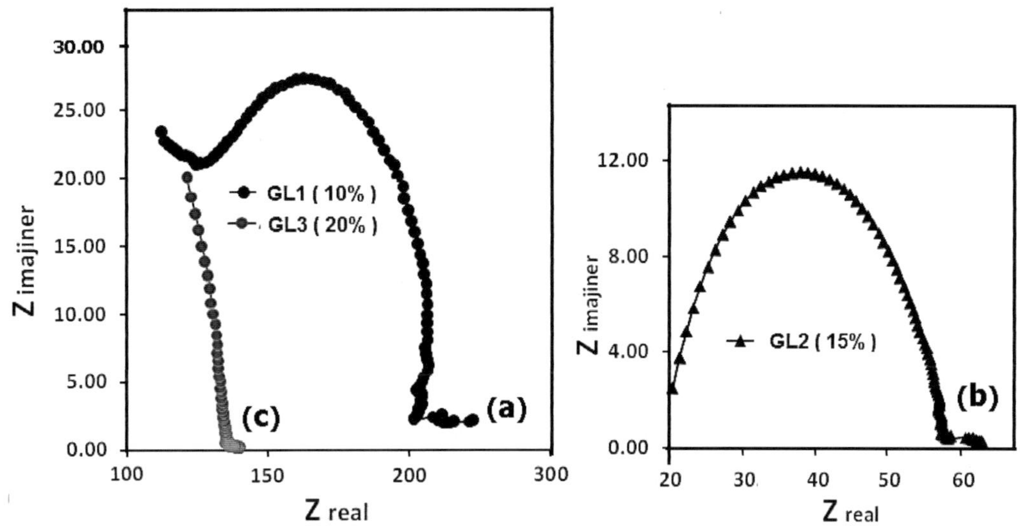

Figure 4. Impedance of PVDF coated anode (a) 10 weight % (GL1), (b) 15 weight % (GL3) and (c) 20 weight % (GL2).

At 10 weight % coated anode (GL1 sample) and 20 weight % (GL3 sample), PVDF has a high viscosity and density, making it difficult to interact with graphite in the lower layer. Consequently the impedance curve shows the nature of graphite only, as shown in Figure 4 (a) and 4(c). Figure 4(b) shows the impedance curve of a sample of 15% PVDF coated anode (GL2 sample). Interaction between the molecules of PVDF and the interaction of the PVDF molecule with graphite reaches the maximum state level so that the impedance value on the curve shows the intrinsic nature of the PVDF molecule. From the measurement results obtained that the best PVDF polymer electrolyte coating was obtained with a concentration of 15 weight %, namely in the GL2 sample.

Conductivity curves are obtained by plotting the frequency log values with conductivity log values. Conductivity values obtained from the equation:

$$\sigma = (1) / R \times t / A \quad (1)$$

with R = resistance, A = the surface area of the anode used and t = thickness of the anode layer. From equation (1) it can be seen that the presence of PVDF in the composite affects the conductivity value (σ) of the anode, where the increase in PVDF concentration is proportional to the increase in conductivity, σ. Measurable conductivity is ionic conductivity of polymer electrolytes because electrolyte ions move on the surface of the anode.

Figure 5 shows the conductivity curves of standard anode layer samples, G1, G2 and G3. The sample measurement results show that the standard conductivity is obtained around $\sigma = 10^{-3}$ S.cm^{-1}. The addition of 10 weight % PVDF electrolyte (Figure 6a) produces a conductivity value, $\sigma < 10^{-3}$ S.cm^{-1}, as in Figure 5 (b). The value of $\sigma > 10^{-3}$ S.cm^{-1} slightly increases with the addition of 15 weight % PVDF (Figure 5c), and then decreases to $\sigma << 10^{-3}$ S.cm^{-1} with an increase in PVDF content up to 20 weight % by weight (Figure 5). In this condition, the sample conductivity reaches a maximum at 15 weight % PVDF addition concentration.

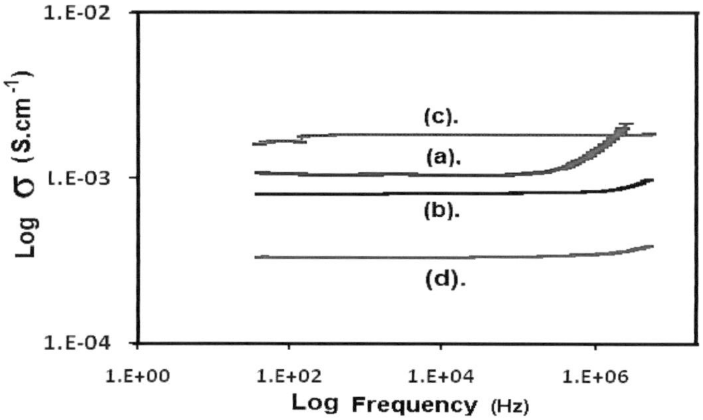

Figure 5. The conductivity curve of non-coated anode from (a) Standard, (b) G1, (c) G2 and (d) G3 samples.

From Figure 5, it can be seen that the conductivity non-coated anode (G3) has slightly higher value to the $LiFePO_4$ standard anode which is liquid $LiPF_6$ electrolyte. This is because the spread of solid electrolytes in PVDF polymer occurs evenly between graphite and carbon molecules. Graphite and carbon have high electronic conductivity values. The presence of electrolyte distribution in the anode also increases electron travel so that its electronic conductivity also increases.

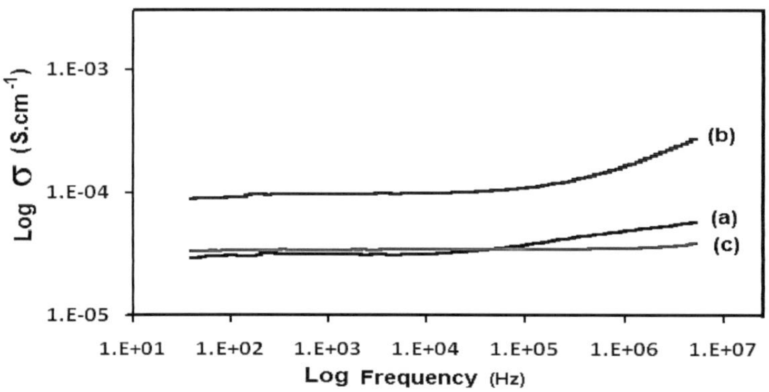

Figure 6. The conductivity curve of PVDF coated anode of (a) GL1, (b) GL2 and (c) GL3 samples.

Figure 6 shows the coated anode conductivity curve with PVDF. From the measurement results obtained the conductivity value of GL1 samples in order $\sigma < 10^{-4}$ $S.cm^{-1}$ (Figure 6a). Then the conductivity of the layer increases to $\sigma = 10^{-4}$ $S.cm^{-1}$ with an increase in the PVDF layer by 15 weight % (Figure 6b). The conductivity value of the anode layer drops again to $\sigma < 10^{-4}$ $S.cm^{-1}$ with the addition of PVDF layer concentration up to 20 weight % (Figure 6b). Based on the curve in Figure 6, it can be seen that the overall conductivity value of the anode after coupled with PVDF is 10, 15 and 20 weight % (each sample GL1, GL2, GL3) is obtained in order, $\sigma = 10^{-4}$ $S.cm^{-1}$ in the GL2 sample. The value of σ is lower than the value of σ in the sample G2. In this sample group, the maximum anode conductivity was obtained at a PVDF concentration of 15 weight %, $\sigma = 10^{-4}$ $S.cm^{-1}$ in the GL2 sample.

3.2. Characterization with SEM-EDS

Observation non-coating anode layer of G2 sample and cross-section of GL2 sample morphology was described in Figure 7. It appears that the surface of the anode layer is composed of granules in the form of a granular plate with varying sizes ranging in size from 3μm to 10μm in length and width. The granules are composed of elements carbon (C of 92 atomic %) and fluorine (F of 5 atomic %), while the element copper (Cu of 3 atomic %) comes from the base. The white colour distributed almost throughout the surface of the anode layer is an element of F. Figure 7(b) shows the morphology of the cross section of the G2 sample. It can be seen that the sample is composed of Cu layer as a base, graphite + PVDF and PVDF layer with 15 weight % content variation. The results of enumeration with EDS in GL2 samples, which were shot in the Graphite + PVDF layer obtained content of elements C (72.24 atomic %), F (20.60 atomic %), O (5.36 atomic %) and Cu (1.81 atomic %). In the GL2 sample, the estimated thickness of graphite + PVDF above the Cu layer is about 45μm, with the thickness of the PVDF polymer layer having a thickness in a range smaller than 4μm.

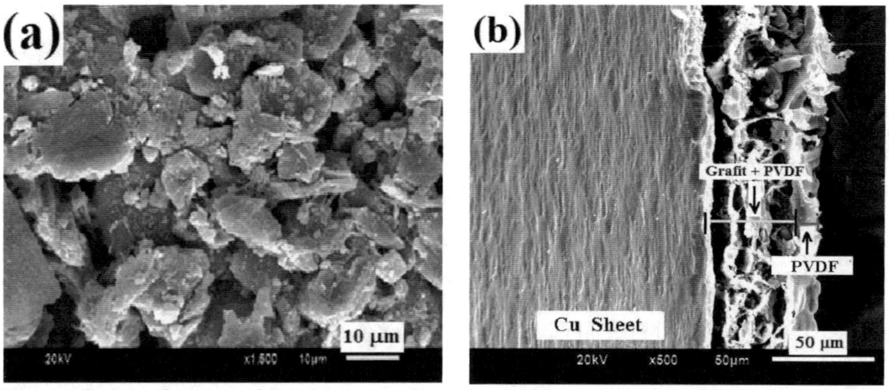

Figure 7. Morfology of (a) Non-coating anode layer (G2 sample) and (b) Cross section of GL2 sample. The thickness of cross section of GL2 sample (graphite + PVDF about 45μm) and the thickness of PVDF polimer less then 4μm.

4. Conclusion

Anode sheet synthesis for applications of lithium-free batteries and coated with Polyvinylidene fluoride polymer (PVDF) has been done. PVDF polymer coating is given with a content of 10, 15 and 20 weight % by weight. The results of the measurement can be concluded that the variation of PVDF electrolyte concentration affects the intrinsic properties of the anode. Conductivity value, σ is obtained in the order between 10^{-2} - 10^{-3} S.cm^{-1} for variations in PVDF content around 15weight % in the G2 sample. Adding PVDF polymer coating over the anode layer on the GL2 sample with a content variation of about 15 weight % causes a conductivity value, σ decreases to be in order 10^{-4} S.cm^{-1}.

Acknowledgment

This research was carried out under Budget Implementation List (DIPA) of Centre for Science and Technology of Advanced Materials – BATAN and a part of Lithium Ion Battery consortium project under the Ministry Research Technology and Higher Education (SINAS 2018).

References

[1] S. Sudaryanto, E. Yulianti, D. Deswita, W. Wahyudianingsih, D. N. P. Anggreyni, and E. Kartini, "Interfacial Properties of Solid Polymer Electrolyte for Lithium-Ion Battery," *IOP Conf. Ser. Mater. Sci. Eng.*, 2019.

[2] Deswita, S. Sudaryanto, H. Djodi, and E. Kartini, "The improving conductivity of LiFePO4 by optimizing the calendaring process," *IOP Conf. Ser. Mater. Sci. Eng.*, vol. 432, no. 1, pp. 0–8, 2018.

[3] A. Purwanto *et al.*, "Improving cylinder-type LiFePO4 Battery Performancevia Control of Internal Resistance," *Mater. Res. Express*, vol. 5, no. 4, 2018.

[4] Q. Si, D. Mori, Y. Takeda, O. Yamamoto, and N. Imanishi, "Silicon-carbon composite

IOP Conf. Series: Materials Science and Engineering **924** (2020) 012032 doi:10.1088/1757-899X/924/1/012032

electrode materials prepared by pyrolysis of a mixture of Manila hemp, silicon powder, and flake artificial graphite for lithium batteries," *Energies*, vol. 10, no. 11, pp. 1–9, 2017.

[5] Q. Cheng, Y. Okamoto, N. Tamura, M. Tsuji, S. Maruyama, and Y. Matsuo, "Graphene-Like-Graphite as Fast-Chargeable and High-Capacity Anode Materials for Lithium Ion Batteries," *Sci. Rep.*, vol. 7, no. 1, pp. 1–14, 2017.

[6] U. Tocoglu, T. Cetinkaya, O. Cevher, M. O. Guler, and H. Akbulut, "Nanostructured silicon thin film electrodes for li-ion batteries," *Acta Phys. Pol. A*, vol. 123, no. 2, pp. 380–382, 2013.

[7] K. Kierzek, "Influence of Binder Adhesion Ability on the Performance of Silicon/Carbon Composite as Li-Ion Battery Anode," *J. Mater. Eng. Perform.*, vol. 25, no. 6, pp. 2326–2330, 2016.

[8] A. M. Grillet *et al.*, "Conductivity Degradation of Polyvinylidene Fluoride (PVDF) Composite Binder during Cycling: Measurements and Simulations for Lithium-Ion Batteries," *J. Electrochem. Soc.*, vol. 163, no. 9, pp. A1859–A1871, 2016.

[9] U. Farooq *et al.*, "Effect of binder and composition ratio on electrochemical performance of silicon/graphite composite battery electrode," *Mater. Lett.*, vol. 136, no. September, pp. 254–257, 2014.

[10] M. J. Kim, C. H. Lee, M. H. Jo, and S. K. Jeong, "Electrochemical Decomposition of Poly(Vinylidene Fluoride) Binder for a Graphite Negative Electrode in Lithium-Ion Batteries," *Mater. Sci. Forum*, vol. 893, no. Cv, pp. 127–131, 2017.

[11] L. Zou, F. Kang, Y. P. Zheng, and W. Shen, "Modified natural flake graphite with high cycle performance as anode material in lithium ion batteries," *Electrochim. Acta*, vol. 54, no. 15, pp. 3930–3934, 2009.

[12] P. P. Prosini, M. Carewska, C. Cento, and A. Masci, "Poly vinyl acetate (PVA) used as a binder for the fabrication of a LiFePO4-based composite cathode for lithium-ion batteries," *Electrochim. Acta*, vol. 150, pp. 129–135, 2014.

IOP Conf. Series: Materials Science and Engineering **924** (2020) 012033 doi:10.1088/1757-899X/924/1/012033

Process Mining with Applications to Automotive Industry

Michael Siek*, Ryan Malik Gunadharma Mukti
Bina Nusantara University Jakarta, Indonesia 11480

*michael.s@binus.edu

Abstract. Process mining as a modeling and analysis tool can be used to improve the business performance by looking at the actual business processes. This paper presents the applications of process mining in automotive industry. Using event log data with timestamps, process mining algorithms, like inductive miner and fuzzy miner were able to automatically generate car manufacturing processes, automatically checking the conformance between the actual processes and the predefined standard ones, and identify and solve any bottlenecks and issues in the car manufacturing processes. A number of car manufacturing issues were considered in this research, such as process delays, stagnant workflow, mismanagement, faulty production and labor insufficiency. The modeling and statistical results show promising leverage of process mining in automotive industry that can lead to the autonomous car manufacturing with abilities for real-time process auditing and reengineering.

Keywords: automated process model generation, real-time process auditing and reengineering, inductive miner, fuzzy miner, identification of process bottleneck

1. Introduction

Mining a business process or process mining often leads to new discoveries and insights into the real processes occurred in many businesses[1]. These findings could explain some detailed reasons why the businesses run stagnate or without any progress. Analyzes of these findings and insights are crucial in order to give some benefits to most of the enterprises, such as detecting the process bottleneck and how to solve this issue [2]. Some bottleneck or interferences in the processes may disrupt the whole of business performance.

As an analysis tool used in process management, a process mining utilizes simulation methods of playing out or playing in of events, which usually occur in many companies' processes and determine what to do in tackling any process problems [3]. The process mining requires event log data as inputs for generating business process model and its further objects for analysis. The event logs are sets of event or process data with timestamps to indicate when a process starts and ends [4]. Some example of events or processes in automotive industry, such as: casting, stamping, engine assembling and quality controlling [5].

The objective of this research is to apply process mining algorithms for automotive industry in the purposes of automated generation of car manufacturing process and automated conformance checking of the actual manufacturing process leading to process auditing and reengineering. The main principles of car manufacturing in Toyota were chosen and utilized for creating production scenarios and event logs, testing process mining algorithms and their resulting process models. There are several scenarios included in the case of Toyota Manufacturing, which visage a number of certain hindrances causing delays and stagnant workflow in the car manufacturing processes. Some hindrances include

IOP Conf. Series: Materials Science and Engineering **924** (2020) 012033 doi:10.1088/1757-899X/924/1/012033

faulty engine manufacture, prolonged wait time of quality control inspection, mismanagement of certain parts that require attention, and faulty production overall [6].

2. Experimental method

2.1. Process mining methodologies

Despite having a widely-used Business Process Model (BPM), we employ Petri net as process model representation as it is more universal [7]. A Petri net is able to visualize how events are transition to one another based on the event logs. Some reliable tools like ProM and RapidProM were utilized for mining manufacturing processes from event log data and checking the conformance of the actual processes with the standard manufacturing processes in Toyota [8]. Two algorithms for process mining like inductive miner and fuzzy miner [9-11] were exploited to obtain a consistent process model results from event logs. In fuzzy model, a rough diagram of process can be generated from event logs with the frequency information of each process transition for producing every individual car. In other words, it allows us to verify how often a particular process travels to the other process[12]. In fuzzy model diagram, the frequency information is depicted by the arrows between events or processes; the thicker the arrow the more frequent the process leading to the next process.

2.2. Event log data collection

Event log data is utilized as the inputs for process mining algorithms to produce car manufacturing processes represented by Petri net. These event logs are specific database employed to identify and check on how the businesses operates and find in case of the company has some unused or unnecessary process [13]. Though most companies may not have event logs, however they have time stamped records of events that are part of the company's business processes. For instance, the events can be taken from the meetings held, reports made, items manufactured, and so forth [14].

Table 1. Fragment of synthetic event log data in automotive industry.

Scenario ID	Case ID	Activity	StartTimestamp	EndTimeStamp	Cost	Resources
1	311656	Casting	2019-05-07 07:45	2019-05-07 09:25	100	Mike
1	311657	Engine production	2019-05-07 09:30	2019-05-07 11:30	200	Pietro
1	311658	Stamping	2019-05-07 07:45	2019-05-07 08:55	50	John
1	311659	Welding	2019-05-07 09:00	2019-05-07 10:00	100	John
1	311660	Painting	2019-05-07 10:05	2019-05-07 11:00	50	James
1	311661	Assembly	2019-05-07 13:15	2019-05-07 15:45	200	Charles
1	311662	Quality Control	2019-05-07 16:00	2019-05-07 18:00	100	Charles
1	311663	Shipping and delivery	2019-05-07 18:05	2019-05-07 19:00	50	Susan
2	311665	Engine production	2019-05-08 08:00	2019-05-08 09:45	200	Pietro
2	311666	Stamping	2019-05-08 08:00	2019-05-08 09:00	50	John
2	311667	Welding	2019-05-08 09:05	2019-05-08 09:50	100	John
2	311668	Painting	2019-05-08 09:50	2019-05-08 10:30	50	James
2	311669	Assembly	2019-05-08 10:35	2019-05-08 11:58	200	Charles
2	311670	Quality Control	2019-05-08 13:00	2019-05-08 14:00	100	Mike
2	311671	Shipping and delivery	2019-05-08 15:00	2019-05-08 15:55	50	Susan

For data collection in this research, we investigated the car manufacturing processes in Toyota, specifically on how the company manufactures its cars, how many cars is possibly produced, how often something can go wrong in each process, and how the issues can be resolved. Subsequently, we created various types of scenarios in the event logs covering all possible standard and non-standard procedures of car manufacturing that could occur within a certain time frame for one production line, with production scenarios with production issues. Using this event log data, we conducted some experiments and testing for automated generation of car manufacturing processes and checking the conformance or compliant of the actual car manufacturing processes with the standard car manufacturing processes [15].

A fragment of synthetic event log data utilized for process mining in automotive industry is listed in Table 1. This data consists of four main attributes: case ID, activity, starting timestamp and ending timestamp. The other attributes are used to create the scenarios for various types of car manufacturing processes. For instance, scenario ID 1 illustrates the "perfect" process scenario from the process of casting until shipping and delivery of the car. In other case, scenario ID 2 indicates an event which the manufacturing plant chooses to re-use an asset rather than casting a new engine. In scenario ID 3, we consider the manufacturing processes and packing of spare parts. Lastly, scenario ID 4 is a scenario in which a certain car manufacturing requires to build a new engine after the first production is considered faulty.

3. Results and discussion

3.1. Automated generation of manufacturing processes

Two process mining algorithms: inductive miner and fuzzy miner, were performed to automatically generate car manufacturing processes from event log data as partly listed in Table 1.

3.1.1. Inductive miner. The automated generation of car manufacturing processes using inductive miner from event log data is depicted as Petri net in Figure 1. The Petri net diagram produced by inductive miner provides us with how the manufacturing processes flow with some information of the time duration and frequency for each process.

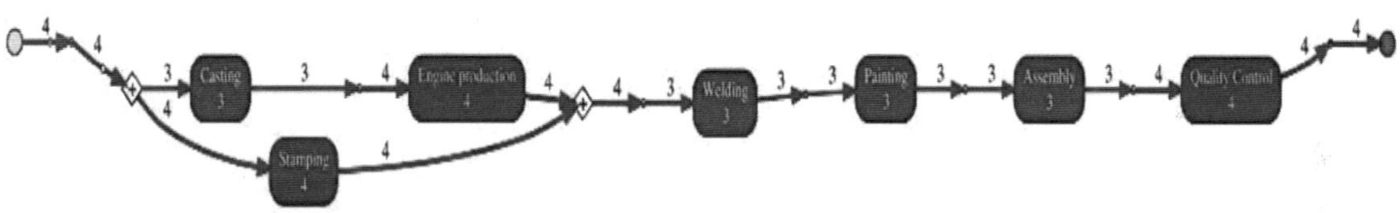

Figure 1. Automated generation of automotive manufacturing processes using process mining induction represented as Petri net.

The green icon on the left indicates the start of the car manufacturing process, while the red icon indicates the end of the car manufacturing process. What comes after the green icon is a '+' icon, this icon indicates that the process is split and will work in parallel with each other. For this case, it can be seen from Figure 1 that the processes that are working in parallel is casting, stamping, and engine production. It is also shown that after casting is completed; the process continues to engine production; while stamping ends with its own route. Once they finish, the process ends back to a new '+' icon indicating that the parallel sequences come together and follow by further process to be executed. Some subsequent processes include welding, painting, assembly, quality control, and end process. These manufacturing processes look simple since we hide some process paths for better understanding of the generated car manufacturing processes from the used event log data.

In addition to Petri net diagram automatically generated by inductive miner from event log data, we can check some anomalies in the car manufacturing processes as illustrated in Figure 2. It is a dotted chart for the event logs indicated by index traces versus timestamps of the events or processes with the information of each process's time duration and frequency. Using this chart, we can find the manufacturing process issues and analyze the abnormal processes in comparison to the standard processes in car manufacturing. In more detailed example, we can obviously see that on May 17th the production line appears to take longer than the rest of the production scenarios. With closer inspection, we can look that after conducting a quality control, there is a need to recast a part for the car, therefore insinuating that the previous casting caused an error in the car production.

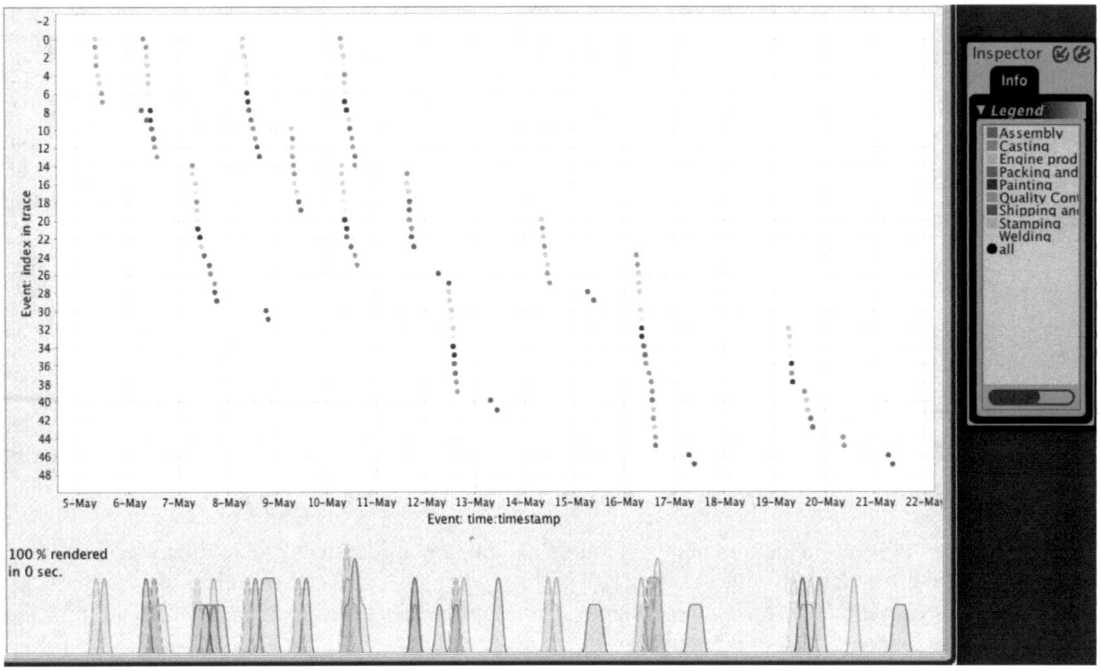

Figure 2. Event: index in trace versus time: timestamp depicting some variations on time duration and frequency of the car production processes.

Another important dotted chart is the one indicated by event name versus time stamps as illustrated in Figure 3. This chart allows us to check how often a particular activity in car manufacturing need to be performed in regards to all scenarios. From this dotted chart, we can see the list of events that have occurred in both the process and event log, with the timestamp placed as the handle to separate the events based on its frequency and occurrence. It becomes apparent in this dotted chart that the least frequent event that executed in the process is 'Packing and Vanning', occurring only 3 times within the entire processes

3.1.2. Fuzzy miner. The other process mining algorithm utilized in this research is fuzzy miner. In similar case to inductive miner algorithm, we used event log data to automatically generate fuzzy process model as depicted in Figure 4. In the fuzzy process model, the frequency information of a particular process is indicated by an arrow connecting between events or processes; the thicker the arrow the more frequent the process leading to the next process.

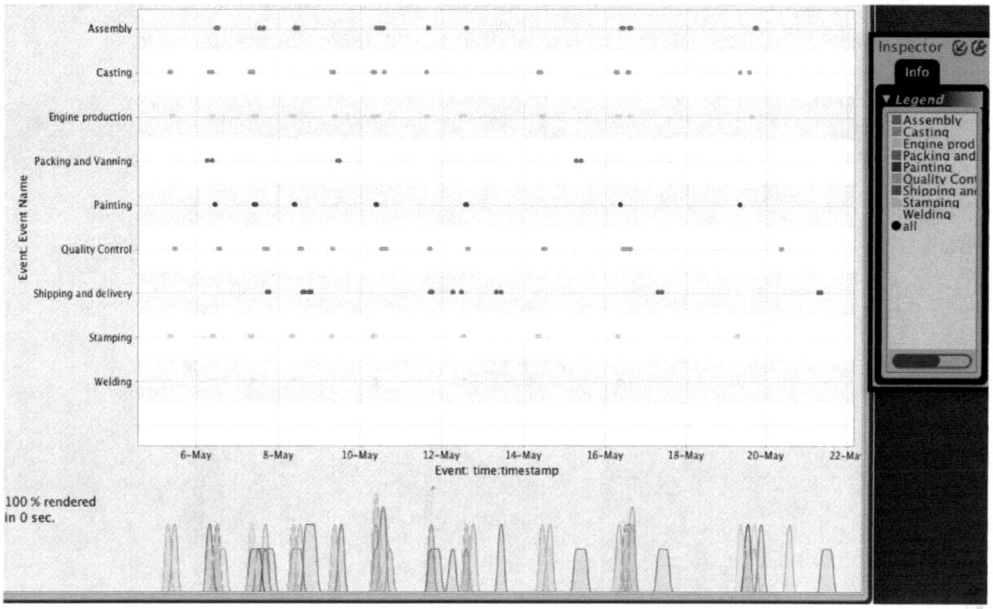

Figure 3. Event: event name versus time: timestamp comparing production time durations of each process for every single car production.

It can be inferred that the car manufacturing processes can continue progressively. How often a certain process follow the standard predefined process indicates that there is high conformance between the actual and standard car manufacturing procedures. The current process does not need to impact any drastic changes since the processes indicated here still follow the standard ones. In other aspect, there are some arrow connections in the fuzzy model diagram that loop back onto it. The most likely reason is due to the timestamps and the sequence of activities overlapped in the event log data. In case of some processes do not continue to any other process, the fuzzy model creates a loop in which it ends at its own process once it completes.

Table 2. Statistical evaluation measures to indicate the alignments.

	Average	Max.	Min.	Std. Deviation	#Case Value of 1.00
Raw Fitness Cost	2.75	5	0	2.22	0
Calculation Time (ms)	2,378.50	2,677	2,031	274.59	0
Num. States	2,771.00	4,813	1,244	1,531.30	0
Move-Model Fitness	0.97	1	0.92	0.04	2
Trace Fitness	0.94	1	0.9	0.04	1
Move-Log Fitness	0.97	1	0.93	0.03	1
Trace Length	42	48	30	8.49	0
Queued States	7,806.50	9,574	6,516	1,416.13	0

3.2. Automated conformance checking of manufacturing processes

For conformance checking, we utilize some statistical evaluation measures to indicate the process alignments, such as: raw fitness cost, move-model fitness, trace fitness and move-log fitness. Table 2 lists all of these statistical evaluation measure values that indicate how the actual processes to the standard car manufacturing processes conform. The move-model fitness of 0.97, trace fitness of 0.94 and move-log fitness 0.97 indicate reliable conformance or alignment, whereas the raw fitness cost of

2.75 did not provide good indicator of conformance. The latter is due to the fact that the number of cases for every scenario in event log data is insufficient for the process mining.

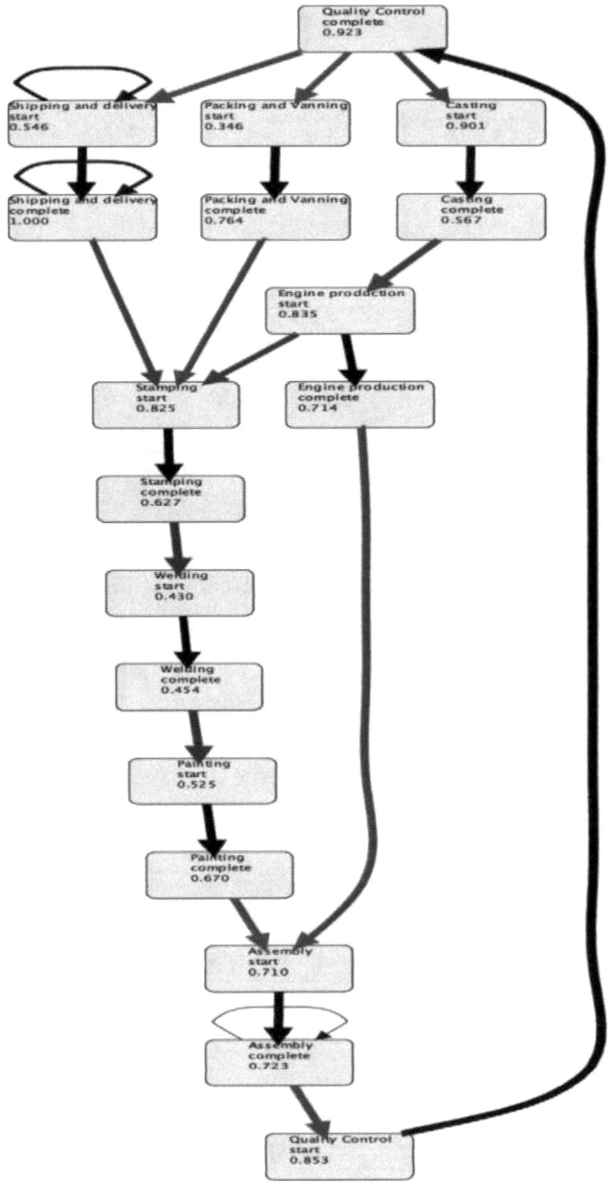

Figure 4. Automotive manufacturing processes generated by fuzzy-based process mining from event log data.

There are also some other statistics to measure the speed of process model generation, number of states, trace length and queued states. It can be inferred from these alignment statistics that the generated Petri net took time on average of 2,3 milliseconds to create with an average trace length of 42. As the alignment statistic shows, there are 7,806 queued states on average, which explains why the the Petri net diagram sometimes looks scrambled and messy. Figure 5 shows the conformance checking of automotive manufacturing processes generated from event log data, represented as Petri net.

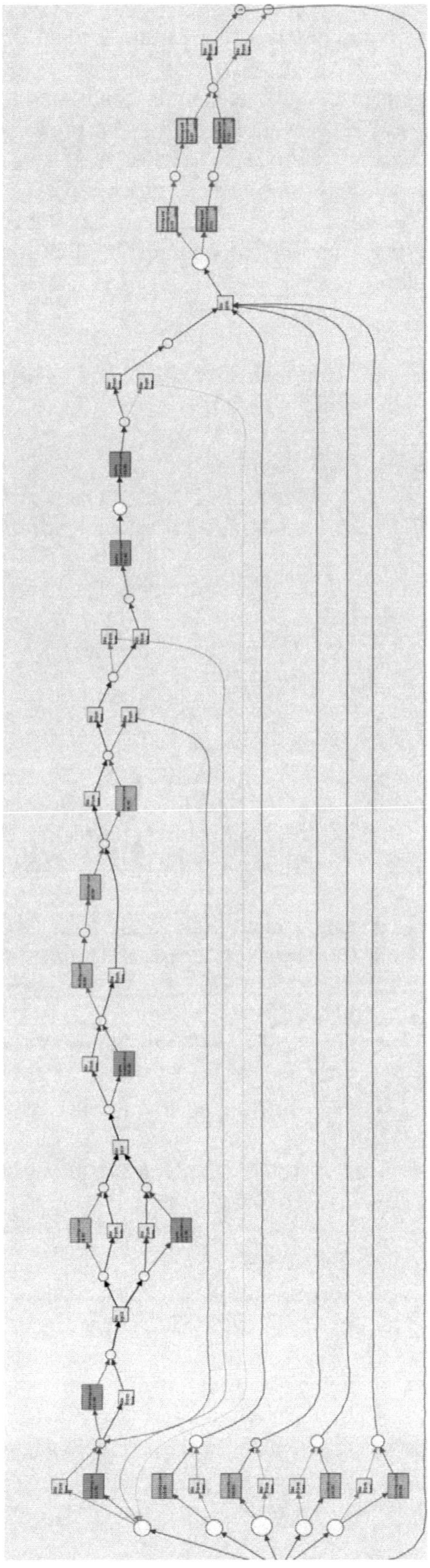

Figure 5. Conformance checking of automotive manufacturing processes from event log data.

IOP Conf. Series: Materials Science and Engineering **924** (2020) 012033 doi:10.1088/1757-899X/924/1/012033

4. Conclusion

The applications of process mining in automotive industry are quite straightforward. There are at least three applications: automated generation of car manufacturing processes, automated conformance checking of the actual car manufacturing processes to the standard ones, and to identify and solve any bottleneck in the manufacturing processes. The leverages of process mining in automotive industry can lead to the autonomous manufacturing with abilities for real-time process auditing and reengineering. A number of car manufacturing issues such as process delays and stagnant workflow due to mismanagement, faulty production and labor scarcity can be immediately detected by the process mining system. This indicates process mining can be a reliable tool towards the automated car manufacturing technology in the near future.

Acknowledgment
This research has been conducted and funded through a research grant program supported by Bina Nusantara University with a number of 033/VR.RTT/IV/2019.

References

[1] W. M. van der Aalst, H. A. Reijers, A. J. Weijters, B. F. van Dongen, A. A. De Medeiros, M. Song, *et al.*, "Business process mining: An industrial application," *Information Systems,* vol. 32, pp. 713-732, 2007.
[2] W. M. van der Aalst, "Process mining: Discovery, conformance and enhancement of business processes," ed: Springer-Verlag Berlin, 2011.
[3] S. Park and Y. S. Kang, "A study of process mining-based business process innovation," *Procedia Computer Science,* vol. 91, pp. 734-743, 2016.
[4] W. M. van der Aalst, *Process Mining. Data science in action*: Springer-Verlag Berlin, 2016.
[5] Toyota. (2019). *How we manufacture*. Available: http://www.toyotaindonesiamanufacturing.co.id/manufacturing/how-we-manufacture?lang=id
[6] Á. Valencia-Parra, B. Ramos-Gutiérrez, Á. J. Varela-Vaca, M. T. Gómez-López, and A. G. Bernal, "Enabling Process Mining in Aircraft Manufactures: Extracting Event Logs and Discovering Processes from Complex Data," in *Proceedings of the Industry Forum at BPM 2019*, 2019, pp. 166-177.
[7] S. J. Leemans, E. Poppe, and M. T. Wynn, "Directly follows-based process mining: Exploration & a case study," in *2019 International Conference on Process Mining (ICPM)*, 2019, pp. 25-32.
[8] W. M. van der Aalst, A. Bolt, and S. J. van Zelst, "RapidProM: mine your processes and not just your data," *arXiv preprint arXiv:1703.03740,* 2017.
[9] D. P. Solomatine and M. B. Siek, "Modular learning models in forecasting natural phenomena," *Neural networks,* vol. 19, pp. 215-224, 2006.
[10] R. Mans, W. M. van der Aalst, and H. Verbeek, "Supporting Process Mining Workflows with RapidProM," in *BPM (Demos)*, 2014, p. 56.
[11] S. M. Tedjojuwono, "Virtual Lines Sensors for Moving Object and Vehicle Counters," *IAENG International Journal of Computer Science,* vol. 44, 2017.
[12] D. Chapela-Campa, M. Mucientes, and M. Lama, "Mining frequent patterns in process models," *Information Sciences,* vol. 472, pp. 235-257, 2019.
[13] E. González López De Murillas, H. A. Reijers, and W. M. Aalst, "Connecting databases with process mining: a meta model and toolset," *Software and Systems Modeling (SoSyM),* vol. 18, pp. 1209-1247, 2019.
[14] S. van Zelst, "Process mining with streaming data," *Technische Universiteit Eindhoven,* 2019.
[15] Á. J. Varela-Vaca, J. A. Galindo, B. Ramos-Gutiérrez, M. T. Gómez-López, and D. Benavides, "Process mining to unleash variability management: discovering configuration workflows using logs," in *Proceedings of the 23rd International Systems and Software Product Line Conference-Volume A*, 2019, pp. 265-276.

Electrical Properties Study of Solid Polymer Electrolyte based on PolyCaproLactone/$LiCF_3SO_3$ Added Plasticizer

Evi Yulianti*, Sudaryanto, Deswita, and Wahyudianingsih
Center for Science and Adveanced Materials, BATAN
Puspiptek Serpong, Tangerang 15314

*yulianti@batan.go.id

Abstract. The electrical properties of solid polymer electrolyte based on biodegradable polymer PolyCaproLactone (PCL)/$LiCF_3SO_3$ with PC plasticizer addition have been studied. Polymer electrolyte membranes were prepared by solution casting technique by varying propylene carbonate (PC) composition (0-60) % weight. The membranes were characterized by the Electrical Impedance Spectroscopy at room temperature to measure its ionic conductivity and other electrical properties such as dielectric constant and tangent loss. The PCL/$LiCF_3SO_3$ system containing 20% weight PC plasticizer exhibits the highest conductivity of 4.92×10^{-4} S/cm at room temperature. The dielectric constant shows decreases with frequency increasing and it depends on dissociation of ions in the polymer electrolyte system. The tangent loss suggests that there are relaxing dipoles in the polymer electrolyte that shift toward higher frequency region. The membrane microstructure also investigated by XRD and the result showed that the PC plasticizer made the PCL/$LiCF_3SO_3$ system become more amorphous

Keywords : PCL/$LiCF_3SO_3$, polymer electrolyte, propylene carbonate (PC), electrical properties

1. Introduction

Polymer electrolytes represent an attractive class of solid state materials that exhibit the benefit of various desirable properties such as safety, high flexibility, mechanical stability, multiple charging/discharging cycle and ease fabrication into desired shapes and sizes and their capability to form electrode-electrolyte contact. The solid polymer electrolyte (SPE) is an alternative substitute for the liquid electrolyte in the electrochemical device such as battery. In the battery cell SPE can also separate the two electrodes electronically to prevent short-circuit. Solid polymer electrolytes are essentially polymer salt complexes formed by incorporation salts in a polymer matrix containing heteroatoms such as O, N, S, F etc as Lewis base which serves to coordinate the cations. The solid polymer electrolytes have many advantages to apply in many devices such as in lithium ion batteries, fuel cells, supercapacitors, solar cells, and electrochromic display devices [1,2].

The most polymer electrolyte studied is poly(ethylene oxide) (PEO) with various inorganic salts dissolved in its matrix. PEO has demonstrated its good performance as a solid polymer electrolyte, however the high degree of crystallinity of PEO restricts its use in battery. A few efforts have already made on natural polymers such as chitosan, cellulose acetate, starch [3-5] and synthetic polymers such as poly(vinyl alcohol) (PVA), poly(vinyl chloride) (PVC) [6,7] to obtain the new polymer electrolytes for their application in various electrochemical devices. However there are some disadvantages of these films including poor mechanical properties and low conductivity.

Various approaches have been undertaken by many researcher to achieve higher conductivity and improve the mechanical properties of polymer electrolyte, such as (1) formation cross link networks, (2) polymers blending [8], (3) addition of inorganic filler such as SiO_2, TiO_2 and ZrO_2 [9-12], and (4) plastization [7,13]. In plastization, the low molecular weight plasticizer with high dielectric constant such as ethylene carbonate, propylene carbonate etc are added to the polymer electrolyte. Among of them, plastization is the most effective ways to increase the ionic conductivity. The incorporation of plasticizer will increase ionic conductivity by enhancement the amorphous phase of the polymer electrolyte and increase the flexibility and release the mobile charge carriers due to ion dissolution effect. In addition, plasticizer can increase volume within electrolyte system and decreases viscosity by making the ion mobility became easier [14,15].

In this research, PolyCaproLactone (PCL) is used as the base polymer for studies on lithium conducting membrane. PCL is a synthetic biodegradable polymer that widely used in biomedical application because this polymer is nontoxic and has an excellent level biocompatibility. PCL can make the complex compound by coordinating cations in the salt with the carbonyl group and a Lewis base (ester oxygen). PCL also has low glass transition temperature (-60°C) that allows polymer chain to make segmental motion at room temperature, so that the ions could hop from one complexation site to another easily. A few researchers have attempted to make PCL as polymer electrolyte for battery and supercapacitor application by complexing with various salts [16-18]. In this paper we will report the effect of plasticizer addition to the electrical properties of polymer electrolyte PCL incorporated with $LiCF_3SO_3$

2. Experimental method

2.1. Materials and methodology

Polymer Polycaprolactone (PCL) with molecular weight 80.000, Lithium Triflat ($LiCF_3SO_3$) 99.995% salt and propylene carbonate (PC) as plasticizer were purchased from Sigma Aldrich and used without any purification. Tetrahydrofuran (THF) (p.a) as solvent was purchased from Merck.

Polycaprolactone membrane were prepared by dissolving polycaprolactone (PCL) beads in tetrahydofuran and stirred up for 3 hours. The lithium salt $LiCF_3SO_3$ was added to the PCL solution (in weight percent) as much as 20% and stirred up for 2 hours. Then PC plasticizer was added to the solution by varying composition (0, 10, 20, 40, 60) % by weight. The mixture was stirred continuously until homogenous and then poured into a petry dish. The membrane is formed by slow evaporation in a vacuum oven for ± 72 hours at 50°C. After that membranes were stored in a desiccators for 2 days to remove any remaining solvent [1].

2.2. Characterization

The impedance measurement is done using HIOKI LCR Hi-Tester Model 3532-50 controlled by computer at room temperature. The membrane was placed between the electrodes and impedance values measured with a range of frequencies from 42 Hz to 5 MHz with AC amplitude of 1.0 Volt. In the plot, an imaginary impedance, Z_i at vertical axis versus real impedance, Zr at horizontal with same scale, and the bulk resistance (R_b) of polymer electrolyte is obtained by the intercept of semicircle with the x-axis (Z_r) [14]. Hence, The ionic conductivity (σ) of polymer electrolyte membrane can be calculated by using equation (1) :

$$\sigma_{dc} = \left(\frac{1}{Rb}\right) x \left(\frac{d}{A}\right) \tag{1}$$

Where R_b is bulk resistance, d is the thickness of membrane and A is the surface area of contact [19].

IOP Conf. Series: Materials Science and Engineering **924** (2020) 012034 doi:10.1088/1757-899X/924/1/012034

The microstructural studies were analyzed at room temperature by Empyrean PANalytical XRD (45 kV, 40 mA) with Cu-Kα radiation (λ = 1.5418Å) and scattering range angle 2θ = (5-60)° with a step size 0.02 deg/step.

3. Results and Discussion

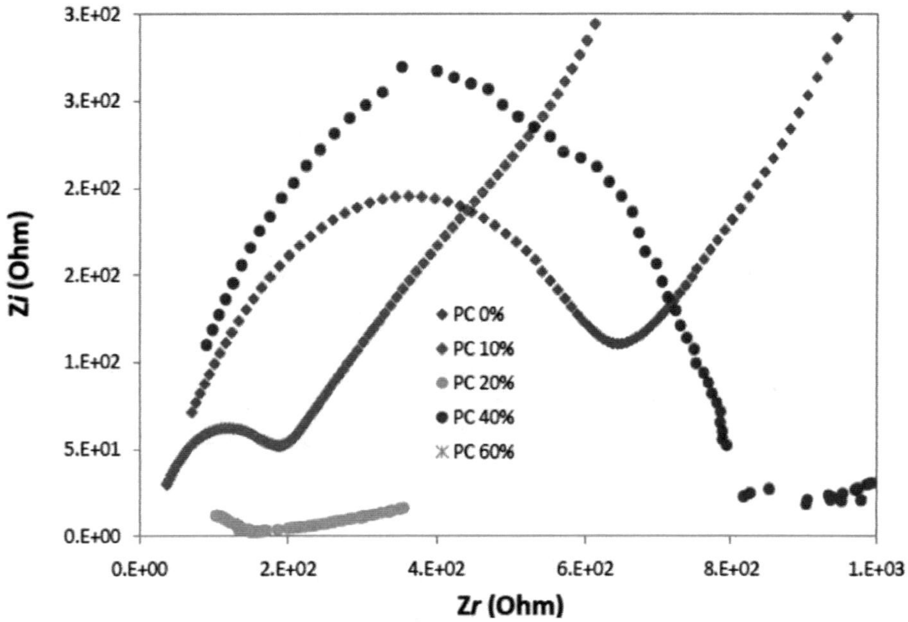

Figure 1. Impedance plot of complex PCL/LiCF$_3$SO$_3$ and PC plasticizer with varying composition at ambient temperature.

Figure 1 only shows the impedance plot for the highest conductivity sample, and Cole-Cole plot of sample with PC 60% could not be seen because the value is out of range of the graph. Figure 1 shows a typical cole-cole plot for complex PCL/LiCF$_3$SO$_3$ with various concentration of PC plasticizer. The complex impedance basically consist of (i) a title spiked, (ii) a depressed semicircle or (iii) a combination of depressed semicircle and title spike. From the plot, it can be observed that these polymer electrolyte mostly show a combination of depressed semicircle in high frequency region and an inclined spike at lower frequencies [20].

Figure 2 show ionic conductivity of all samples as a function of PC concentration. The ionic conductivity of membrane PCL/LiCF$_3$SO$_3$ (PC 0%) has value 1.4 x 10^{-5} S/cm. The PC plasticizer addition to complex PCL/LiCF$_3$SO$_3$ as much 10 w.t% and 20 w.t% can increase the ionic conductivity by one order and get maximum value of 4.92 x 10^{-4} S/cm by 20 w.t% plasticizer addition. The increase of ionic conductivity can take place because PC plasticizer could lessen the Coulomb force between cation and anion in the lithium salt that is why more salt could dissociate becames free mobile ions. In addition, the plasticizer addition can create more pathway for ion conduction and also can increase ionic mobility [7,14]. The plasticizer addition made the polymer electrolyte system more amorphous, hence reduces the energy barriers which enhances the mobility of the ion. However, the ionic conductivity decreased further than 40 w.t% PC plasticizer due to the accumulation of plasticizer cluster, which can block the ions transport within the polymer matrix. Shaima et.al. reported that while EC plasticizer was added to CuI/PVA nanocomposite, the ionic conductivity would increase from 1 x 10^{-7} to 1 x 10^{-5} S/cm [14]

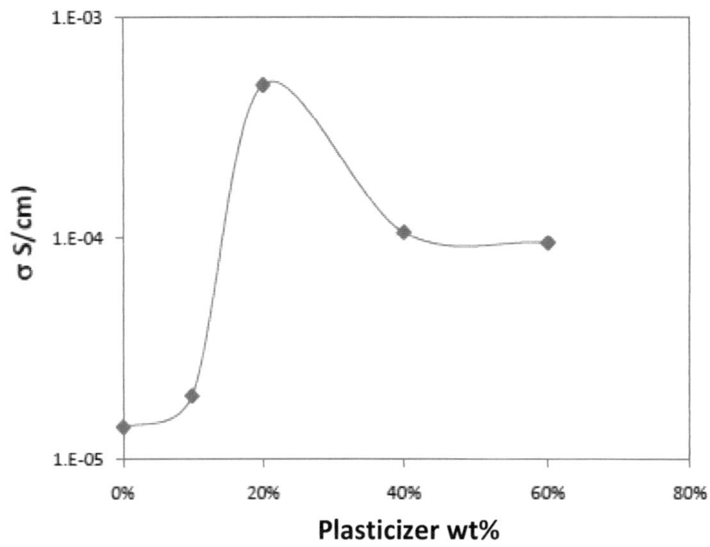

Figure 2. Ionic Conductivity of PCL/$LiCF_3SO_3$ as a function of PC plasticizer concentration.

The best way to describe the ionic conductivity behaviour of polymer electrolyte is by studying its permittivity (dielectric properties). Dielectric constant (ε_r), charge carrier (n) and the dissociation energy (U) are associated to the equation :

$$\eta = \eta_0 \exp\left(-U/(\varepsilon_r kT)\right) \tag{2}$$

From the equation it can be described that, when the dielectric constant (ε_r) increases, the number of charge carrier increases. In polymer electrolytes, ions are the charge carrier where the dielectric constant represents the stored charge in material. While dielectric loss (ε_i) is a measure of energy losses to move ions and align dipoles when the polarity of dielectric field reverses rapidly. From the impedance measurement, ε_r *(ε')* and ε_i *(ε")* can be calculated using the following equation [21]:

$$\varepsilon' = \frac{Z'}{\omega C_o(Z'^2 + Z''^2)} \tag{3}$$

$$\varepsilon'' = \frac{Z''}{\omega C_o(Z'^2 + Z''^2)} \tag{4}$$

Where $C_0 = \varepsilon_0 A/t$ where ε_0 is the permittivity of the free space, A is the area of the sample and t is the thickness of the sample and ω is equal to $2\pi f$ which f in frequency taken from the data in Hz [20].

The frequency dependence of dielectric function of solid materials reflected the dynamic response of the constituents of the solid. Figure 3(a) and (b) show the frequency dependence of the dielectric constant (*ε'*) and the dielectric loss (*ε"*) corresponding to PCL/$LiCF_3SO_3$ polymer electrolyte added PC plasticizer. All sample show similar characteristic in which the ε' is higher at lower frequency due to orientation of polar groups, ions and space charges polarization at these frequencies. In the low frequency region is observed dispersion with a high value of ε' and ε" which is attributed to the dielectric polarization effect Even so, when the frequency is increased, dipoles are not succeed to make parallel with the applied field and they are oscillating begins to lag behind and the ε' start decrease [22,23].

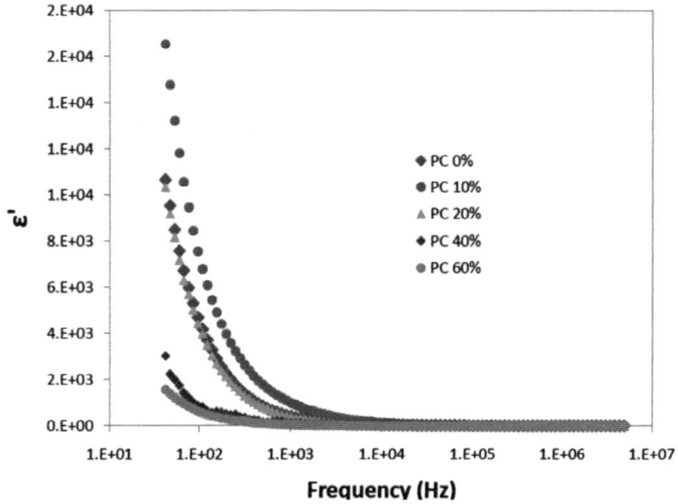

(a)

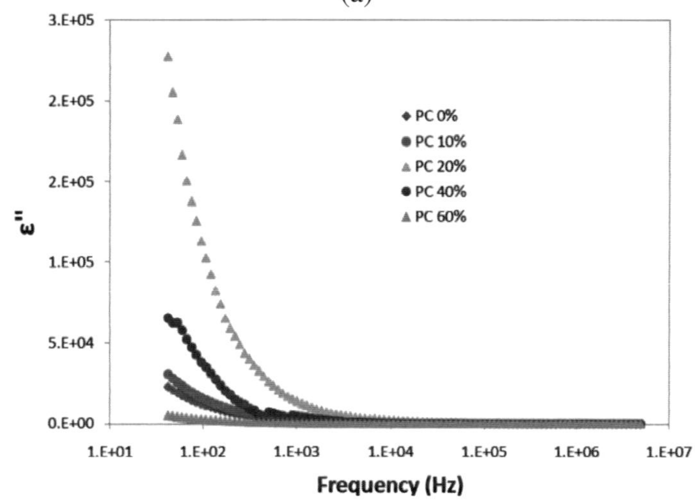

(b)

Figure 3. Frequency dependence of (a) the dielectric constant (ε') and (b) the dielectric loss (ε") corresponding to PCL/$LiCF_3SO_3$ polymer electrolyte added PC plasticizer.

More over the dielectric loss (ε") becomes very large at lower frequencies due to free charge motion within the solid materials. The larger value of dielectric loss at low frequencies is occured as a consequence of the existence of mobile charges within the polymer backbone beside the interfacial polarization at PCL and $LiCF_3SO_3$ interfaces. At higher frequencies region, dielectric constant decreases rapidly and becoming frequency independent. This is because the charge carriers and dipoles in the polymer chain find it hard to translate and orient, respectively, according to the direction of the applied field. At higher frequency, the electric field periodic reversal occurs so fast that there is no excess ion diffusion in the direction of the field [24,25].

Propilene carbonate that has high dielectric constant is able to dissociate more lithium salt into cations and anions that cause an increasing in number density of mobile ions. This indicates that the increase in conductivity is due to the increase in the concentration of mobile ions. The polarization is due to the charge accumulation decrease, leading to the decrease in the value of ε' and ε". Yusof et al.

also reported that the more plasticizer was added, the more salt dissociate become free ions, therefore increased the stored charge in the solid electrolyte [25].

Figure 4 (a) and (b) depict the variation of real part, M' and imaginary part, M'' of electric modulus respectively corresponding to PCL/$LiCF_3SO_3$ polymer electrolyte added plasticizer. The equations for real electrical modulus M' and imaginary modulus M" can be expresses as follow:

$$M' = \varepsilon' / (\varepsilon'^2 + \varepsilon''^2) \tag{5}$$
$$M'' = \varepsilon'' / (\varepsilon'^2 + \varepsilon''^2) \tag{6}$$

At low frequency regions, both M' and M" tend towards zero values. The long tail at lower frequency region is attributed to the large capacitance associated with the electrodes. The curves show that both of M' and M" increase as the frequencies increase, the dispersions are found at higher frequencies. The presence of the peaks in the modulus at higher frequencies for all polymer complexes implies that the polymer electrolyte membranes are ionic conductor [26].

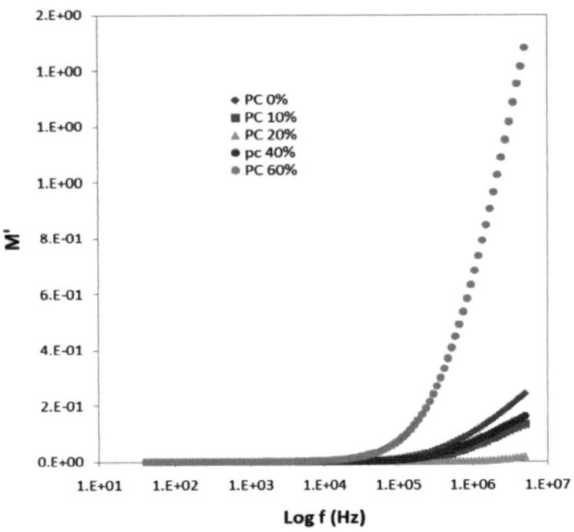

(a)

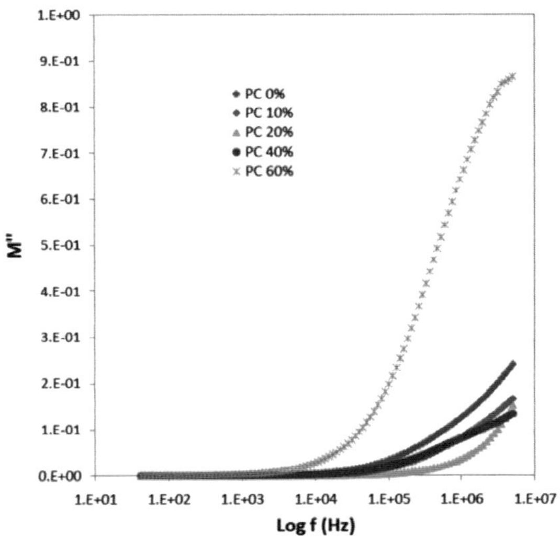

(b)

Figure 4. Frequency dependence of (a) real part of Modulus (M') and (b) imaginary part of Modulus (M") corresponding to PCL/$LiCF_3SO_3$ polymer electrolyte added PC plasticizer.

The observed dispersion is mainly due to conductivity relaxation spreading over range frequencies and indicates the presence of a relaxation time. The absence of peak in M' diagram is due to the fact that M' in complex dielectric modulus is equivalent to ε'. In Figure 4(a) is shown that the highest conductivity sample (20 w.t% PC) has shifted to higher frequency and has the lowest M' value results from the increase in the mobility of the polymer segment and charge carriers with the plasticizer addition. As mentioned before plasticizer addition made polymer electrolyte more amorphous that make the orientation of the charge carriers and molecular dipoles become easier.

The M" curves of the ion conducting electrolyte exhibit a relaxation peak in high frequency region corresponding to the ionic conductivity relaxation time. In Figure 4(b) there are no peak of M" can be observed that is due to the experimental frequency limitation. So it couldn't be determined the exact value of ionic conductivity relaxation frequency (f_σ) and ionic conduction relaxation time (τ_σ) [20,26].

Figure 5 depicts the variation in tangent loss as a function of frequency of PCL/$LiCF_3SO_3$ membrane for different concentration of PC plasticizer and measured at room temperature. The loss spectra characterized by peak appearing at high frequency region for both plasticizer free PCL/$LiCF_3SO_3$

and plasticized PCL/LiCF_3SO_3 samples. This suggests the existence of dipole dipole relaxation in all the samples. Relaxation occurred as a result of ionic charge carriers within polymer materials obeying the change direction of the applied field. From Figure 5 it is observed that the peak frequency shifted towards higher frequency as the plasticizer concentration increases until 20 w.t%. The peak shift towards higher frequency indicates quicker relaxation time. The peak intensity suggests the easiness of ion movement within polymer matrix. Beyond 40 w.t% plasticizer addition the tangent δ peaks shift towards lower frequency. This due to the excessive plasticizer will accumulate to make cluster that make ion movement become slower. On addition of plasticizer there is increasing in the amorphous phase in the PCL membrane as confirmed by XRD data below. The small and mobile PC molecule can accelerate the segmental motion by increasing the existing free volume. Thus, the relatively fast segmental motion coupled with mobile ions enhances the transport properties on plasticization [15,27].

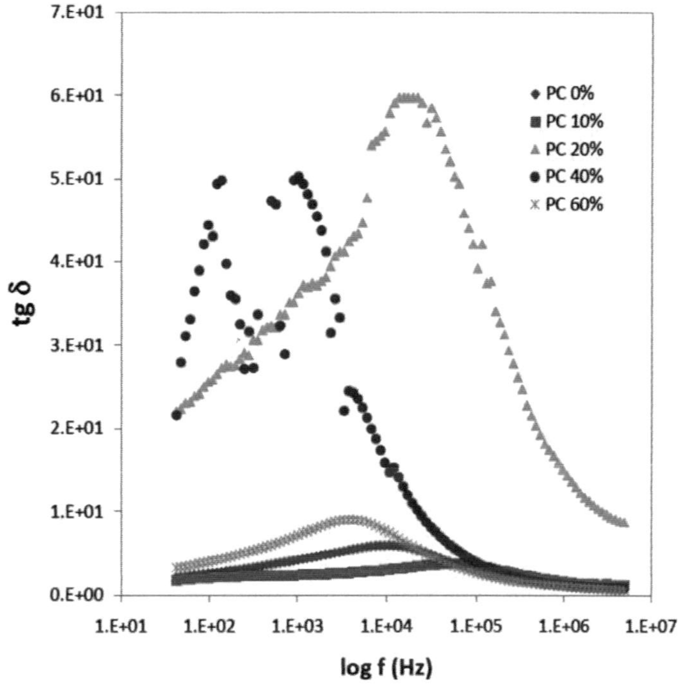

Figure 5. Frequency dependence of tangent δ corresponding to PCL/LiCF_3SO_3 polymer electrolyte with different concentration of PC plasticizer.

Figure 6 depict the XRD pattern of PCL/LiCF_3SO_3 complexes polymer electrolyte with various compositions of PC plasticizer measured at room temperature. There are no difference XRD pattern with the previous research by Yulianti [28]. There are three sharp diffraction peaks in pure PCL and these peaks seem at $2\theta = 21.2$, 22.1 and 23.79 angles which are plane of (110), (111), and (200) of the orthorhombic crystal system respectively. These sharp peaks are recognized as the crystalline phase of PCL, which comes from the arrangement of polymer side chains caused by the intermolecular interaction between PCL chains through the hydrogen bonding. Besides to these crystalline peaks, there is a small continuous spectrum centered on $2\theta = 21.0°$ that confirms the film is semicrystalline consisting of both crystalline and amorphous phases.

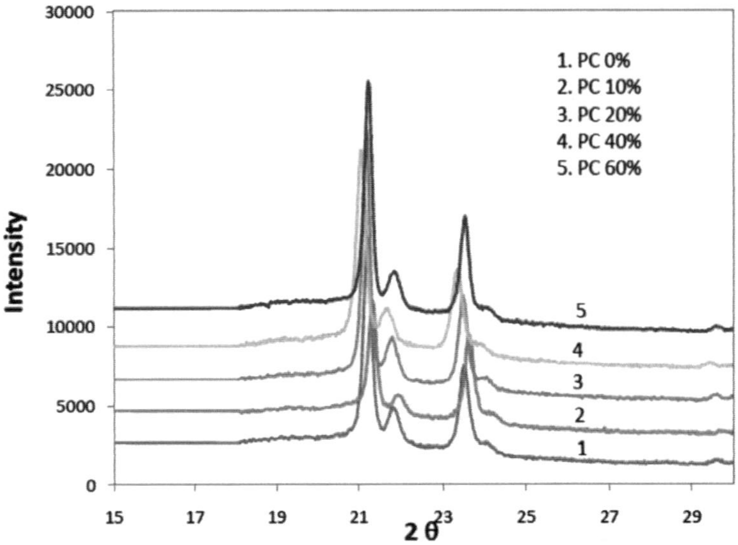

Figure 6. XRD pattern of PCL/$LiCF_3SO_3$ complexes with PC plasticizer addition at different composition .

The addition of PC plasticizer to the polymer complexes PCL/$LiCF_3SO_3$ have not much changed in the PCL X ray diffraction pattern. The addition of 10 wt.% plasticizer has not changed the PCL diffraction pattern significantly, there are slightly broadening of the three peaks. The addition of 20% plasticizer and more causes the diffraction peaks intensity decreasing and the diffraction peaks became broader. This indicates that there is phase transformation from crystalline into an amorphous phase due to polymer plasticization leads to polymer chains separation followed by the structure rearrangement. The more the amorphous phase is formed the conductivity of the electrolyte polymer will increase. The plasticizer addition can create more free volume in polymer matrix because swelling effect so that Li ions transport become easier. The result inferred from XRD study is in good agreement with conductivity measurements as described previously.

4. Conclusion

Solid polymer Electrolyte based on PolyCaproLactone/$LiCF_3SO_3$ complexes with PC plasticizer addition have been success prepared by solution casting method. The electrical properties of membrane polymer electrolyte consist of ionic conductivity, dielectric permittivity, electrical modulus and tangent loss have been studied to investigate the effect of plasticizer addition. The PCL/ $LiCF_3SO_3$ complex with 20 w.t% PC plasticizer addition has the highest ionic conductivity value of 4.92×10^{-4} S/cm, increase by one order of magnitude from 1.40×10^{-5} S/cm for unplasticized membrane. The dielectric permittivity study shows that the PC plasticizer addition proposes that number of charge carriers influences the increase in conductivity significantly.

References

[1] C. P. Fonseca, F. Cavalcante, F. A. Amaral, C. A. Zani Souza, and S. Neves, "Thermal and Conduction Properties of a PCL-biodegradable Gel Polymer Electrolyte with LiClO 4 , LiF 3 CSO 3 , and LiBF 4 Salts," *Int. J. Electrochem. Sci*, vol. 2, pp. 52–63, 2007.

[2] N. H. Ahmad and M. I. N. Isa, "Ionic Coductivity and Electrical Properties of Carboxymethyl Cellulose - NH 4 Cl Solid Polymer Electrolytes," *J. Eng. Sci. Technol.*, vol. 11, no. 6, pp. 839–847, 2016.

[3] E. Yulianti, A. K. Karo, and L. Susita, "Synthesis of Electrolyte Polymer Based on Natural Polymer Chitosan by Ion Implantation Technique," *Procedia Chem.*, vol. 4, pp. 202–207,

IOP Conf. Series: Materials Science and Engineering **924** (2020) 012034 doi:10.1088/1757-899X/924/1/012034

2012.

[4] N. I. Harun, R. M. Ali, A. M. M. Ali, and M. Z. A. Yahya, "Conductivity studies on cellulose acetate – ammonium tetrafluoroborate based polymer electrolytes," *Mater. Res. Innov.*, vol. 15, pp. 168–172, 2011.

[5] F. N. Jumaah, N. N. Mobaraka, A. Ahmad, N. Ramli, F. N. Jumaah, N. N. Mobaraka, and A. Ahmad, "Characterization of ɽ -carrageenan and its derivative based green polymer electrolytes Characterization of Ų -Carrageenan and Its Derivative Based Green Polymer Electrolytes," *AIP Conf. Proceeding*, vol. 768, no. 2013, 2014.

[6] S. Rajendran, M. Sivakumar, and R. Subadevi, "Li-ion conduction of plasticized PVA solid polymer electrolytes complexed with various lithium salts," *Solid State Ionics*, vol. 167, no. 3–4, pp. 335–339, 2004.

[7] M. Y. A. Rahman, A. Ahmad, T. K. Lee, Y. Farina, and H. M. Dahlan, "Effect of Ethylene Carbonate (EC) Plasticizer on Poly (Vinyl Chloride)-Liquid 50% Epoxidised Natural Rubber (LENR50) Based Polymer Electrolyte," *Mater. Sci. Appl.*, vol. 2, no. July, pp. 818–826, 2011.

[8] M. F. Z. Kadir, Z. Aspanut, S. R. Majid, and A. K. Arof, "FTIR studies of plasticized poly(vinyl alcohol)-chitosan blend doped with NH4NO3 polymer electrolyte membrane," *Spectrochim. Acta - Part A Mol. Biomol. Spectrosc.*, vol. 78, no. 3, pp. 1068–1074, 2011.

[9] T. Winie, N. S. M. Hanif, C. H. Chan, and A. K. Arof, "Effect of the surface treatment of the TiO2 fillers on the properties of hexanoyl chitosan/polystyrene blend-based composite polymer electrolytes," *Ionics (Kiel).*, vol. 20, no. 3, pp. 347–352, 2014.

[10] S. B. Aziz and Z. H. Z. Abidin, "Ion-transport study in nanocomposite solid polymer electrolytes based on chitosan: Electrical and dielectric analysis," *J. Appl. Polym. Sci.*, vol. 41774, p. n/a–n/a, 2014.

[11] S. Navaratnam, K. Ramesh, and W. J. Basirun, "Investigation of ion conducting behaviour of composite chitosan based polymer electrolytes," *Mater. Res. Innov.*, vol. 15, no. 1, pp. 184–186, 2011.

[12] Sudaryanto, E. Yulianti, and Patimatuzzohrah, "Structure and properties of solid polymer electrolyte based on chitosan and ZrO2 nanoparticle for lithium ion battery," in *AIP Conference Proceeding*, 2016, vol. 1710, pp. 020003–1.

[13] M. A. Ramlli, M. N. Chai, and M. I. N. Isa, "Influence of Propylene Carbonate as A Plasticizer in CMC-OA Based Biopolymer Electrolytes: Conductivity and Electrical Study," *Adv. Mater. Res.*, vol. 802, pp. 184–188, 2013.

[14] S. A. Mohamed, A. A. Al-Ghamdi, G. D. Sharma, and M. K. El Mansy, "Effect of ethylene carbonate as a plasticizer on CuI/PVA nanocomposite: Structure, optical and electrical properties," *J. Adv. Res.*, vol. 5, no. 1, pp. 79–86, 2014.

[15] S. K. Deraman, N. S. Mohamed, and R. H. Y. Subban, "Conductivity and electrochemical studies on polymer electrolytes based on poly vinyl (chloride) - ammonium triflate-ionic liquid for proton battery," *Int. J. Electrochem. Sci.*, vol. 8, no. 1, pp. 1459–1468, 2013.

[16] H. J. Woo and A. K. Arof, "Vibrational studies of flexible solid polymer electrolyte based on PCL-EC incorporated with proton conducting NH4SCN," *Spectrochim. Acta - Part A Mol. Biomol. Spectrosc.*, vol. 161, no. March, pp. 44–51, 2016.

[17] H. J. Woo, C.-W. Liew, S. R. Majid, and a. K. Arof, "Poly(-caprolactone)-based polymer electrolyte for electrical double-layer capacitors," *High Perform. Polym.*, vol. 26, no. 6, pp. 637–640, 2014.

[18] K. Sownthari, "Synthesis and characterization of an electrolyte system based on a biodegradable polymer," *Express Polym. Lett.*, vol. 7, no. 6, pp. 495–504, 2013.

[19] M. Prabu, S. Selvasekarapandian, A. R. Kulkarni, G. Hirankumar, and C. Sanjeeviraja, "Conductivity and dielectric studies on LiCeO2," *J. Rare Earths*, vol. 28, no. 3, pp. 435–438, 2010.

[20] S. Misenan, "Conductivity , Dielectric And Modulus Studies of Methylcellulose-NH4TF Eurasian Journal of Conductivity , Dielectric And Modulus Studies of Methylcellulose-NH 4

TF Polymer," no. December, 2018.

[21] Y. Chernyak, "Dielectric Constant, Dipole Moment, and Solubility Parameters of Some Cyclic Acid Esters," *J. Chem. Eng. Data*, no. 1, pp. 416–418, 2006.

[22] Y. M. Yusof, N. A. Majid, R. M. Kasmani, H. A. Illias, and M. F. Z. Kadir, "The Effect of Plasticization on Conductivity and Other Properties of Starch/Chitosan Blend Biopolymer Electrolyte Incorporated with Ammonium Iodide," *Mol. Cryst. Liq. Cryst.*, vol. 603, no. 1, pp. 73–88, 2014.

[23] P. Dhatarwal and R. J. Sengwa, "Effects of PEG plasticizer concentrations and film preparation methods on the structural , dielectric and electrical properties of PEO – PMMA blend based plasticized solid polymer electrolyte films," vol. 55, no. January, pp. 7–18, 2017.

[24] M. F. Shukur, Y. M. Yusof, S. M. M. Zawawi, H. a Illias, and M. F. Z. Kadir, "Conductivity and transport studies of plasticized chitosan-based proton conducting biopolymer electrolytes," *Phys. Scr.*, vol. T157, p. 014050, 2013.

[25] Y. M. Yusof, H. A. Illias, and M. F. Z. Kadir, "Incorporation of NH4Br in PVA-chitosan blend-based polymer electrolyte and its effect on the conductivity and other electrical properties," *Ionics (Kiel).*, vol. 20, no. 9, pp. 1235–1245, 2014.

[26] S. Fadli, M. Zamri, F. A. Latif, A. Malik, M. Ali, N. Kamaluddin, and F. Hadip, "Ionic conductivity and dielectric properties of LiBF 4 doped PMMA / ENR 50 filled acid modified SiO 2 electrolytes," *Procedia Technol.*, vol. 15, pp. 849–855, 2014.

[27] E. Yulianti, Sudaryanto, H. Jodi, Deswita, and Mashadi, "Li Ion Conducting Polymer Based on Polymer Polyvinylidene Fluoride and Li Triflate," *J. Sains Mater. Indones.*, vol. 15, no. 1, pp. 24–28, 2013.

[28] E. Yulianti, "Study of solid polymer electrolyte based on biodegradable polymer polycaprolactone," vol. 15, no. 3, pp. 467–471, 2019.

Electrochemical Impedance Spectroscopy Study of $LiFePO_4$/Li during Discharging Process

Sudaryanto*, Yustinus Purwamargapratala, Evi Yulianti, Wagiyo Honggowiranto, and Evvy Kartini
Center for Science and Technology of Advanced Materials-BATAN
Building 71, PUSPIPTEK Area, Serpong, South Tangerang, Banten 15314, Indonesia

*dryanto@batan.go.id

Abstract. Electrochemical impedance spectroscopy (EIS) study has been applied to investigate the electrochemical properties of $LiFePO_4$/Li half-cell at initial state (before and after charge/discharge process), and a certain dept of discharge (DoD). The impedance spectra were collected over the frequency range from 42 Hz to 5 MHz. The EIS spectra of $LiFePO_4$/Li half-cell show the appearance of semicircles in which the diameter increased with DoD. An equivalent circuit was constructed to each spectrum in order to elucidate to electrochemical properties changes of $LiFePO_4$/Li half-cell battery as function of DoD. The information of bulk and interfacial resistances are obtained by modeling with the equivalent circuits according with the Nyquist plots of each spectrum. The experiment results show the increase in the resistance of $LiFePO_4$/Li half-cell with the DoD.

Keywords: Electrochemical impedance spectroscopy, Lithium ion batteries, Depth of discharge, $LiFePO_4$

1. Introduction

Lithium-ion batteries (LIB) are the most promising power system, since they exhibit higher operating voltage and energy density. Therefore, many intensive studies are being performed by many researchers, in order to develop a new battery system and to improve its performance [1].

Lithium iron phosphate ($LiFePO_4$) or LFP (lithium ferrophosphate), an olivine-type cathode material, has great potential application in lithium-ion batteries because of having high theoretical capacity (170 $mAhg^{-1}$), low manufacturing cost, good thermal stability, high reversibility Li-ion intercalation/de-intercalation, and high safety, as well as environmental friendly [2-4]. However, the $LiFePO_4$ has poor rate capability due to poor electronic conductivity (10^{-9} - 10^{-10} S cm^{-1}) and low Li^+ diffusion rate (10^{-4} - 10^{-12} cm^2 s^{-1}). Therefore, many studies have been conducted on the material preparation and characterization as well as the electrochemical performances of $LiFePO_4$ in order to improve their electronic and ionic transport properties [2,5-7].

The electrochemical properties of battery system can be studied by several techniques such as using cyclic voltammetry (CV), charge/discharge (CD), and electrochemical impedance spectroscopy (EIS). Recently, many researches used EIS technique, to study the electrode materials because it can explain the relationship between the micro-structure and the electrochemical properties [8]. For example, EIS had been used to study the reversibility of Li-ion intercalation/de-intercalation processes of $LiFePO_4$ [2,9,10]. However, most studies of eletrochemical performance using EIS were conducted using full cell system. In other hands, a half-cell battery where metalic lithium foil is used as counter electrode (anode), the lithium foil is large enough, so that it can be considered almost does not affect the EIS behavior [8]. Thus, it can simplify the model when discussing the experimental data of EIS spectra.

In this study, EIS study of $LiFePO_4$/Li half-cell was conducted in order to understand the effect of charge/discharge process on the electrochemical performance of lithium ion batteries. The effect of charging was studied by measuring EIS of the battery cell before and after charging/discharging. While the effect of discharging was studied by taking the EIS spectra from battery cell with various depth of discharge (DoD). The DoD was defined as the percentage of battery capacity that has been discharged expressed as a percentage of maximum capacity. Equivalent circuit model was used to analyze the EIS spectra.

2. Experimental method

2.1. Cell battery preparation

In this study, a commercially available $LiFePO_4$ powder obtained from MTI was used as cathode active material. The powder was mixed with conductive carbon (acetylene black) and Polyvinylidene fluoride (PVDF) dissolved in N-Methyl-2-pyrrolidone (NMP) in the weight ratio of 91:4:5 to form slurry. The slurry was then coated on aluminum foil in a certain thickness using doctor blade. After solvent evaporation, the cathode sheet was pressed and dried at 120 °C under vacuum oven for 24 h. CR2032-type coin cells were assembled in a glove box (Vigor, China) with H_2O content was controlled below 1 ppm. Metallic lithium foil 1.0 cm^2 was used as anode. The electrolyte was commercially availabe 1.0 M $LiPF_6$ dissolved in a mixture of ethylene carbonate (EC), ethylmethyl carbonate (EMC) and dimethyl carbonate (DMC) (1:1:1, v/v) and Celgard 2400 polyethylene was used as the separator.

2.2. Cell characterization

The performance of the prepared coin cell was examined using the charging and discharging activities (CD) and electrochemical impedance spectroscopy (EIS). A BST8-3 8 Channel Battery Analyzer used to examine the charge/discharge performance and to prepare the sample at a certain depth of discharge (0-100 % DoD). The measurement was conducted within cut-off voltages of 2.5-4.2V and charge/discharge rate of 0.1C. Here C is theoretical capacity, calculated based on cathode ($LiFePO_4$) active material weight. The DoD sample preparation was performed after 2nd cycle charge/discharge process. The impedance characteristic was studied by using a computer controlled impedance analyzer (HIOKI LCR Hi-Tester Model 3532, Japan) with an amplitude of the AC signal of 1V in the frequency range of 42 Hz to 5 MHz. Before collecting the EIS data, the sample was kept at least 1 hour after charging and discharging up to certain DoD for equilibration.

3. Results and discussion

Figure 1 shows the 1st and 2nd charge and discharge of the assembled $LiFePO_4$ coin half-cell. It was observed that the discharged capacity at 1st cycle was 1.43 mAh. Considering weight of the active material ($LiFePO_4$) inside the coin (12.06 mg) the specific capacity of the coin cell battery was 116 $mAhg^{-1}$ or about 68% of the theoretical specific capacity of $LiFePO_4$ (170 $mAhg^{-1}$). In other hands, the 1st charging capacity was only 1.05 mAh. It means that the battery has charge efficiency 135% at the 1st cycle. The charge capacity or Coulombic efficiency (CE) as a battery parameter to monitor the magnitude of side reactions, has been of great interest in recent years [9]. The charge/discharge efficiency is the ratio of the energy can be taken out from a battery (discharging capacity) divided by the energy can be put in (charging capacity)[9]. Generally, efficiency is a dimensionless number between 0 and 1.0, or 0% to 100%. Efficiencies may not exceed 100%. However, in this study the charge efficiency at 1st cycle was exceed 100% as shown in Figure 2. This can occur because of the formation of solid electrolyte interphase (SEI), which forms in the first charge/discharge cycle, then decreases significantly in the second cycle, and decreases further in the third cycle and so on. Generally, SEI formation is highest in the 1st charge/discharge cycle because pure anodes and cathodes previously did not form a passive layer which electronically isolated electrodes from electrolytes [11]. However, it is depending on negative-to-positive capacity ratio, surface area of particles, operation conditions, etc.[12]. In this study, the charge capacity at the 2nd cycle was 1.73 mAh and the discharge capacity was 1.38 mAh, thus the charge efficiency became 78%. After 2nd cycle and upto 10th cycle the charge efficiency was then stable at 98 – 100 % as shown in Figure 2. Therefore, we considered that after 1st cycle, the SEI at anode was relatively stable, so that further cycling, resulting in much lower SEI formation.

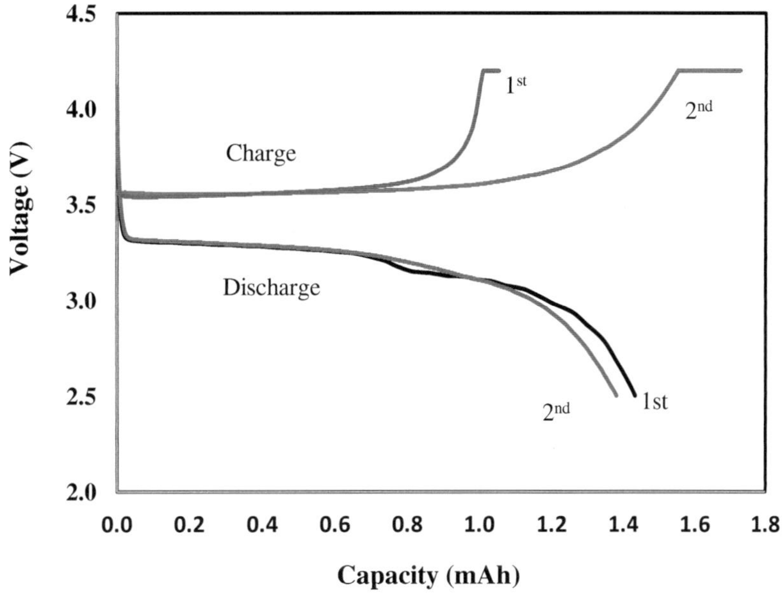

Figure 1. The 1st and 2nd charge/discharge profile of $LiFePO_4$ coin cell.

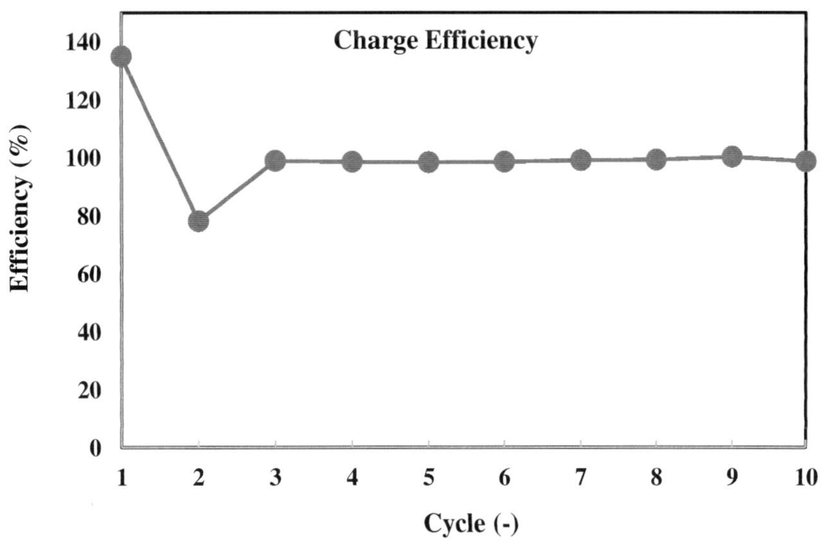

Figure 2. The charge efficiency of $LiFePO_4$ coin cell as function of cycle number.

IOP Conf. Series: Materials Science and Engineering **924** (2020) 012035 doi:10.1088/1757-899X/924/1/012035

During charge and discharge processes, chemical reactions occur at both sides of electrode. The reaction (1) occurs at the cathode or working electrode (iron phosphate)

$$Li^{+} + e^{-} + FePO_4 \leftrightarrow LiFePO_4 \quad , \tag{1}$$

and the reaction (2) occurs at the anode or counter electrode (lithium foil)

$$Li^{+} + e^{-} \leftrightarrow Li\ . \tag{2}$$

During discharge, lithium ions carry the current from the anode to the cathode and the delithiation happens at the anode and lithiation occurs at the cathode. During charging, the reverse happens when an overvoltage is applied. The lithiation and delithiation process may affect the impedance properties of the battery. Therefore we measured the impedance characteristic of $LiFePO_4$/Li half-cell before and after charge/discharge process as shown in Figure 3.

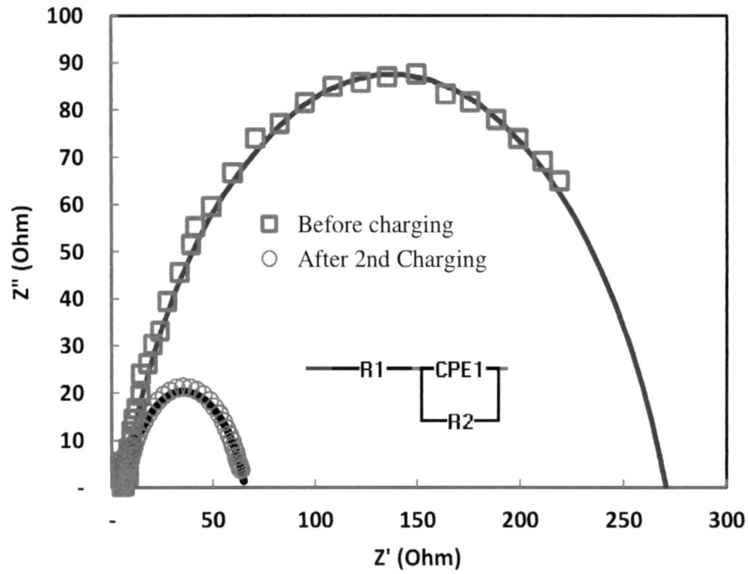

Figure 3. Nyquist Plot of $LiFePO_4$/Li half-cell impedance experimental data and their equivalent circuits.

Figure 3 displays the EIS spectra of $LiFePO_4$/Li half-cell before applied initial charging and after 2nd charging. The solid lines indicate the simulation curve when an equivalent circuit as shown in the figure is employed. The Nyquist plots show the appearance of semicircle for both before and after charging. However, the Nyquist plots at before charging are different from that at after 2nd cycle. The semicircle at the after charging is much smaller. This result shows the same tendency with previous report by Li Wang *et al.* [8]. The difference in radius between the before and after charging is revealing that the charge/discharge process changed the resistance of the battery. In order to investigate the changes in the EIS spectra of $LiFePO_4$/Li half-cell due to charging process, an equivalent circuits are employed as displayed in the Figure 3. The lithium foil as anode is large enough, so it can be considered almost does not affect the EIS behavior. Therefore, the total impedance could be regarded as the intrinsic resistance (Rs) which is related to the combined resistance of electrolyte and electrode material, charge transfer resistance (Rct) which is originated from the electronic and ionic resistances at the electrode-electrolyte interface, and the double-layer capacitance (Cdl) [8]. In the model of equivalent circuits as shown in Figure 3, R1 and R represent Rs and Rct, respectively. The phase element (CPE) has been introduced in the equivalent circuit instead of double-layer capacitance Cdl. Though the equivalent circuit simulated by the electrochemical impedance software is different from the previous research [8], it can match the measurement results totally. The fitting results are shown by solid line in Figure 3 and the values of R1 and R2 are summarized in Table 1.

Table 1. Nyquist Analysis of $LiFePO_4$/Li half-cell before and after charging.

	Before charging	**After 2nd charging**
R1 (Ohm)	4.6	6.1
R2 (Ohm)	265.9	59.1

Table 1 displays the variation of R1 (Rs) , and R2 (Rct) with charging process, respectively. It can be found that Rct decreases after charging process, revealing that the charge/discharge process enhance the transference performance of lithium-ion in the electrolyte. There is no obvious differences of Rs at before and after charging since it come from the intrinsic resistance of the electrolyte and electrode.

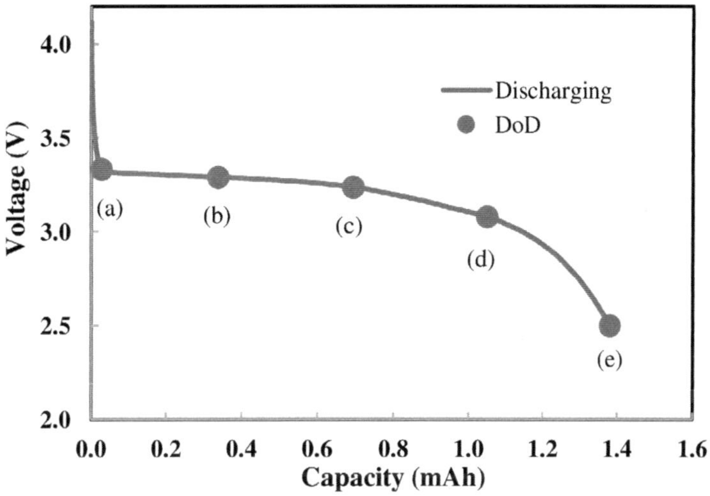

Figure 4. The discharging process (solid line) and the discharging stop position (circle mark) to obtain a certain DoD (a) (0%) (b) 25% (c) 50% (d) 75% and (e) 100% of $LiFePO_4$/Li half-cell.

In order to study the EIS spectra of $LiFePO_4$/Li half-cell during charging process, we measured the EIS of the battery cell with various DoD as shown in Figure 4. As mentined before, DoD indicates the percentage of the battery that has been discharged relative to the overall capacity of the battery. Practically the samples were prepared by discharging the battery at a certain time with considering the time used to discharge until lower voltage cut off. For example, when the total discharging time is 10 hours, to obtain 25% DoD the battery is discharged for 2.5 hours at the same condition. After keeping for at least 1 h at room temperature, and measuring the EIS, the sample then fully charged and used for other DoD sample with similar manner. The EIS spectra of $LiFePO_4$/Li half-cell at each DoD are shown in Figure 5.

Figure 5 shows the Nyquist plots of EIS spectra of $LiFePO_4$/Li half-cell at various DoD. The solid lines indicate the simulation curve when equivalent circuits are employed. The Nyquist plots show the appearance of semicircle for all DoD. However, the semicircles are different one and others in the radius. The radius of the semicircle increased with the DoD. In order to investigate the changes in the EIS spectra of $LiFePO_4$/Li half-cell due to discharging process, equivalent circuits are employed to fit the experimental data. The circuits used same with that for Figure 3, consisted of the intrinsic resistance (R1), charge transfer resistance (R2) and the double-layer capacitance which represented by CPE. It is clear that the equivalent circuit simulated by the electrochemical impedance model (solid line) are perfectly matched with the measurement results. The values of R1 and R2 are summarized in Table 2.

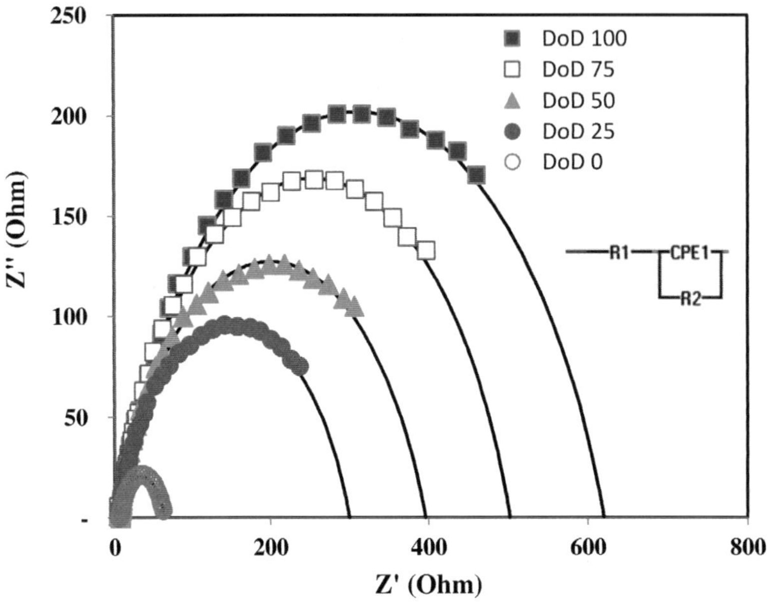

Figure 5. Nyquist Plot of $LiFePO_4$/Li half-cell impedance experimental data and their equivalent circuits.

Table 2 demonstrated the variation of R1, and R2 with various DoD. It is clear that there are no significant differences of R1 at entire DoD. R1 is the intrinsic resistance (Rs) which is related to the combined electrolyte and electrode material resistances, thus it may not be affected by charge/discharge process. While the R2 increased with the DoD. The higher the DoD the higher the R2. Here R2 is charge transfer resistance (Rct) which is originated from the electronic and ionic resistances at the electrode/electrolyte interface. The increase in Rct, revealing that the discharge process decreased the transference performance of lithium-ion in the electrolyte. In general, internal resistance of battery cell includes (a) ohmic resistance, (b) charge transfer resistance or active resistance, and (c) diffusion resistance known as concentration resistance [13,14]. Hence the increase in charge transfer resistance, enhance the battery internal resistance. In other word, internal resistance of battery growth along with the discharging process.

Table 2. Nyquist Analysis of $LiFePO_4$/Li half-cell with various DoD.

DoD	0	25	50	75	100
R1 (Ohm)	6.1	4.5	4.2	4.3	4.5
R2 (Ohm)	59.1	295.2	391.0	498.9	616.0

4. Conclusion

EIS measurements of assembled $LiFePO_4$/Li half-cell have been conducted at initial state (before and after charge/discharge process), and a certain depth of discharge (0-100%, DoD). Entire EIS spectra of $LiFePO_4$/Li half-cell show the appearance of semicircles. The radius of semicircle decreased with charging and increased with discharging processes. The EIS spectra can be fitted totally with an equivalent circuit, composed of the intrinsic resistance, charge transfer resistance and the double-layer capacitance which is represented by CPE. The charge/discharge did not affect the the intrinsic resistance significantly but affected the charge transfer resistance. It was found that the charge transfer resistance decrease with charging and increased with discharging. Since charge transfer resistance is a part of internal resistance, it was clear that the dischaging or capacity loss increased the internal resistance of the $LiFePO_4$/Li half-cell.

Acknowledgment
This study was carried out under DIPA Budget of Center for Science and Technology of Advanced Materials – BATAN number 3446.001.001. The authors would like to thank researchers as well as technicians at Center for Science and Technology of Advance Materials, BATAN, for their supports.

References

[1] B. Scrosati and J. Garche, "Lithium batteries: Status, prospects and future," *J. Power Sources*, vol. 195, no. 9, pp. 2419–2430, 2010.
[2] C. Park, S. Bin Park, S. H. Oh, H. Jang, and W. Il Cho, "Li Ion Diffusivity and Improved Electrochemical Performances of the Carbon Coated LiFePO 4," vol. 32, no. 3, pp. 836–840, 2011.
[3] B. Wu, Y. Ren, and N. Li, "$LiFePO_4$ Cathode Material", in *Electric Vehicles-The Benefits and Barriers,* Dr. Seref Soylu (Ed.), Sanghai, InTech,, 2011, pp. 199-216. Available from: http://www.intechopen.com/books/electric-vehicles-the-benefits-and-barriers/lifepo4-cathode-material
[4] W. Honggowiranto, Sudaryanto, E. Kartini, and A. Purwanto, "Electrochemical performance of LiFePO4cylinder cell battery," *AIP Conf. Proc.*, vol. 1710, no. February 2016, pp. 1–9, 2016.
[5] C. Hu *et al.*, "Mg Doping of LiFePO 4 by co-Precipitation for Lithium Ion Batteries," vol. 5, pp. 1457–1463, 2010.
[6] K. Hong *et al.*, "Preparation of $LiFePO_4$ Using Chitosan and its Cathodic Properties for Rechargeable Li-ion Batteries," vol. 30, no. 8, 2009.
[7] J. Hu *et al.*, "Doping Effects on Electronic Conductivity and Electrochemical Performance of LiFePO 4," vol. 25, no. 3, pp. 405–409, 2009.
[8] L. Wang *et al.*, "Electrochemical Impedance Spectroscopy (EIS) Study of," *Int. J. Electrochem. Sci.*, vol. 7, no. 2012, pp. 345–353, 2012.
[9] F. Yang, D. Wang, Y. Zhao, K. L. Tsui, and S. J. Bae, "A study of the relationship between coulombic efficiency and capacity degradation of commercial lithium-ion batteries," *Energy*, vol. 145, pp. 486–495, 2018.
[10] Y. Chen and J. Chen, "A Study of Electrochemical Performance of $LiFePO_4/C$ Composites Doped with Na and V," vol. 7, pp. 8128–8139, 2012.
[11] S. J. An, J. Li, Z. Du, C. Daniel, and D. L. Wood, "Fast formation cycling for lithium ion batteries," *J. Power Sources*, vol. 342, no. February 2018, pp. 846–852, 2017.
[12] S. J. An, J. Li, C. Daniel, D. Mohanty, S. Nagpure, and D. L. Wood, "The state of understanding of the lithium-ion-battery graphite solid electrolyte interphase (SEI) and its relationship to formation cycling," *Carbon N. Y.*, vol. 105, pp. 52–76, 2016.
[13] L. P. Fan and J. J. Li, "Overviews on internal resistance and its detection of microbial fuel cells," *Int. J. Circuits, Syst. Signal Process.*, vol. 10, pp. 316–320, 2016.
[14] R. Weber, A. J. Louli, K. P. Plucknett, and J. R. Dahn, " Resistance Growth in Lithium-Ion Pouch Cells with LiNi 0.80 Co 0.15 Al 0.05 O 2 Positive Electrodes and Proposed Mechanism for Voltage Dependent Charge-Transfer Resistance ," *J. Electrochem. Soc.*, vol. 166, no. 10, pp. A1779–A1784, 2019.

Study Performance of $LiFePO_4$/Graphite Cylindrical Pouch Cell

Wagiyo Honggowiranto*, Evvy Kartini and Agus Sudjatno**
Science and Technology Center for Advanced Materials, National Nuclear Energy Agency, Puspiptek Area, Serpong, South Tangerang 15314, Indonesia

*wagiyo@batan.go.id;
**kartini@batan.go.id

Abstract. Cylindrical cell Lithium ion battery with a pouch casing has been made at the Integrated Battery Laboratory, BATAN. LiFePO4 double sided coated aluminum foil with a thickness of 180 μm is used as a cathode sheet. The anode sheet is made from two-sided artificial graphite coated on copper foil with a thickness of 197 μm. The composition, crystal structure, of the coated LiFePO4 was measured by XRD. Battery cells are built by rolling thin layers of cathode, separator, and anode material into cylindrical shape using a rolling machine. The cylindrical cell is inserted into an aluminum bag and then sealed at 175oC. Cylindrical cell-bag inserted into the Glove Box then filled with ~ 2.5 ml of LiPF6 electrolyte liquid. A vacuum sealing machine is used to seal the remainder of the bag set at 175 oC. The performance of lithium ion batteries is characterized by using a battery analyzer. LiFePO4 shows a release capacity of 250.00 mAh, with a specific capacity of 124.10 mAh/g in the 1st cycle and 100.20 mAh/g after 100th cycle, at a rate of 0.3C. The cell shows good performance after 100 cycles with 80.74% retention.

***Keywords*:** $LiFePO_4$; Graphite, Lithium ion batteries , cylinder cell, pouch cell

1. Introduction

One of the most important power sources worldwide is the rechargeable lithium-ion batteries, since their variety applications such as in mobile phones, laptop computers, digital cameras, electrical vehicles and hybrid electrical vehicles. One cell of rechargeable lithium-ion battery contains several components, namely cathode, anode, separator, electrolyte, current collector and connector. The most important components is the cathode material because relating to the performance of the batteries. The most well-known cathode materials, layered $LiCoO_2$ and $LiNiO_2$, spinel $LiMn2O_4$, have been studied extensively [1-3]. $LiCoO_2$ is being used as the cathode material for commercial lithium-ion batteries, but it is toxic and expensive.

The cathode materials of lithium transition metal phosphates $LiMPO_4$ (M = Mn, Fe, Ni, and Co) with an ordered olivine-type structure, has a high theoretical specific capacity (170 mAh/g) [4-7]. However, it is difficult to attain its full capacity because its electronic conductivity is very low, which leads to an initial capacity loss and poor rate capability, and diffusion of Li+ ion across the $LiFePO_4/FePO_4$ boundary is slow due to its intrinsic character. The electronic conductivity of $LiFePO_4$ is ~10^{-9} S cm^{-1}, which is much lower than those of $LiCoO_2$ (~10^{-3} S cm^{-1}) and $LiMn_2O_4$ (2×10^{-5}-5×10^{-5} S cm^{-1}). There are two methods to improve the electronic conductivity. One method is to introduce conductive additives including a carbon coating [8] and supervalent cation doping [3]. Another method is to control the particle size by optimizing the synthesis conditions [6]. $LiFePO_4$ can be synthesized by different methods such as a solid-state reaction [9], a sol-gel method [10], a co-precipitation method, and a hydrothermal method [11,12], and so forth. Most of the studies are related to the improvement of the cathode materials, and performance of the battery, but only in half cell or coin cell.

In 2012, the Science and Technology Center for Advanced Materials, National Nuclear Energy Agency, with the support from the Ministry Research and Technology, Indonesia opened an Integrated Battery Laboratory (miniature pilot plant for battery line production) located in Serpong, Indonesia [13-15]. In this laboratory, the scientists have the opportunity to study all relevant process steps of line production of batteries in form of coin, cylinder or pouch cells [16,17].

The aim this study is to analyst cyclic performance of $LiFePO_4$/graphite occurred in cylindrical pouch cell by using Battery analyser [18,19].

2. Experimental method

2.1. Electrode manufacturing

The first step of battery production line was the electrode manufacturing. The raw materials for electrode processing are commercial powders $LiFePO_4$ as cathode materials, and artificial graphite as anode materials. All cathode and anode powder are commercial from MTI Co. Aluminum foil and copper-foil are used as current collectors. To form slurry, the cathode $LiFePO_4$powders were mixed with artificial carbon and polyvinylidene fluoride (PVDF) that dissolved in N-methyl-2-pyrrolidinone (NMP). The weight ratio of active materials with artificial carbon and PVDF was 80:15:5. About 10 ml slurry was mixed in stainless crucible attached to vacuum mixing equipment. It was mixed for about 30 minutes with a certain speed. Mixing time and energy input are process parameters by means of which particle stability and degree of dispersion can be defined. Similar processes were done for the artificial graphite slurry, but without carbon black.

2.2. Electrode characterization

In order to quantify these electrode sheets in early stage of production, the Al-foil coated with $LiFePO_4$/C (LFP/C) and Cu-foil coated with artificial graphite were characterized by x-ray diffraction. The crystal structure of the $LiFePO_4$/C sheet and artificial graphite sheet were characterized by X-ray diffraction (XRD) using a PANalytical EMPYREAN Series-2, with Cu Kα, scan range from $10^o$to 80°. All those characterizations were done at the Integrated Battery Laboratory, BATAN [8,14].

2.3. Assembly of cylinder pouch cell

$LiFePO_4$ double sided coated aluminum foil with a thickness of 180 μm is used as a cathode sheet. The anode sheet is made from artificial two-sided graphite coated on copper foil with a thickness of 197 μm. Both aluminum and copper foil each function as a collector of positive and negative currents, respectively. Battery cells are built by rolling thin layers of cathode, separator, and anode material into cylindrical shape using a rolling machine. The cylindrical cell is inserted into an aluminum bag and then sealed at 175°C. Negative and positive connectors made of nickel and aluminum are each welded at the top of aluminum and copper. Cylindrical cell-bag inserted into the Glove Box then filled with ~ 2.5 ml of $LiPF_6$ electrolyte liquid. A vacuum sealing machine is used to seal the remainder of the bag set at 175 °C.The c-pouch cell was left for 12 hours in glove box, before taken out for testing its electrochemical cell performance.

In the present work, aluminum case was chosen, since it was simpler, lighter and the dimension could be customized. The product was named as *a cylinder pouch cell or C-pouch cell*, a new form of battery cell.

3. Results and Discussions

3.1. X-ray diffraction of $LiFePO_4$ and graphite sheet

The performance of cathode $LiFePO_4$ after coating, drying and calendaring are analysed by using the x-ray diffraction, as shown in Figure 1.

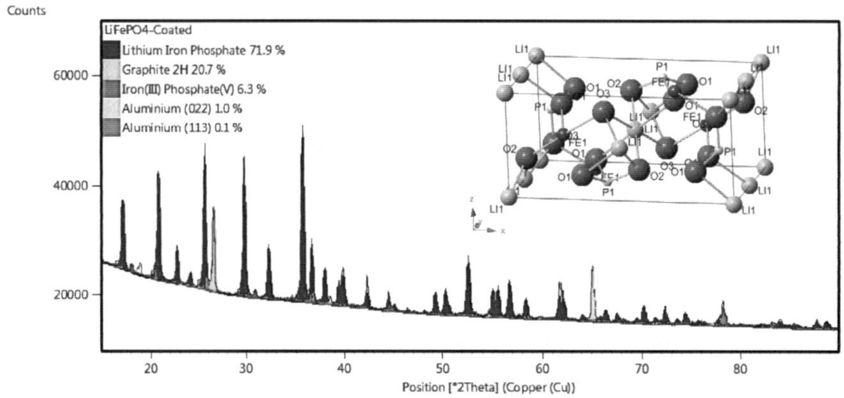

Figure 1. X-ray diffraction pattern of $LiFePO_4$/coated on Al-foil.

Figure 1 shows the x-ray diffraction data of $LiFePO_4$/coated on Al-foil that has been refined using the High-score plus program [14]. The major phase of ~71.9% was from the Lithium Iron Phosphate, $LiFePO_4$, with the olivine structure and space group (No.) *P n m a (62)*. The lattice parameters a, b, and c are 10.322(3) Å, 6.003(2) Å, and 4.691(2) Å, respectively. Another peaks observed were corresponding to graphite, iron phosphate and aluminum phases. To form slurry, 85% $LiFePO_4$, 10% PVDF and 5% carbon black were mixed together in the vacuum mixer. It is known that the $LiFePO_4$ has low ionic conductivity (~10^{-9} S/cm), therefore in order, to increase its electronic conductivity, the addition of carbon black was needed as carbonaceous sources. Due to this reasons the graphite peaks were observed in the x-ray data [16]. The iron phosphate might appear from the dissociation of lithium hydroxide from $LiFePO_4$ during vacuum mixing or coating. Another Al peaks related to aluminum foil current collector, where $LiFePO_4$ was coated on it.

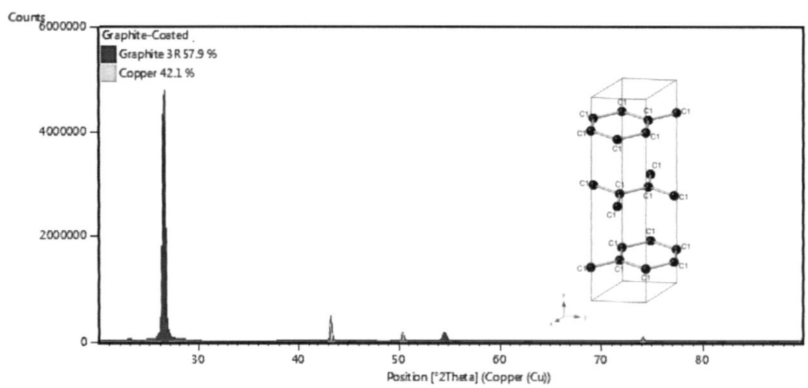

Figure 2. X-ray diffraction pattern of Graphite-coated on Cu-foil.

Figure 2 showed x-ray diffraction pattern of the artificial graphite-coated on Cu foil. The data was compared to the standard graphite material. It was clear that the diffraction peaks corresponding to the graphite ~57.9% and copper 42.1%. The composition of graphite was rather similar with copper caused by the overlapping peak at 2θ~ 26.2559°. The graphite had a space group (No.) P 63/m m c (194), with the lattice parameters a, b, and c are 2.4603(3) Å, 2.4603(3) Å, and 6.726(1) Å, respectively. From crystal structure refinement, the reliability factors Rwp, Re, and X2 are 19.65156, 0.50427 and 1518.70400, respectively. It should be mentioned here that all the preparation was done open air, not in glove box. This was due to larger production of electrode sheets for cylinder cells. Similar refinements have been performed to the electrolyte material Li_3PO_4 [14,16].

3.2. Electrochemical testing

Performance of C-pouch cell was analyzed by using a BTS-8 MTI Battery analyzer. The step setting is shown in table 1. It is important to perform initial charging before doing all the charge discharge cycling. After rest for 1 minute, it was charged with a constant current (CC_Chg) 100 mA until the voltage reached 4 Volt. Then it was kept in this charge at contact voltage (CV_Chg) until the current decreased to 5mA, then it was discharged with a constant current (CC-DChg) 100 mA until the voltage reached 2 Volt and rest for 5 minutes. These steps were cycled for hundredth times to determine its capacities, stability and Coulombic efficiency. The electrochemical performance was also performed for other types of coin and cylinder cells [15,20]. Figure 3 shows the Volt-Current-Time curve during the cycling.

Table 1. Step setting for electrochemical performance of cylinder pouch cell.

		Step setting scheme		
StepNum	**StepType**	**Time(mm:ss)**	**Vol(V)**	**Cur(mA)**
1	Rest	1:00		
2	CC-Chg		4	100
3	CV-Chg		4	5
4	CC-DChg		2	100
5	Rest	5:00		
6	Cycle	Start Step: 1	Cycle Total: 100	
7	Stop			

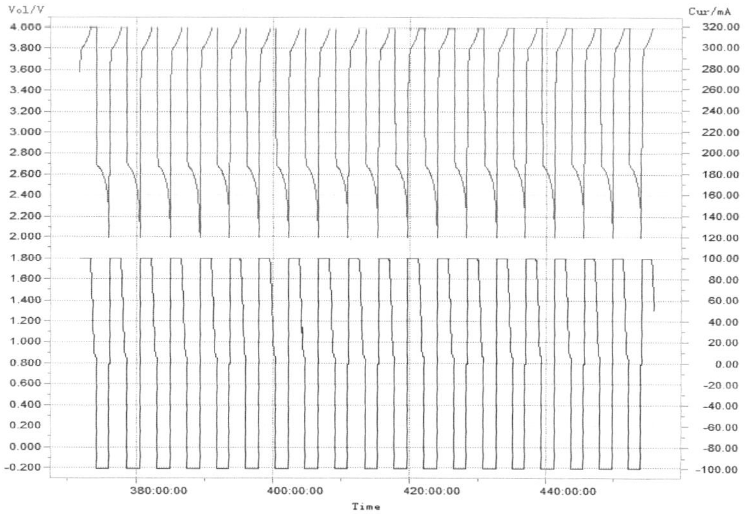

Figure 3. Charge-discharge curve , voltage and current versus time of C-pouch cell $LiFePO_4$/Graphite.

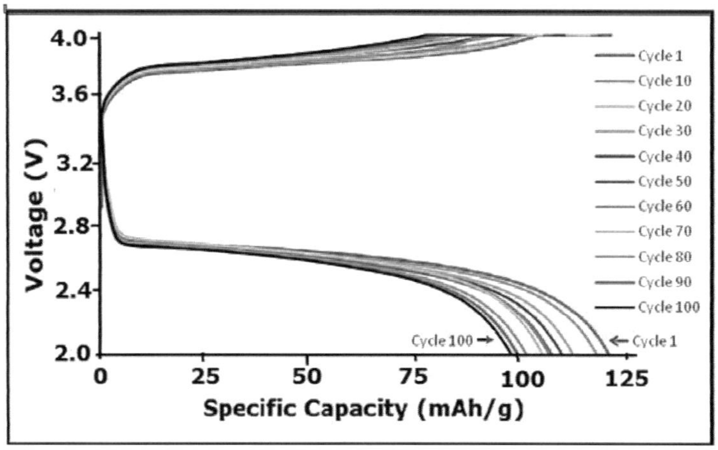

Figure 4. Charge discharge curves of $LiFePO_4$/Graphite Cylindrical Pouch cell at various cycles at 0.3C.

Figure 4 shows the charge discharge curves of LFP/Graphite cylindrical pouch cell with various cycles from 1st to 100th cycles. The discharge capacities for the first and 100th cycles were ~124.10 mAh/g and 100.20 mAh/g, respectively. The cycling performance of LFP/Graphite cylinder pouch cell showed the best capacity off all samples at 0.3C-rates, about 124.10 mAh/g with small decreasing after 100th cycles as shown in Figure 4. The results showed that the cylinder pouch showed good performance and a better stability after 100th cycles. The results are comparable to the LFP/C using cylinder 18650 [9].

4. Conclusions

The electrochemical properties of LFP/Graphite cylindrical pouch cell lithium ion battery (c-pouch cell) exhibited a discharge capacity of 250 mAh, with the specific capacity capacities for the first and 100th cycles were ~124.10 mAh/g and 100.20 mAh/g, respectively. The battery showed good performance after 100 cycles at rate 0.3C with rentention of 80.74%. The c-pouch cell can be used as an alternate casing product of lithium ion rechargeable battery.

Acknowledgment

This work is partially supported by the DIPABATAN fiscal year 2018, and INSINAS Consortium 2018 from the ministry Research, Technology and Higher Education.

References

[1] I.A. Bobrikov, A.M. Balagurov, C.W. Hu, C.H. Lee, T.Y. Chen, S. Deleg, D.A. Balagurov, Structural evolution in LiFePO4-based battery materials: In-situ and ex-situ time-of-flight neutron diffraction study, J. Power Sources. 258 (2014) 356–364.

[2] H. Xu, J. Sun, L. Gao, Hydrothermal synthesis of LiMnO2 microcubes for lithium ion battery application, Ionics (Kiel). 19 (2013) 63–69.

[3] W. Honggowiranto, Sudaryanto, E. Kartini, A. Purwanto, Electrochemical Performance of LiFePO 4 Cylinder Cell Battery, in: AIP Proceeding "6th Nanosci. Nanotechnol. Symp. (NNS2015)," 2016: pp. 030042–1–7.

[4] B. Jin, H. Gu: Preparation and characterization of LiFePO4 cathode material by hydrothermal method , Solid State Ionics, 2008.

[5] X. Guo, Y. Zhang, H. Xiang, Synthesis and Electrochemical Property of Flowerlike LiFePO 4

by Poly (ethylene glycol) -assisted Hydrothermal Process, Chin.J.Chem.Phys. 26 (2013) 337–340.

[6] Z. Liu, L. Sun, W. Yang, J. Yang, S. Han, D. Chen, Y. Liu, X. Liu, The synergic effects of Na and K co-doping on the crystal structure and electrochemical properties of Li4Ti5O12 as anode material for lithium ion battery, Solid State Sci. 44 (2015) 39–44.

[7] C.P. Supriadi, E. Kartini, W. Honggowiranto, K.T. Basuki, Synthesis and Characterization of Carbon Material Obtained From Coconut Coir Dust by Hydrothermal and Pyrolytic Processes, IJTech 2017. 8 (2017) 1470–1478.

[8] K. Fei-yu, M. a Jun, L.I. Bao-hua, Effects of carbonaceous materials on the physical and electrochemical performance of a LiFePO 4 cathode for lithium-ion batteries, New Carbon Mater. 26 (2011) 161–170.

[9] Z. Guo, Z. Chen, High-temperature capacity fading mechanism for LiFePO4/graphite soft-packed cell without Fe dissolution, J. Electroanal. Chem. 754 (2015) 148–153.

[10] P. Manikandan, P. Periasamy, R. Jagannathan, Sol–gel synthesis and impedance characteristics of networked nanocrystalline olivine cathode for Li-ion full cells, J. Mater. Chem. A. 1 (2013) 15397.

[11] C. Sato, E. Hoshi, K. Uematsu, T. Ishigaki, K. Toda, M. Sato, Hydrothermal Reaction by Microwave Heating for Synthesis of LiFePO 4 as Cathode Material, Trans. Mater. Res. Soc. Japan. 35 (2010) 397–400.

[12] A.Z. Syahrial, B. Priyono, A.H. Yuwono, E. Kartini, H. Jodi, Johansyah, Synthesis of Lithium Titanate (Li4Ti5O12) through Hydrothermal Process by using Lithium Hydroxide (LiOh) and Titanium Dioxide (TiO2) Xerogel, Int. J. Technol. 3 (2016) 392–400.

[13] E. Kartini, Emerging battery research in Indonesia: The role of nuclear applications, in: B.A. Cunningham, O. Catherine, G. Shohini (Eds.), AIP Proceeding "5th Int. Conf. Women Physics, Waterloo, Canada, 5-8 August 2015"., AIP Publishing LLC, 2015: pp. 100001–2.

[14] E. Kartini, M. Manawan, Solid Electrolyte For Solid-State Batteries : Have lithium-ion batteries reached their technical limit , in: 6th Nanosci. Nanotechnol. Symp., 2016: pp. 1–6.

[15] E. Kartini, A. Purwanto, S. Sudaryanto, H. Honggowiranto, R.N.S. Rofika, Indonesian consortium of lithium ion battery for solar street lamp, in: IOP Conf. Ser. Mater. Sci. Eng., 2018: pp. 1–6.

[16] N.I.P. Ayu, E. Kartini, L.D. Prayogi, M. Faisal, Crystal structure analysis of Li 3 PO 4 powder prepared by wet chemical reaction and solid-state reaction by using X-ray diffraction (XRD), Ionics (Kiel). (2016) 1051–1057.

[17] Deswita, Sudaryanto, H. Djodi, E. Kartini, The improving conductivity of LiFePO 4 by optimizing the calendaring process, in: IOP Conf. Ser. Mater. Sci. Eng., 2018: pp. 1–8.

[18] N. Sofyan, G.T. Setiadanu, A. Zulfia, E. Kartini, Effect of Different Calcination Temperatures and Carbon Coating on the Characteristics of LiFePO4 Prepared by Hydrothermal Route, IJET. 9 (2017) 3310–3317.

[19] A.Mulyono, T. Mustika, H. Sulaikan, E. Kartini, Development of battery performance data acquisition system for monitoring battery performance inside solar cell system, in: IOP Conf. Ser. Mater. Sci. Eng., 2018: pp. 1–8.

[20] W. Honggowiran, E. Kartini, Characterization of LiFePO4 Cathode by Addition of Graphene for Lithium Ion Batteries, in: AIP Proceeding, 2016.

IOP Conf. Series: Materials Science and Engineering **924** (2020) 012037 doi:10.1088/1757-899X/924/1/012037

Synthesize & Characterization of $Li_3PO_{4(0.5)}LiI_{(0.25)}LiCl_{(0,25)}$ Solid Electrolyte for Lithium Ion Battery

Gde Paksi Raganata[1*], Evvy Kartini[2], Heri Jodi[2], Wahyudianingsih[2], Agus Sudjatno[2], and Philips Nicolas Gunawidjaja[1]**
[1]Department of Physics, Faculty of Information Technology and Sciences, University of Parahyangan, Jl. Ciumbuleuit No. 94, Kota Bandung 40141, Indonesia
[2]National Agency of Nuclear Energy, Jl. Kawasan Puspiptek, Kota Tangerang Selatan 15310, Indonesia

*paksiraganatag@gmail.com ;
**kartini@batan.go.id

Abstract. Lithium phosphate (Li_3PO_4) is a promising candidate for solid electrolytes. However its ionic conductivity still stands at 3×10^{-7} S/cm which is very low compared to liquid electrolytes which are contained within widely distributed lithium ion batteries at 10^{-3} S/cm. This requires a solution to increase its conductivity which is the addition of dopants. The proposed dopants are lithium iodide (LiI) and lithium chloride (LiCl) which is mixed with the Li_3PO_4, creating the new solid electrolyte system $Li_3PO_{4(0.5)}LiI_{(0.25)}LiCl_{(0,25)}$. The mixture had an ionic conductivity of 8×10^{-4} S/cm observed through electrochemical impedance spectroscopy (EIS). The crystal structure observation using x-ray diffraction identified the Li_3PO_4 belongs to the Hermann-Mauguinn symmetrical space group of P m n b.

Keywords : Li_3PO_4, LiI, LiCl, solid electrolyte, ionic conductivity

1. Introduction

Solid electrolytes have been developed extensively as an alternative to liquid electrolytes. The liquid ones are deemed more risk because of the flammability of most of its materials [1]. Aside of that, due to the oxidation happening inside the liquid electrolytes which results in the batteries itself expanding. This may cause the internal structure of liquid batteries to change, resulting in electrolyte moving through the separator which may cause short circuit from connection between Anode and Cathode [2]. Lithium phosphate (Li_3PO_4) is a promising network former for solid electrolytes as reported by Evvy et al in 2016 which was created by mixing lithium hydroxide (LiOH) and phosphoric acid (H_3PO_4) via wet chemical reaction and lithium oxide (Li_2O) and phosphorus pentoxide (P_2O_5) via solid state reaction (melt-quenching) [3-5]. The highest ionic conductivity obtained through the study was via solid state reaction which measured was measured at 3×10^{-7} S/cm.

The meager ionic conductivity requires an additive to exceed the ionic conductivity of liquid electrolytes which measures at 10^{-3} S/cm [6] hence the addition of dopant salts. These salts increase ionic conductivity of network formers by enriching the lithium content in which they only react statically inside the electrolyte [7]. Examples of dopants include lithium iodide (LiI), silver iodide (AgI), lithium chloride (LiCl), and silver chloride (AgCl). Research about dopants have been conducted extensively throughout the years. Makhsun et al in 2012 added AgI and LiI which increased the conductivity of

$AgPO_3$ to 10^{-2} S/cm [8]. Kaus et al in 2009 added LiI into Li_3PO_4 which increased its conductivity to 10^{-3} S/cm and also did it with several concentration which gave its optimal at 45-50 % [9]. Other experiments also include the work of Tanaka et al in in 1988 about the production of LiCl + Li_2O + TeO_2 glass which has an ionic conductivity of 10^{-6} S/cm [10] and the work of Kharbachi et al in 2020 about the $LiBH_4$-LiCl-P_2S_5 system which with an ionic conductivity measured at 10^{-3} S/cm [11].

In this research, two dopants are chosen which are LiI and LiCl. The wide usage of these materials influenced its selection for the experiment. Their ionic conductivities measured respectively at 10^{-7} S/cm and 10^{-6} S/cm [12, 13]. The material is a mixture of Li_3PO_4 which was the product of solid state reaction from the research of Nur et al [4], LiI, and LiCl with a molar ratio of 2:1:1 which will be labeled $Li_3PO_{4(0.5)}LiI_{(0.25)}LiCl_{(0,25)}$. The material is created via wet chemical reaction which is grinding using mortar and pestle. Properties of the materials are also examined to be used as comparation between other researches. Electrical property will be measured using electrochemical impedance spectroscopy (EIS)[14] and the crystal structure will be observed using x-ray diffraction (XRD) [15]. The addition of these dopants is expected to increase the base conductivity of Li_3PO_4 from 3×10^{-7} S/cm to compete with or even exceed those of liquid electrolytes.

2. Experimental method

2.1. Experimental Diagram

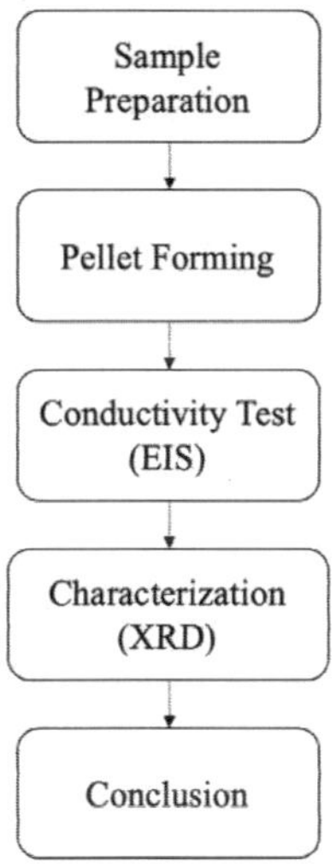

Figure 1. The flow chart of experimental diagram of the research.

Figure 2. The $Li_3PO_{4(0.5)}$ $LiI_{(0.25)}$ $LiCl_{(0,25)}$ mixture in the form of powder.

The experiment is first done by mixing all the ingredients into one homogenous powder (all three components are in powder form). The raw materials used are Li_3PO_4, LiI (MERCK), and LiCl (Sigma ALDRICH). In the sample preparation, the materials are weighed together according to the molar ratio of 2:1:1. The materials are then grinded using a mortar and pestle. For the electrochemical impedance spectroscopy (EIS), the materials must be shaped into an easily measured symmetrical form hence the forming of pellet using a portion of the mixture. The remaining powder is then used for characterization using x-ray diffraction (XRD).

IOP Conf. Series: Materials Science and Engineering **924** (2020) 012037 doi:10.1088/1757-899X/924/1/012037

2.2. Sample preparation

The materials (Li_3PO_4, LiI, and LiCl) in the form of powder are first weighed using a digital scale according to its molar ratio. A total of 30 g of material is used. Using the relation between mass ratio and molar ratio, the mass of each component within the mixture can be determined, via equation (1).

$$w_i = x_i \frac{M_i}{\Sigma_j x_j M_j} \tag{1}$$

With w_i being the mass ratio, x_i the molar ratio of the weighed component, M_i the molar mass of the component in g/mol, and $\Sigma_j x_j M_j$ as the sum of the molar mass of all the components multiplied by its molar ratio.

Table 1. The mass of each component within the mixture.

Component	Mass (g)
Li_4PO_4	17.03
LiI	9.84
LiCl	3.11

The whole mixture is then grinded together for four hours using a mortar and pestle in room temperature until it becomes homogenous. After the mixing, the sample is placed within a porcelain crucible and dried frequently using an oven at 50°C in between processes to remove water particles (due to LiI and LiCl being highly hygroscopic).

2.3. Pellet Forming

For the EIS, the sample must be shaped into small cylindrical pellets. The process is done by putting the powder into a mold. The mold is then pressed vertically using a hydraulic presser. After the pressing, the pellet is then re-dried in the oven.

2.4. Electrochemical Impedance Spectroscopy (EIS)

The pellets are then measured to obtain its diameter and thickness and Ag paste is applied on it. The test uses a HIOKI 3532-50 LCR HiTESTER LCR meter and is run through a measuring frequency of 5 Hz to 5000000 Hz. During the LCR test, the pellet is placed between two electrodes and a plot of its conductance is obtained. Using the formula in equation (2) of conductivity, its ionic conductivity can be obtained which is as follows:

$$\sigma = G\left(\frac{h}{A}\right) \tag{2}$$

where σ is the conductivity in S/cm, G is the conductance in S, h is the thickness of the pellet in cm, and A is the surface area of the pellet in cm^2. The result is then plotted into two graphs. The first graph is the Cole-Cole plot to observe its impedance, and the second is the conductivity to measuring frequency graph. The conductivity graph is put into logarithmic scale to observe the obtain its conductivity according to the Jonscher Power Law:

$$\sigma_{total} = \sigma_{DC} + A\omega^n \tag{3}$$

Where σ_{total} is the total conductivity in S/cm, σ_{DC} is the DC conductivity, and $A\omega^n$ is the AC conductivity with n being the power constant ($0 < n < 1$).

2.5. X-Ray Diffraction (XRD)

The XRD characterization in this experiment uses a Malvern Panalytical Type Empyrean diffractometer. The peaks are observed from a 2θ value of 10° to 80°. The peak values from this test would then be processed using Rietveld refinement to discover its new lattice information. The peaks would then be plotted into intensity to 2θ graph.

3. Results and discussion

3.1. Electrochemical Impedance Spectroscopy (EIS)

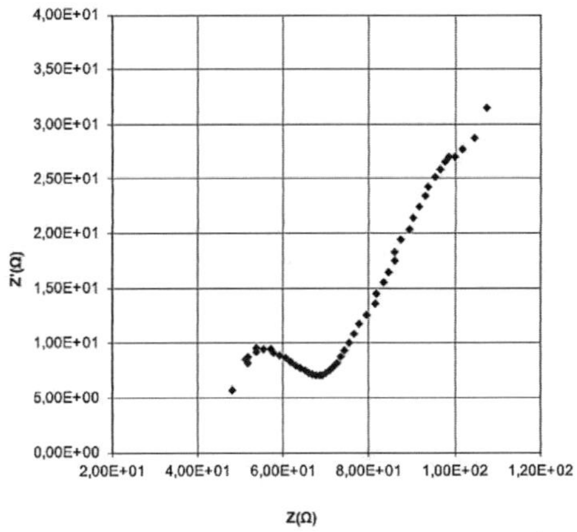

Figure 3. Cole-Cole plot of the $Li_3PO_{4(0.5)}LiI_{(0.25)}LiCl_{(0,25)}$ mixture.

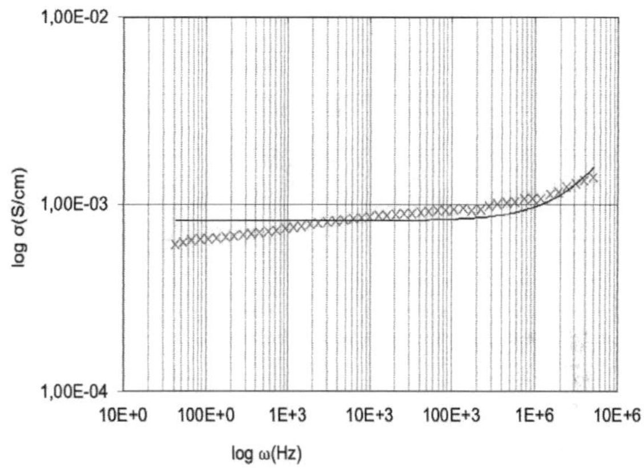

Figure 4. Conductivity graph of the $Li_3PO_{4(0.5)}LiI_{(0.25)}LiCl_{(0,25)}$ mixture.

Table 2. Impedance and DC conductivity value of $Li_3PO_{4(0.5)}LiI_{(0.25)}LiCl_{(0.25)}$ mixture.

R (Ω)	**σ (S/cm)**
4.83	8×10^{-4}

The EIS test shows that the $Li_3PO_{4(0.5)}LiI_{(0.25)}LiCl_{(0,25)}$ mixture had an ionic conductivity in the value of 8×10^{-4} S/cm. Compared to the ionic conductivity of the Li_3PO_4 created by Evvy et al. [3], which had an ionic conductivity of 3×10^{-7} S/cm, the addition of dopants have increased the conductivity by several orders. The conductivity however still stands lower compared to the works of Kaus et al and Makhsun et al which electrolyte systems are able to reach 10^{-3} S/cm or even 10^{-2} S/cm which even exceeds those of liquid electrolytes. The work of Makhsun et. al [8] however involves the process of melt quenching which is heating the material to achieve a more amorphous structure first, hence the higher conductivity.

3.2. X-Ray Diffraction (XRD)

The crystal structure observation which uses XRD and is analyzed using Rietveld refinement gave the information of the crystal structures of each of the components. The Li_3PO_4 used belonged to the Hermann-Mauguinn symmetry space group P m n b [16], the LiI belonged to the F m -3 m space group [17], and the LiCl also belonged to the F m -3 m space group [17]. The crystal structure of Li_3PO_4 made by Evvy et al using solid state reaction also belonged to the P m n b space group which means its structure remain unchanged because the materials are only mixed as a composite. The lattice parameters for the Li_3PO_4 are a = 4.8600 Å, b = 6.0700 Å, and c = 10.2599 Å, for LiI are a = 10.3415 Å, b = 5.5693 Å, c = 6.6374 Å, and for LiCl are a = 3.8300 Å, b = 3.8300 Å, and c = 3.8300 Å.

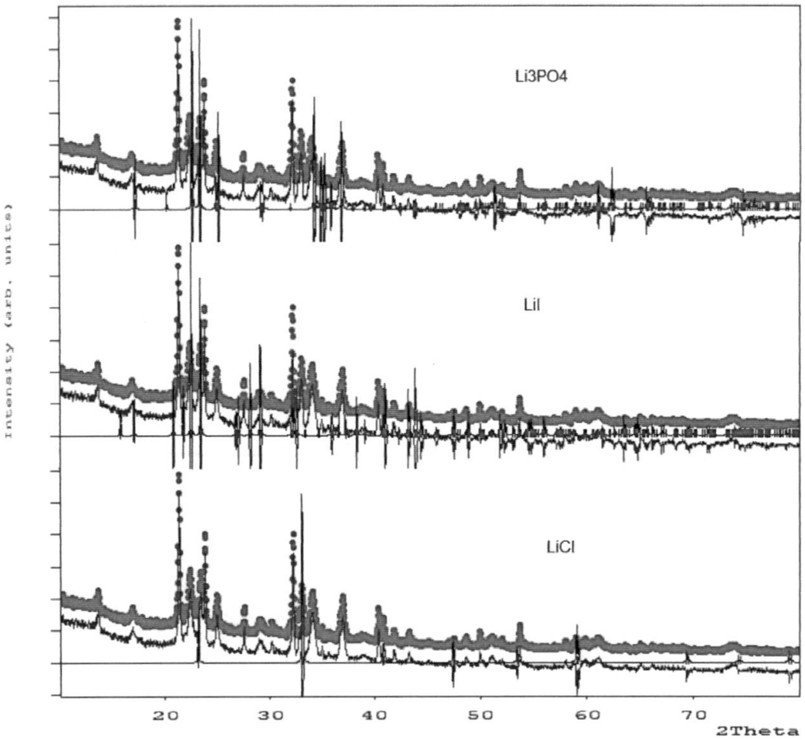

Figure 5. X-ray diffraction results of the $Li_3PO_{4(0.5)}LiI_{(0.25)}LiCl_{(0,25)}$ mixture with each graph representing a phase detection of Li_3PO_4, LiI, and LiCl (represented by the green tick marks below the graph).

Table 3. Rietveld refinement result of the $Li_3PO_{4(0.5)}LiI_{(0.25)}LiCl_{(0,25)}$ mixture.

Formula		**Li_3PO_4**	**LiI**	**LiCl**
Space Group		**P m n b**	**F m -3 m**	**F m -3 m**
Lattice Parameters	a (Å)	4.8600	10.3415	3.8300
	b (Å)	6.0700	5.593	3.8300
	c (Å)	10.2599	6.6374	3.8300
	α (°)	90	90	90
	β (°)	90	90	90
	γ (°)	90	90	90
R_{exp} (%)		4.75	4.75	4.76
R_{wp} (%)		95.7	94.2	98.9
X^2		405	392	432

4. Conclusion

Based on the results of the experiment, it is concluded that the addition of LiI and LiCl as dopants for the Li_3PO_4 solid electrolyte did increase its ionic conductivity. The $Li_3PO_{4(0.5)}LiI_{(0.25)}LiCl_{(0,25)}$ system achieved an ionic conductivity of 8×10^{-4} S/cm, which is close to that of liquid electrolytes conductivity which stands at 10^{-3} S/cm.. This value has been increasing about three orders of magnitudes from the Li_3PO_4 ionic conductivity which stands at 3×10^{-7} S/cm. The crystal structure observation revealed that the Li_3PO_4 still belongs to the same space group of P m n b as reported by Evvy et. al,[5] due to only being mixed together with the dopants as a composite. For further research, it is suggested to make the composite with different concentration which can be related to changes in ionic conductivity and to and to find the optimal portion of dopants within Li_3PO_4.

Acknowledgment
Author would like to thank the National Agency of Nuclear Energy, Indonesia for supporting with perspective and experimental facilities.

References

[1] M. S. Whitingham, "Lithium Batteries and Cathode Materials," *Chem Rev*, vol.104, pp. 4271-4302, 2004
[2] J. L. Souquet and M. Duclot, "Batteries: Glassy Electrolytes," *Elsevier*, vol 1, 2001
[3] E.Kartini, M.Manawan, M.Avdeev, and M.F.Collins, "Neutron diffraction study on Li_3PO_4 solid electrolyte for lithium ion battery", Physica B: Condensed Matter 551 (2018) 320-326 https://doi.org/10.1016/j.physb.2017.11.079
[4] Nur I. P. Ayu, Evvy Kartini, Lugas D. Prayogi, Muhamad Faisal, and Supardi "Crystal Structure Analysis of Li_3PO_4 Powder Prepared by Wet Chemical Reaction and Solid- State Reaction by using X-Ray Diffraction," *Ionics*, vol. 25, Jan. 2016"
[5] E.Kartini, V.Yapriadi, H.Jodi, M.Manawan, "Solid electrolyte composite Li4P2O7–Li3PO4 for lithium ion battery",.Progress in Natural Science: Materials International 2020 (in press), https://doi.org/10.1016/j.pnsc.2020.01.020
[6] D. Linden and T. B. Reddy, "Handbook of Batteries," *McGraw-Hill*, vol. 3, 2001
[7] J. L. Souquet and M. Duclot, "Batteries: Glassy Electrolytes," pp. 457-462, 2001
[8] Makhsun, T. Sakuma, and E. Kartini, "Characteristic of $(AgI)_{0,44}(LiI)_{0,22}(AgPO_3)_{0,34}$ Ionic Conductor Prepared by Melt Quenching Method," *Jurnal Sains Materi Indonesia*, 2012
[9] N. H. Kaus and A. H. Ahmad, "Conductivity Studies and Ion Transport Mechanism in $LiI-Li_3PO_4$ Solid Electrolyte," *Ionics*, vol 15, pp. 197-201, 2009
[10] K. Tanaka, T. Yoko, H. Yamada, and K. Kamiya, "Structure and Ionic Conductivity of $LiCl-Li_2O-TeO_2$ Glasses," *Journal of Non-Crystalline Solids*, vol. 103, pp. 250-256, 1988
[11] El Kharbachi, Abdel, et al, "Pseudo-ternary LiBH4· LiCl· P2S5 system as structurally disordered bulk electrolyte for all-solid-state lithium batteries," *Physical Chemistry Chemical Physics*, 2020
[12] C. C. Liang, "Conduction Characteristics of the Lithium Iodide-Aluminum Oxide Solid Electrolytes," *Journal of the Electrochemical Society*, vol. 120, pp. 1289-1292, 1973
[13] M. B. Mogensen and E. Henneso, "Properties and Structure of the LiCl-Films on Lithium Anodes in Liquid Cathodes," *Acta Chimica Slovenica*, vol. 3, pp 519-534, 2016
[14] "LCR Measurement Primer," *IET Labs Inc.*, vol. 1, 2018
[15] L. M. Kyu, "X-Ray Diffraction for Materials Research," *Apple Academic Press Inc.*, vol. 1, 2016
[16] J. Zemann, "Die kristallstruktur von lithiumphosphat," Acta Crystallographica, vol. 13, pp. 863-867, 1960
[17] H. Ott, "Die Raumgitter der Lithiumhalogenide," Physikalische Zeitschrift, vol. 24, pp. 209-213, 1923

IOP Conf. Series: Materials Science and Engineering **924** (2020) 012038 doi:10.1088/1757-899X/924/1/012038

The Future of All Solid State Battery

Evvy Kartini[*1] and Carla Theresa Genardy[2]
[1] Center for Science and Technology for Advanced Materials, National Nuclear Energy Agency, Indonesia
[2] Concord College Shrewsbury, UK

*evvy.kartini@gmail.com

Abstract. All Solid State Battery, is a type of battery which uses no liquid electrolyte, instead, it uses solid electrolyte (alternative to lithium-ion battery). This type of battery still has not been commercially used, but will soon be used in electric vehicles. The demand for the production of solid state battery is due to the advantages that arise from the safety issue. Solid electrolytes are the solid that exhibits ionic conductivities about 10^{-3} S/cm, and it conducts ions between the cathode and anode in the lithium ion battery. Various solid electrolytes based on phosphate glasses have been developed. This article reviews the developments of solid electrolytes and its applications.

Keywords : Solid state battery, solid electrolyte, superionic conducting glass, ionic conductivity

1. Introduction

There are numerous types of rechargeable batteries, including rechargeable Li-ion batteries (LIBs). LIBs have several advantages, such as, long life cycle, high power density, low self-discharge property, high gravimetric, and volumetric energy. Currently, lithium-ion batteries (LIB) are widespread and promising candidates for future application, such as electric vehicles and energy storage [1]. Nonetheless, they suffer from raw materials availability, safety concerns, and limited energy storage capacity. State-of-the-art lithium-ion cells consist of two porous electrodes (anode and cathode) and a separator, as depicted in Fig. 1 current collector, which consist of the active material, conductive agents, and binder [2]. The ion transfer requires a liquid electrolyte which is mainly composed of aprotic organic solvents and a conductive salt.

Many of the issues that current LIBs are facing can be traced back to this liquid electrolyte. Safety concerns, in fact, arise from the flammability of the solvents. Side reactions of the solvents and the conductive salt lead to capacity fading and aging [2]. All solid-state batteries (ASSB), in contrast, are not only inherently safer due to the lack of flammable organic components, but also offer the potential for a dramatic improvement of energy density. Instead of a porous separator soaked with liquid electrolyte, ASSBs use a solid electrolyte, which acts as electrical insulator and ionic conductor at once (Fig. 1(b)). The use of a compact solid electrolyte acting as a physical barrier for lithium dendrites also enables the use of lithium metal as the anode material [3]. All solid-state batteries (ASSB), in contrast, are not only inherently safer due to the lack of flammable organic components, but also offer the potential for a dramatic improvement of energy density. Instead of a porous separator soaked with liquid electrolyte, ASSBs use a solid electrolyte, which acts as electrical insulator and ionic conductor at once (Fig. 2).

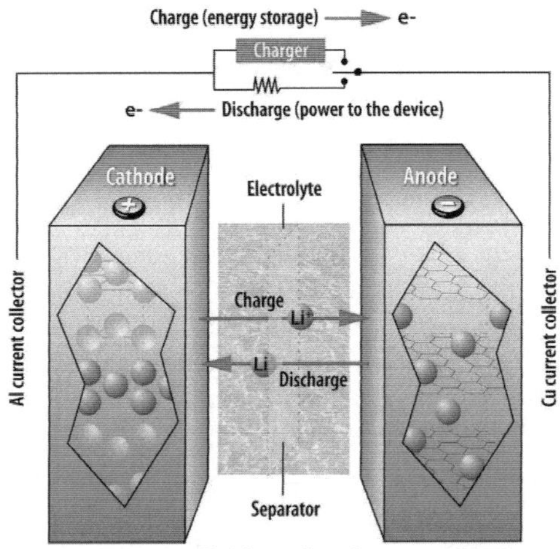

Figure 1. How the lithium-ion battery (LIB) works.

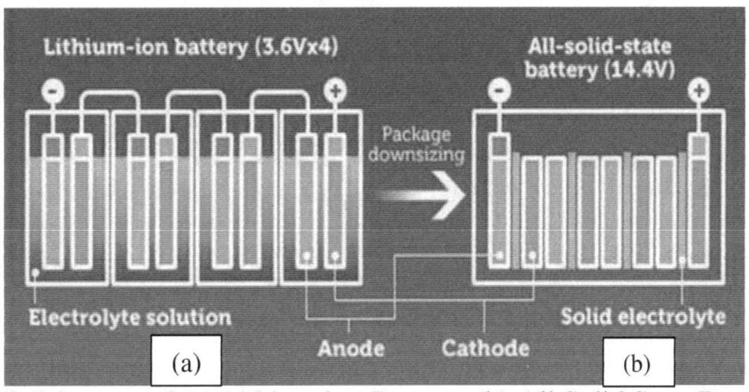

Figure 2. Diagram of (a) Lithium ion Battery; (b) All Solid State Battery [1].

2. Advantages of Solid State Batteries

2.1 Advantages of Solid State Batteries over Lithium-Ion Batteries.
There are several advantages of all solid-state battery over lithium ion batteries. It is non-flammable. In Lithium Ion batteries, the ions that flow from the cathode to the anode may form dendrites on the surface of the cathode, which can pass through the liquid electrolyte and to the anode causing a short circuit [3]. When a short circuit is formed, the battery will get hotter and might explode. Solid State Batteries solved this case due to the use of solid electrolyte which prevents the formation of dendrites. It has a higher energy density, due to is attainable by direct-series-stacking of battery cells. It reduces battery size, especially when thin-film of solid electrolyte is used. It has longer lifespan and higher capacity: solid state battery enables use of lithium metal anode which increases cell voltage and capacity.

2.2 Challenges of using solid state batteries:
Finding a solid material that is conductive enough to be used in large batteries, e.g. polymer or ceramic (solid electrolyte has to have high ionic conductivity). To obtain material that has ionic conductivity,

superionic materials are used, which are materials suitable for creating solid electrolytes as they allow ions to move quickly and freely through their crystal structure.

Ionic conductivity is typically low, which prevents fast charge and discharge. Forming a stable conductive interface between solid electrolyte and electrode is difficult. Low power due to high solid electrolyte-electrode resistivity. The poor electronic conductivity can cause electron beam radiation damage.

2.3 Different types of solid electrolytes used in solid state batteries:

2.3.1. Glass-based electrolyte. There are several glass-based electrolytes: (i) Simplifies fabrication of the battery cell as it allows them to plate the alkali metals and strip them on both electrodes. Hence, preventing the formation of dendrites; (ii) Uses sodium instead of lithium. Sodium is cheaper as it can be easily extracted from seawater; (iii) More environmentally friendly and can operate in extreme temperatures unlike lithium.; (iv) Absence of grain boundaries; (v)Flexibility of size and shape at satisfactory cost; (vi) Can be prepared over a wide range of composition. Glass-based electrolyte consist of glass former (i.e.P_2O_5), glass modifier (i.e.Ag_2O, Li_2O) and salt-doping (i.e. AgI, LiI). The most well known study superionic conducting glasses are the silver and lithium system, respectively, AgI-$AgPO_3$, LiI-$LiPO_3$, and the mixed AgI-Ag_2S-$AgPO_3$, AgI-LiI-$LiPO_3$,etc.[4-8]. Usually, the superionic glass was prepared by mixing appropriate amounts of AgI, $AgNO_3$, $NH_4H_2PO_4$, heating up $100^oC/h$ to 600^oC within 6 hrs, casting and quenching into liquid-N_2. It will resulted a clear yellowish glass, while the undoped $AgPO_3$ is a clear transparent glass [4-11].

2.3.2 Lithium-based electrolyte. Lithium based electrolyte, has several advantageous, namely (i) Ultrafast charging; (ii) A superionic conductor, hence can allow ions to move quickly through it (high ionic conductivity); (iii) Very small internal resistance, hence can produce greater power; (iv) Cells can stack together without interference; (v) Can retain their charges for long period of time, hence has long lifespan. The lithium phosphate, Li_3PO_4 has been chosen as one good candidates for solid state battery [9-10]. Its application in the thin film battery has been well studied. However, the limitation of low ionic conductivities, made modification of this material by doping with LiI, AgI or mixed salts [12-13].

2.3.3 Sulfide electrolyte. It has been investigated that $Li_{10}GeP_2S_{12}$ is one of the best solid ionic conductor (12mS/cm). However, it is expensive as Germanium is used. Germanium can be replaced with tin, but it will reduce the ionic conductivity of the substance, unstable, and toxic. The best solid ionic conductor is $Li_{9.54}Si_{1.74}P_{1.44}S_{11.7}Cl_{0.3}$ (25mS/cm). It is amendable to fast charging due to the very high ionic conductivity and can sustain high temperature [14].

2.3.4 Oxide electrolyte. Proven to be better than sulfide electrolyte as it is thermally and chemically stable, does not require a controlled environment for processing. More environmental friendly as it does not produce toxic gases, unlike sulfide electrolyte that produces hydrogen sulfide which can cause acid rain. The common one used is C-LLZO, for use in high-voltage batteries. It is because C-LLZO is electrochemically inert over wider voltage. The problem with C-LLZO is that it is not thin enough because to slot a sloid electrolyte between anode and cathode requires dense thin films (less than 50μm thick).

3. Results and discussion

Figure 3(a) and 3(b) show examples of superionic conducting glass AgI-$AgPO_3$ and the series of AgI-Ag_2S-$AgPO_3$, respectively. The ionic conductivity has been increased by three orders of magnitudes, for AgI-$AgPO_3$~10^{-13} S/cm from the $AgPO_3$ ~10^{-7} S/cm. There are several Influence of factors governing behavior of solid electrolytes grain interior atomic configuration is on ionic conductivity. Impacts of grain boundaries Behavior of solid electrolyte-electrode interferences. The phenomena have been described by Aniya et al, on the Bond fluctuation model shown in Fig.4 [14]. In this model, it is assumed that the structure consist of two regions: the glass network and the doped salt. By the insertion of salt, the network fluctuates and the distance, d, increases, thus increase the ionic conductivity.

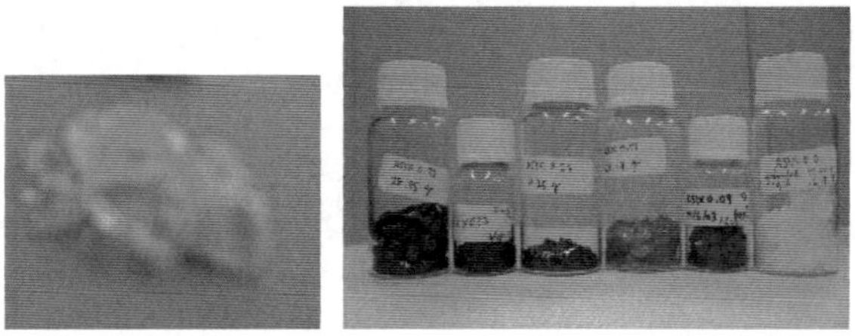

(a) (b)

Figure 3. Superionic conducting glass AgI-$AgPO_3$ and the series of AgI-Ag_2S-$AgPO_3$ [4,6].

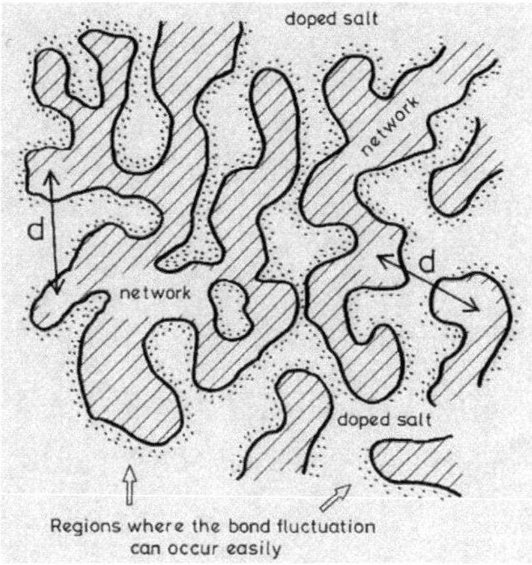

Figure 4. Bond fluctuation model shown in Fig.4 [15].

4. Conclusion

It is concluded that the research and development of on LIB moving forward. This paper has summarized the early development of solid electrolyte for all solid state battery. The research on solid electrolyte is growing due to safety reason. The future of solid state battery will be replacing the current commercial lithium ion battery which based on liquid electrolyte. In the future, all solid state battery will be very safety and efficient battery. This is because of the solid electrolyte has high ionic conductivity, and also it has high melting point. The use of solid state battery in electric cars removes all the cooling elements, which makes the battery pack lighter. Hence, the car will have a longer range and smaller battery size. The application, though is still not commercially available, but its prospect is promising to replace the current lithium ion battery.

Acknowledgment

This work is partly financially supported by the Ministry Research Technology / National Research and Innovation Agency (BRIN), through the Rispro Mandatory LPDP 2020 and the National Nuclear Energy Agency, Indonesia.

.

References

[1] B. Scrosati and J. Garche, "Lithium batteries: Status, prospects and future," *J. Power Sources*, vol. 195, pp. 2419–2430, 2010.

[2] Joscha Schnell, Till Günther, Thomas Knoche et.al., All-solid-state lithium-ion and lithium metal batteries – paving the way to large-scale production, *Journal of Power Sources 382,* pp. 160–175, 2018

[3] Sveinbjörnsson, D. Þ., Vegge, T., Norby, P., & Mogensen, M. B. 2014. Design and Characterisation of Solid Electrolytes for All-Solid-State Lithium Batteries. Department of Energy Conversion and Storage, Technical University of Denmark.

[4] E. Kartini, S.J. Kennedy, K. Itoh, T. Kamiyama, M.F. Collins, S. Suminta, "Anion effect on the structure of Ag_2S-$AgPO_3$ superionic glasses", *Solid State Ionics* 167 (1–2), pp. 65–71, 2004

[5] E. Kartini, M.F. Collins, "Neutron scattering and thermal measurements on the superionic-conducting Ag_2S–$AgPO_3$ glass system", *Solid State Ionics* 314, pp.633–636, 2002.

[6] E. Kartini, M. Arai, F. Mezei, M. Nakamura, M. Russina, "Structure and dynamics on superionic conducting phosphate glasses by neutron scattering", *Phys. B Condens. Matter*. 385–386, pp. 236–239 (2006).

[7] E. Kartini, S.J. Kennedy, K. Itoh, T. Fukunaga, S. Suminta, T. Kamiyama, "Characterization of the intermediate-range order in new superionic conducting AgI–Ag_2S–$AgPO_3$ glasses by neutron diffraction", *Appl. Phys. A*, 1238 (1–3), pp. 1236–1238, 2002.

[8] E. Kartini, M. Manawan , M. F. Collins, M. Avdeev, 2017, "Neutron diffraction study on Li_3PO_4 solid electrolyte for lithium ion battery", *Physica B: Condensed* 551, pp. 320–326, 2018

[9] N.I.P. Ayu, E. Kartini, L.D. Prayogi, M. Faisal, Supardi, "Crystal structure analysis of Li_3PO_4 powder prepared by wet chemical reaction and solid-state reaction by using X-ray diffraction (XRD)", Ionics (Kiel) 353, pp.1–7, 2016

[10] E. Kartini, T.Y.S. Panca Putra, I. Kuntoro, T. Sakuma, K. Basar, O. Kamishima, J. Kawamura, "Recent studies on lithium solid electrolytes $(LiI)_{0.3}(LiPO_3)_{0.7}$ for secondary battery", *J. Phys. Soc. Jpn.* 79, pp. 54–58, 2010.

[11] E. Kartini, M. Nakamura, M. Arai, Y. Inamura, K. Nakajima, T. Maksum, W. Honggowiranto, T.Y.S.P. Putra, "Structure and dynamics of solid electrolyte $(LiI)_{0.3}(LiPO_3)_{0.7}$", *Solid State Ionics* 262, pp. 833–836, 2014.

[12] Evvy Kartini, Valentina Yapriadi, Heri Jodi, Maykel Manawan, Cipta Panghegar, Wahyudianingsih, Solid electrolyte composite Li4P2O7–Li3PO4 for lithium ion battery, Progress in Natural Science: Materials International, https://doi.org/10.1016/j.pnsc.2020.01.020

[13] D.P. Singh, K. Shahi, K.K. Kar, "Scaling behavior and nearly constant loss effect in AgI–$LiPO_3$ composite glasses", *Solid State Ionics* 231, pp. 102–108, 2013.

[14] M. Tatsumisago, M. Nagao, A. Hayashi, "Recent development of sulfide solid electrolytes and interfacial modification for all-solid-state rechargeable lithium batteries", *J. Asian Ceram. Soc.* 1 (1)17–25, 2013.

[15] M. Aniya, J. Kawamura**, "**Medium range structure and activation energy of ion transport in glasses" *Solid State Ionics* 154, pp. 343, 2002

AUTHOR INDEX